Coating and Drying Defects

SPE MONOGRAPHS

Injection Molding

Irvin I. Rubin

Nylon Plastics

Melvin I. Kohan

Introduction to Polymer Science and Technology: An SPE Textbook

Edited by Herman S. Kaufman and Joseph J. Falcetta

Principles of Polymer Processing

Zehev Tadmor and Costas G. Gogos

Coloring of Plastics

Edited by Thomas G. Webber

The Technology of Plasticizers

J. Kern Sears and Joseph R. Darby

Fundamental Principles of Polymeric Materials

Stephen L. Rosen

Plastics Polymer Science and Technology

Edited by M. D. Baijal

Plastics vs. Corrosives

Raymond B. Seymour

Handbook of Plastics Testing Technology

Vishu Shah

Aromatic High-Strength Fibers

H. H. Yang

Giant Molecules: AN SPE Textbook

Raymond B. Seymour and Charles E. Carraher

Analysis and Performance of Fiber Composites, 2nd ed.

Bhagwan D. Agarwal and Lawrence J. Broutman

Impact Modifiers for PVC: The History and Practice

John T. Lutz, Jr. and David L. Dunkelberger

The Development of Plastics Processing Machinery and Methods

Joseph Fred Chabot, Jr.

Organic Coatings: Science and Technology. Volume I: Film Formation, Components, and Appearance

Zeno W. Wicks, Jr., Frank N. Jones, and S. Peter Pappas

Organic Coatings: Science and Technology. Volume 2: Applications, Properties, and Performance

Zeno W. Wicks, Jr., Frank N. Jones, and S. Peter Pappas

Coating and Drying Defects: Troubleshooting Operating Problems

Edgar B. Gutoff and Edward D. Cohen

Coating and Drying Defects

Troubleshooting Operating Problems

EDGAR B. GUTOFF

EDWARD D. COHEN

With Chapter 10 authored by
GERALD I. KHEBOIAN

A WILEY-INTERSCIENCE PUBLICATION

JOHN WILEY & SONS

New York • Chichester • Brisbane • Toronto • Singapore

This text is printed on acid-free paper.

Library of Congress Cataloging in Publication Data:

Gutoff, Edgar B.
Coating and drying defects : troubleshooting operating problems /
by Edgar B. Gutoff and Edward D. Cohen.
p. cm. — (SPE monographs)
Includes index.
ISBN 0-471-59810-0 (alk. paper)
1. Coatings—Defects. 2. Drying. I. Cohen, Edward D.
II. Title. III. Series.
TP156.C57G88 1995
667′.9—dc20 94-21972

Printed in the United States of America

10 9 8 7 6 5 4 3 2 1

To our wives, Hinda, Luiza, and Grace,
whose support and encouragement made this work possible

Contents

Preface

This book is intended to serve as a guide to help solve the various types of problems that occur in the coating and drying of films on continuous sheets or webs. It is intended for manufacturing and quality control personnel, the operators and supervisors who are directly involved in the production of coated products, and for the engineers and scientists who are involved in the design and production of coated products. The material presented is nonmathematical and should, we hope, be easy to understand. The main focus is on eliminating the many types of coating defects that, unfortunately, do occur in a coater. These defects must be eliminated to produce a satisfactory high-quality product at a competitive price. The difference between a successful and an unsuccessful process can be how rapidly defects are eliminated, how rapidly problems are solved, and how permanent is the cure. Other types of problems that coating personnel typically encounter in running a coater, such as controlling the drying, are also discussed.

Coated products are pervasive in our economy. They cover a wide range of standalone products as well as being key components in many others. Coated products vary from the painted metal in automobiles and appliances, to the photographic films for color and for medical x-rays, to the coatings used to produce electronic printed circuits, to the magnetic coatings that store information in computers and in music and video recorders, and to the coated catalysts in the converters used to reduce pollution in automobiles. These coatings all have a common basis, which is the replacement of air on a substrate with a liquid to give the final product having the desired properties. The goal of all these processes is to produce the desired coating at a competitive cost and yield.

The rapid advances in coating technology coupled with rapid development of product formulation technology has led to many new products that are more uniform and have a low defect level, resulting in higher yields and lower costs. Quality levels that were acceptable only a few years ago are no longer acceptable and in future will be even more demanding. Environmental needs and the resulting economic considerations will also become more stringent and will require that a much higher fraction of the starting materials end up as usable product, not scrap to be disposed of by burial in landfill or by incineration.

Although painted surfaces suffer from many of the same defects that occur in coated continuous webs and substrates, this book does not cover paint film defects as such. It also does not cover defects in the very thin films produced by vacuum deposition techniques. It does cover defects in most coated webs. For all coated products, standards are established that specify the characteristics of the coating solutions, the coating process operating conditions, and the final chemical and physical properties of the coating. Unfortunately, the reality of the coating process is that these standards are often not achieved and product is often produced which is, therefore, defective. There are many process steps where defects and nonuniformities can be introduced into the coating process. These defects can have a variety of causes, and many defects that appear different can have similar origins. Conversely, many defects that seem to be similar can have different causes. Ongoing diligence and effort is required to minimize these defects. Troubleshooting to eliminate defects, nonuniformities, and waste is an ongoing function of all personnel, from the operators to the research and development scientists and engineers who design the products and processes.

In our previous book, *Modern Coating and Drying Technology* (VCH Publishers, New York, 1992), we described the basics of the coating processes. It was intended as an introduction to the coating process. Its purpose was to answer the question that we had encountered: How do we learn about the coating process? This book is intended to answer the question: How do I eliminate typical operating problems that occur in a coater, such as a specific defect in the product I am now making, and how do I ensure that defects do not return? This book is intended to be a troubleshooting guide. It provides a description of the troubleshooting process and problem-solving techniques, and descriptions of common problems and typical solutions. Our approach is to present in one volume all the tools needed to troubleshoot defects and the means to eliminate defects, as well as brief descriptions of the major coating processes. A methodology is presented to guide one from the start of the troubleshooting process, when the defect is first detected, to the end of the process, where the mechanism for the formation of the defect is defined and the defect is eliminated.

The ideal state in any coating process is to run the process so that defects are never formed and problems are never encountered. Therefore, some guidelines for defect prevention will be given. Where possible, specific information on the major defect categories is also presented. We focus on the use of instruments and statistical tools, not on the mathematical derivations and the theory of their operation. Some references are given to the supporting theory for those who want to go to the literature for greater detail.

The web coating process consists of several major unit operations:

- Compounding, mixing, and dispersing the coating formulations, including binder components, solvents, coating aids, dyes, matte agents, and active ingredients, such as silver halides, iron oxides, and so on
- Obtaining the support to be coated on, such as polyester film or paper
- Coating the formulation on the support
- Transporting the coated support
- Drying the coating
- Converting the coated web into the final size and form for use by customers
- Shipping the product to customers or to storage

We focus on the steps of the process up to the final slitting operation. Although it is recognized that defects may be introduced there, this book does not cover that operation.

While all coated products are different in terms of their formulation and many different coating processes are used, the underlying science is similar. Many defects in different products have similar causes and similar cures. The principles developed from the elimination of bubbles in a low-viscosity barrier layer apply also to the coating of a low-viscosity photographic layer.

A wide variety of different coating application methods can apply a coating to a web; however, the successful processes are those that are defect-free over a wide range of operating conditions. In industrial environments, coating personnel spend a significant amount of their time in eliminating defects and trying to make the process defect-free. We have observed that while coating personnel may be trained in the basic sciences, there is very little formal training in troubleshooting or problem solving, even though it is one of the main functions of industrial personnel. The basic procedures and tools used to troubleshoot or to problem-solve are similar for a wide variety of different defects and problems.

EDGAR B. GUTOFF
EDWARD D. COHEN

October 1994

Acknowledgment

We wish to acknowledge with gratitude our many colleagues, coworkers, and friends whose help and contribution to this field enabled us to write this book. More specifically, we wish to acknowledge the assistance of Luiza Cohen in editing part of this book, and the suggestions of Dr. Semyon Kisler of Polaroid on Chapter 8, of Dr. José E. Valentini of DuPont on Chapter 7, and of Edwin A. Chirokas of Polaroid on Chapter 4. While acknowledging this help, the authors alone are responsible for every statement in this book. We also wish to thank Dr. Edward G. Denk of Polaroid for suggesting that we write this book.

The authors would appreciate being notified of any errors that may be found in this book.

E.B.G.
E.D.C.

About the Authors

Edgar B. Gutoff is a Consulting Chemical Engineer (194 Clark Rd., Brookline, MA 02146, phone 617-734-7081), specializing in the coating and drying of continuous webs. He also presents on-site coating seminars. He received his B.Ch.E. from City College of New York and his M.S. and Sc.D. in Chemical Engineering from MIT. In 1988, after 28 years at Polaroid, he left to do consulting. Since 1981 he has been a part-time Lecturer and Adjunct Professor at Northeastern University, and in 1994 became an Adjunct Professor at Tufts University. He organized the first biennial AIChE Coating Symposium in 1982, organized a coating course in 1990 given at these symposia and in alternate years at the University of Minnesota, and organized a course in Coating and Drying Defects at the 1994 Coating Symposium. Dr. Gutoff is a Fellow of the AIChE, and in 1994 received the John A. Tallmadge Award for Contributions in Coating Technology from the AIChE and also the Fellow Award from the Society for Imaging Science and Technology for long-term contributions to the engineering aspect of photographic coatings. With Ed Cohen he edited *Modern Coating and Drying Technology*, and with Paul Frost he coauthored *The Application of Statistical Process Control to Roll Products*.

Edward D. Cohen is a DuPont Fellow (E. I. DuPont de Nemours and Co., Central Science and Engineering, Bldg. 328, Box 80328, Wilmington, DE 19880-0328, phone 302-695-3921). He holds a Ph.D. from the University of Delaware and B.S. in Ch.E. from Tufts University. In his 30 years as a technical professional he has worked in both R&D and manufacturing. His research interests include coating process development, drying of photographic films, polyester base development, film defect mechanisms, and image analysis techniques for characterizing coated films. Currently he is a Research Leader in the DuPont Corporate Coating Technology Center. He chaired the 6th International Coating Process Science and Technology Symposium sponsored by the AIChE. He also is an Adjunct Professor at the Center for Interfacial Engineering at the University of Minnesota and teaches continuing education courses for the AIChE. He has several publications and is coeditor of *Modern Coating and Drying Technology*.

Gerald I. Kheboian, author of Chapter 10 (24 Fiske St., Worcester, MA 01602, phone 508-799-6038) holds an Assoc. E.E. from Wentworth Institute and a B.S.B.A. from Clark University. He consults in the field of web handling machinery and drives and tension controls that are applied to the web handling industry, and gives presentations and seminars on these subjects. His professional career began with 15 years at Rice Barton Corporation of Worcester, MA, where he was responsible for roll coaters, flying splice unwinds, color kitchens, and drives and controls. Jerry next joined Polaroid Corporation as a Production Engineer, and eventually became Technical Manager for Drives and Controls for Process Machinery and Chemical Systems. Since 1991 he has been chairman of the TAPPI Drives Short Course for paper machines and winders. Jerry has written a number of papers concerning the horsepower selection and control of coaters and mechanical factors of drive component selection. His extensive experience as a design engineer, startup engineer, and problem solver has led him into the field of consulting and manufacturers' representative for web processing machinery and industrial drives.

CHAPTER 1

Troubleshooting or Problem-Solving Procedure

In this chapter we discuss the problem-solving methodology and procedures that are needed to troubleshoot coating process defects. We present some guidelines for a systematic approach for identifying and defining the defect, for determining its cause, and for eliminating it. The basic assumption is that there is a defect or problem in a coated product on a specific coater and it must be removed. The problem can be a physical defect, such as chatter, ribbing, or streaks, or it can be an operational problem, such as the web not tracking properly in the coater and causing creases, or wet film exiting the dryer. The problem can also include production needs such as increasing the coater line speed, reducing solvent emissions from the film, or improving the performance properties of the product. The advantage to this approach is that some of the standard problem-solving techniques and principles can be used and adapted to the specific needs of the coating or drying process (Kepner and Tregoe, 1965; Brown and Walter, 1983; Pokras, 1989; Rubinstein, 1975; Kane, 1989; Branin and Huseyin, 1977).

A basic assumption here is that a structured process is the most effective method of solving the particular problem that is encountered. An unstructured approach has several disadvantages. It can result in duplication of effort from the lack of a clear focus. The problem can take a long time to solve if the problem solver jumps to conclusions as to the cause without the necessary supporting data. The first thought is often assumed to be correct, and then experiments are run that do not solve the problem. The next suggested cause is assumed to be correct, and it too may fail. Many possible causes can be tested before the correct answer is found. It also frequently occurs that the problem is never truly solved and the factors that cause it may go away on their own or by an accidental change. The problem will then probably reoccur, and the entire procedure will have to be repeated. In addition, most coating processes are sophisticated and involve many parameters. Some of these are known and understood and can be treated scientifically. Some are unknown or are less well understood, and these cause coating to be part art. The defects are often subtle and may result from small process variations or from interactions among several variables. It is essential to have a set procedure to follow to obtain and analyze all the data needed. The use of a formal procedure is more cost-effective than a random shotgun approach.

A systematic troubleshooting procedure may appear to be long and expensive, may strain available resources, and may not appear to be warranted for all situations. However, it is our belief that a formal methodology is effective and should be followed consistently. When necessary it can be adapted to meet specific needs. A major problem for an important customer who is on the just-in-time system and needs product immediately may require a large team that will follow all the steps. On the other hand, a small defect on one side of a web may require a part-time person who will skip some of the steps while still using the basic concepts. Our belief is that we should try to eliminate all defects and to determine the underlying cause of all problems, starting with those of highest priority. Even minor defects with a low loss should be studied. Far too often, when the history of a major loss is assembled, it is seen that the defect started with a low loss frequency that was not treated promptly and eventually built up in severity over a period of time. If the defect had been studied at the first occurrence, subsequent major losses could have been avoided.

The problem-solving process consists of an orderly series of steps. First one defines the problem, then gathers the needed information about the product or process, develops hypotheses as to possible causes, tests the hypotheses, and uses the correct hypothesis to solve the problem. If this sounds very much like the scientific method, well—it is. Since physical process are responsible for the defects, the scientific method is most efficient for solving problems.

The primary focus of this effort is on defects, since they are a major problem in a coater. Table 1-1 lists the steps in this process, and these steps form the general outline for this chapter. In general, the same procedure is used for all coating machines; for a laboratory or pilot coater, the steps are almost the same. The same steps also apply to any type of problem.

1.1 DETECT THE DEFECT

The troubleshooting procedure starts when a defect is first detected or a problem is encountered on the coater. Detection can occur through visual examination by operating personnel or by automatic defect inspection instrumentation. It can be in several locations, such as at the coating applicator, in the dryer, in the windup area of the coater, or upon visual inspection of the coated product.

The use of real-time automatic inspection based on laser reflections or transmission is becoming widespread on many coating lines. These inspection systems detect many defects as the product is coated. The systems give the exact location data on the web and can give information about the size, shape, and frequency of the defect. They are particularly effective in coaters where the web is difficult to view either because of subdued light, as in the case of photographic films, or because the coating machine and the dryers are enclosed. Also, at high coating speeds the nonuniformities can be hard to see. In addition, it is both stressful and boring to examine the coated web physically throughout a shift. Coating machines run continuously, and operating personnel have other functions to perform.

Defects originating in the coater can be also be detected in the finishing or converting area, where the product is cut to size and packaged. Both automatic laser

Table 1-1 Problem-Solving Procedure

1. Detect the defect, recognize the problem.
   - By on-line real-time inspection system.
   - By visual examination in coater.
   - By visual examination in converting area.
   - By customer complaint.
2. Define the defect or the problem.
   - Collect representative samples.
   - Describe the defect but do not name.
   - List and record the characteristic attributes of the defect.
   - Problem statement formulated.
3. Collect the data and analyze the problem.
   - Collect general process information.
   - Collect analytical data on the defect.
   - Collect analytical data on the raw materials.
   - Statistically analyze the process data.
   - Check the standard operating and quality procedures.
4. Analyze the data and identify potential causes of the problem.
   - Brainstorming sessions.
   - Problem-solving techniques.
   - Collect specialized data on process.
   - Conduct statistically designed laboratory experiments.
   - Run computer simulations.
5. Eliminate the problem.
   - Run production coater trials to test hypothesis.
   - This may take several trials, with returns to step 4.
6. Document the results.
   - Collect all data, charts from working sessions, analytical ressults.
   - Prepare report and issue.

inspection and human observation can be used. Typically, representative samples of the product are taken to check the quality of the final product as it is being packaged. The product can also be given use tests to verify the quality. In the case of photographic films, the product is exposed and processed to give the same type of image the customer will see. This can pick up subtle coating defects that are not detected otherwise. Another possible detection location is when the customer uses the product. Although the customer should not receive any poor-quality material, realistically this will occur at some level. A key point to remember is that the location of detection and the location where defect occurs may not be related at all. A spot first seen in the converting area can be caused by problems in the support manufacturing process.

Interestingly, one of the issues that may arise at this point is the lack of agreement by all functional personnel—manufacturing, quality control, research and development, marketing, and management—that there is a problem that needs to be acted on. In fact, some may deny that there is a problem. It cannot be assumed that everyone will recognize that there is indeed a problem. Often, comments such as the following are heard in the coater area:

"We have always seen this."

"There have always been chatter with this product."

"We always get bubbles for the first 2 to 10 rolls."

"We never could get conditions in that zone."

"We get these and always add some surfactant to fix it."

"It will go away in an hour."

"We need to ship product to meet an order."

"The customer will never see this."

Thus the first important step in the troubleshooting process is to ensure that there is agreement that there actually is a problem and that it needs to be solved. The key is communication to all concerned. Use the historical files or check the product and process specification to see whether the product or process is indeed within limits. The use of statistical quality control, ISO standards, and zero-defect goals all should help eliminate the denial phase and ensure agreement on all problems. However, denial is an inherent human reaction and will not go away easily.

Care should be taken at this step to avoid treating the problem and the cause as one step. They are very different steps, and treating them as one will lengthen the troubleshooting procedure and complicate the analysis. It is essential to define the problem rapidly and clearly and to get all to agree that there is a problem to be solved. A particular reason for doing this is that as the problem is being defined it will often turn out that there are several problems that require a solution, not just one. If you pass too quickly to the subsequent steps, this may not become apparent.

Another factor in obscuring a problem is that often gradual changes in a process may occur over a long period. When examined over a short period, these long-time changes can make the process appear to be stable. An example of this is as follows. One of the causes of fog, or high background density in photographic film, is the drying conditions. Thus conditions in the dryer must be optimized and closely controlled to maintain optimum film performance. Figure 1-1 shows typical levels of fog as measured on individual rolls of product for 3 days. On this time scale the process does not have a problem. However, Figure 1-2 shows the performance over 10 weeks; the gradual drift becomes apparent and the problem becomes obvious.

1.2 DEFINE THE DEFECT OR PROBLEM

After detection of the defect, representative samples of the defect should be obtained from a variety of locations and over different times in the process, or else from varying locations within the coated roll. When taking these samples the entire section of the web should be removed and the orientation of the web marked to a reference side of the coater. A large sample normally should be taken, as it is easier to discard an extra few feet of a sample than it is to try and retrieve more material from the scrap system. The samples should be clearly labeled with complete product identification, the time, and the machine being used. The large sample size will ensure that the true defect is seen and that a statistically valid sample is characterized. It is common to

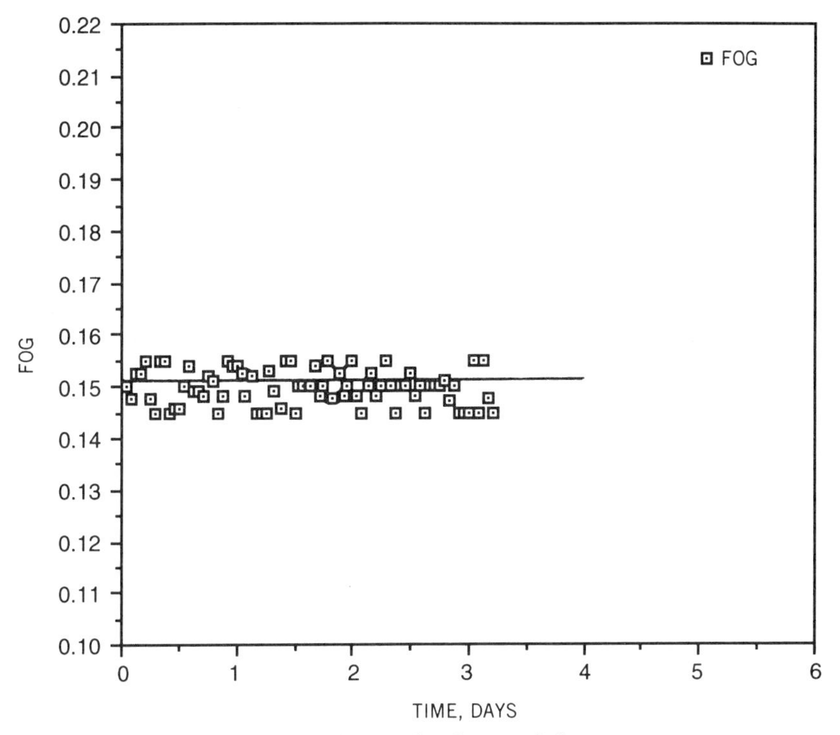

Figure 1-1. Fog levels over 3 days.

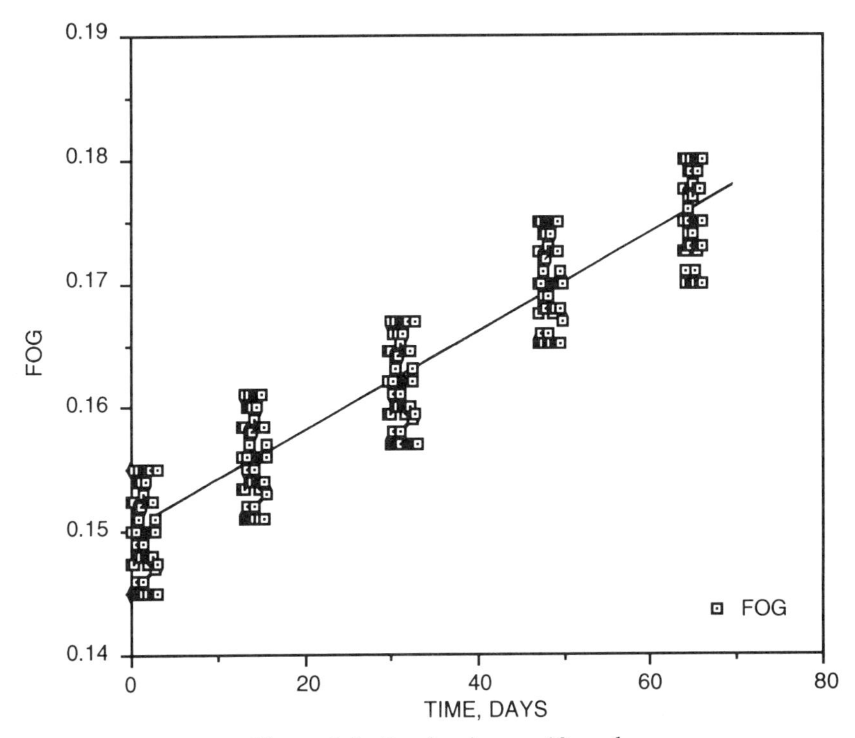

Figure 1-2. Fog levels over 10 weeks.

have several different defects in a coating, and time can be wasted in analyzing a defect that is not the most frequent or the most critical.

A technique that is particularly useful in the later stages of the troubleshooting process is to stop the coater and dry the coating in place. The web should then be marked by the distance from the coating applicator. Once dried, the web can be laid out and analyzed to determine where the defect first appeared. If it is present in the coating solution, it will be visible immediately after coating applicator. If it is caused by dirt in the dryer air system, the first occurrence will be in the dryer. If it is due to contamination on a roll after the dryer, it will not be visible until after that roll. Samples of the support without the coating can be very useful in the troubleshooting procedure and should be taken. Some defects in the final coated product are caused by defects in the support material. A web contaminated with grease will cause defects in the coating on top of it due to wettability problems. Similarly, dirt on the support surface can cause defects. These effects may not be uniform from roll to roll. A sample of the base can help give an insight into the problem. Also, it may not be possible to get a sample of the base later, when it might be needed.

1.3 NAME THE DEFECT

After the sample is obtained, the next step is to name the defect. This is a unique feature of the coating troubleshooting methodology and is critical for its successful implementation. The name may be given to the defect when it is first seen, but whenever it is done it is a key step. While giving a name to a problem is an obvious starting point for any troubleshooting procedure, it is a particularly important consideration for coating and drying defects because these defects often have rather nebulous, hard-to-define characteristics. There can be a wide variety of names for the same defect, and defects that appear similar can have a wide variety of causes. As a result, defects can easily be misnamed and this can have serious consequences. Typically, at the start of the troubleshooting procedure, after the defect is detected, the first observer gives it a name that is descriptive of the defect. A round defect with a clear center that looks as though it had a bubble in its center is a *bubble*, a spot with a center is called *dirt*, transverse direction marks are called *chatter*, a soft nonuniformity in the coating is *mottle*, and so on. In the spray coating of automotive finishes, *pinholes* and *microfoam* both describe the same defect, even though the names imply that the defects are very different. In coating and drying defects there are few tests to quantitatively define a physical nonuniformity such as a bubble or a streak, and there are no universally agreed upon naming conventions or standards as there are with performance properties such as color, hardness, coating weight or coverage, and sensitometry of photographic coatings. This leads to a wide variety of names for defects. It also leads to a variety of defects from many sources having the same name; examples are chatter and streaks, both of which refer to a wide variety of defects from many different causes. This can be seen more easily in Table 1-2, which summarizes some of the widely used common names of defects. These names have been obtained from papers presented at various symposia, from the patent literature, from general publications, and from discussions with many practitioners of the art of coating and drying.

It is our belief that many defects are initially defined incorrectly and that the initial

Table 1-2 Physical Defects

Continuous Defects		
ribbing	tailing streak defects	barring
chatter marks	herringbone	drier bands
streaks	snake air	bead breaks
transverse waves	chatter	weeping
rib pattern	waves	rivulets
wine tears	bands	cascade
Discrete Defects		
bubbles	gel slugs	cracking
thin spots	cell pattern	galaxies
pinholes	chicken tracks	crinkling
blisters	agglomerates	inclusions
Bénard cells	bands	buckling
orange peel	seagulls	air entrainment
picture framing	craters	bubbles with tails
starry night	blotchiness	dirt
fleets	fisheyes	hooks
cavities	mottle	bar marks
cracks	convection cells	shear
cockle	fat edges	randoms
volcanoes		

assignment of a name often implies an incorrect cause. This hampers the subsequent troubleshooting process significantly. This is a more important effect than is generally realized. The phenomenon has been called the *initial name syndrome* in order to point out clearly the importance in the troubleshooting process of correctly identifying a defect before naming it. The initial name syndrome quickly and incorrectly gives a defect a descriptive name that implies both a cause and a specific course of action to eliminate the defect. An incorrect name shifts the focus from the true cause of the defect and results in an ineffective troubleshooting process.

Bubbles are a good example. Once the defect is called a bubble, it implies that air entered the coating solution from a source such as a feed line leak, a poor seal on a pump, too high a mixing speed in the preparation kettle, or air entrainment at the coating stand. Then the initial effort will be focused on examining all these possible causes. If, however, the defect is caused not by an air bubble but by dirt falling onto the wet coating, much effort will be wasted in tracking down possible problems that could not possibly cause the defect.

Another reason for the initial name syndrome is that while defects should be eliminated rapidly on a modern high-speed coater by the operating personnel, the tools needed to analyze defects rapidly are often not readily available. A cursory visual examination, possibly microscopy, along with reference books of typical defects, are widely used to aid in identifying defects, but are not that accurate and can lead to incorrect names. This is particularly true if the operation personnel have other functions that may be of higher priority, such as keeping the coater running. Good preliminary characterizations can go a long way to avoid the initial name syndrome. However, conventional microscopy can be a time-consuming and tedious process, and it requires a skilled person to get good results on a microscope. In addition, manual searching of the past history to correctly identify a defect can also be tedious and often is not very useful. Scanning electron microscopy, EDX (energy dispersive x-ray), and ESCA (electron spectroscopy for chemical analysis) give accurate information but are time consuming. Note also that there is an inherent human tendency, when looking at defect, to associate the defect with a name that implies a cause.

The importance of the initial name syndrome became apparent to us as a result of an ongoing effort in the laboratory to develop improved photographic film test methods based on a low-cost modular digital image analysis system (Cohen, 1993). This digital image system consists of a microscope with resolution capability of 25 to 750× to magnify the defect, a television camera and frame grabber to digitize the image, a computer with software to analyze the image and to calculate quantitative parameters, a high-capacity storage device, and a database to store and retrieve the images.

This equipment was used to improve defect characterization (Cohen and Grotovsky, 1991, 1992, 1993). A motivating factor was the desire by the troubleshooting and product development teams for better and more rapid evaluation technology. Samples were submitted by them and we found that defects could be analyzed rapidly, resulting in much useful data to help in the troubleshooting procedure. As more samples from a variety of sources were analyzed, several observations were made:

1. The vast majority of the defects were submitted with a name already assigned by the first observer.
2. Approximately 50% of these initial names did not describe the defect accurately or correctly.
3. When the digital image analysis results were discussed with the persons submitting the sample, they consistently agreed with our conclusions.

The correct identification after the digital image analysis step significantly improved the troubleshooting process. The causes were more rapidly found, there was less wasted effort and, more important, problems were solved more rapidly. Figure 1-3 shows several examples of defects, all of which were submitted originally as bubbles. These defects are all different, as seen from this analysis, and were not caused by bubbles. The information provided by the digital image analysis led to more rapid discovery of the true causes and elimination of the defects.

The best method of avoiding the initial name syndrome is to avoid giving the defect a poetic or descriptive name that implies a cause at the initial observation. While giving an initial name is a normal tendency for all of us, the temptation should be resisted. Instead, use general terms to describe the defect and its attributes. Instead of calling it a bubble, describe it as series of round defects with no coating in the center. If the defect appears to be chatter, describe it as a series of coating nonuniformities running across the web. Add, for example, that the nonuniformities are light in the center, are $\frac{1}{8}$ in. wide, and are spaced at 2-in. intervals. At least several representative defects should be checked to ensure that the attributes are correct. After that, an innocuous name can be given, but care must be taken that the name does not imply a mechanism.

Although we have no desire to create documents and impose an additional burden

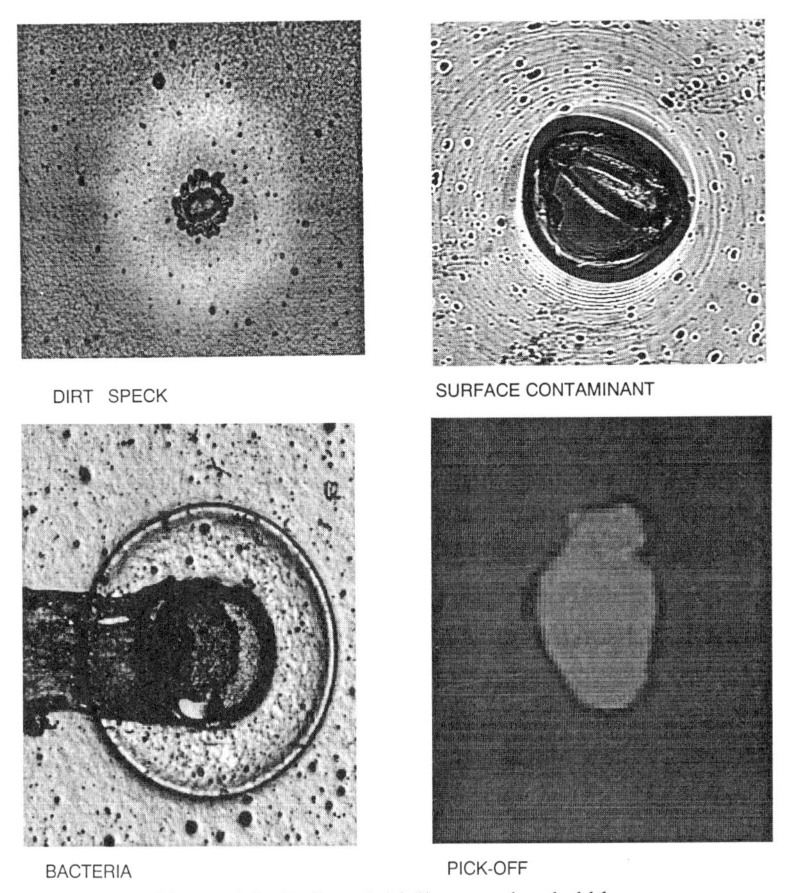

Figure 1-3. Defects initially named as bubbles.

on the operating personnel, at this point it is worthwhile to generate a form to record information gathered so far. This is very helpful for the analytical personnel and for the technical personnel who will become involved in the subsequent steps. Table 1-3 is a filled-in form which contains the essential information that should be available at this stage. This defect is a possible bubble problem.

This discussion was focused on the initial stages of the problem-solving process and its specific application to defects. These same principles should also be applied to the initial stages of any other problems that are encountered in the coater. The problem should be defined as generally as possible and should not be given a name that defines a cause and a cure. At this stage of the process, the problem solver should expand the consideration of the problem and be very inclusive. No possible sources of the defect or problem should be ruled out.

If the coating, for example, is coming out of the coater wet, define the problem as wet coating from the coater. Do not define the problem as being that the coating does not dry. The latter description will focus your efforts on the dryer, whereas the problem may in fact be a heavier coating weight or too high a solvent loading in the coating formulation. If a nonuniform density pattern is seen down the web, define the problem as nonuniform coating density. Do not define it as coating chatter, which

Table 1-3 Defect Initial Data Description

Product Coated

The product being coated was a B-210a vinylidene chloride barrier layer, formulation S-20, lot 45623, on a 1-mil polyethylene terphthalate substrate, S-400. The gravure coater applicator was being used at 400 ft/min.

Defect Description (attributes: size, shape, frequency, appearance, general description, etc.)

Round clusters of spots with no apparent coating in center. They appear to be $\frac{1}{16}$ in. in diameter and occur in clusters.

When and Where Observed

They were first observed on number 5 machine, at 10 p.m. They were first seen at the end of the coater. The initial indications are that they are only on the left side of the web, 2 ft. 6 in. in from the left side of the coater. They were occurring about every 20 ft. After the second roll they were continuous. Subsequent examination showed that they were occurring at the end of the applicator.

Samples Taken

Samples were taken at start, middle, and end of three rolls, Id 050293-2, 050293-5, 050293-9. (We suggest using the date and month as part of the Id. Thus Id 050293-2 means May 29, 1993, roll 2. The location can also be included.)

Operator________________________ Date______________

will cause you to focus on the coating applicator. A broader definition will allow you to look for other causes, such as web vibration, building vibration, substrate quality, and dryer effects.

We also recommend at the conclusion of this step that there be a definition of the problem(s) to which all involved have agreed. This should be written down and kept available to ensure the correct focus of the effort as the team proceeds with the subsequent steps.

1.4 COLLECT THE DATA AND ANALYZE THE PROBLEM

A major requirement for successful problem solving and troubleshooting of defects is to obtain the necessary information about the problem or defect and about the coating process used. Typically, there are many sources of information, and all the information has to be analyzed and integrated to understand the cause of the defect and how to eliminate it permanently. Although all coating processes have some specific characteristics, there is certain basic information that applies to a wide variety of coating techniques and is the basic building block of the data collection process. In this section we focus on the procedure to use in the obtaining of the information.

1.4.1 Process Information

In the next chapter we give more details of some of the special instruments and tools that are needed for this effort. As can be expected, there are many types and sources of information, and the availability will vary from coater to coater within the site, as well as from site to site and company to company. There is some general information needed for the initial analysis that we discuss in this section. The specialized data for certain types of defects are covered in the specific process sections.

First, one should obtain the information and measurements needed to characterize the process both during normal operation and when defects are observed. It is assumed that the coating process is basically as shown in Table 1-4. The process starts with the incoming web and the raw materials for the coating liquids. The next step in the process is compounding or formulating the coating solutions, purifying them, coating them, drying the coating, and then converting the material to the final shape and form. The process does not have to use a continuous web such as rolls of metal or polyester. The coating can also be applied to discrete parts, as in spray-painting cars or dip-coating individual parts.

Each of these unit operation or process steps influences the final coating quality and can cause the eventual formation of defects. Thus process information and data are needed from each step. The data and measurements needed should be made both while the process is running without defects, and while defects are occurring. In most instances a change in a specific parameter, such as viscosity or surface tension, may be more important than the actual level. It is difficult to interpret measurements and data obtained only when defects are occurring because then there is no idea of what the normal state should be. An understanding of the performance and characteristics of the process while it is making good product is as important as while making unacceptable material.

One question that could be asked at this point is why the terms *information* and *measurements* are both used. By *information* we include subjective observations by the people running the process and by those doing the testing, and these observa-

Table 1-4 Coating Process Unit Operations

1. Receive raw materials.
   - Active ingredients
   - Solvents
   - Coating aids
   - Substrate
2. Test raw materials.
3. Compound and formulate coating solution.
4. Purify coating solution.
5. Test coating solution.
6. Coat on substrate.
7. Dry to remove solvent.
8. Cure coating.
9. Inspect for defects.
10. Convert into final form and package.

tions may not be in the form of numerical data such as pH, surface tension, coating weight, and so on. A key effect in running a coating process is to observe all of the process, not just those parts that lead to conventional measurements and data. Subjective observations can and should be used in the troubleshooting procedure. They are often a key part because the cause of the defect may be a result of a change in a property or combination of properties. These observations then lead to a new measurement technique to quantify the property.

A basic rule of troubleshooting is that one can never have too much data. It is important to observe the process and record the key information. There is, however, also a danger of recording too much data and filling many file cabinets with never-used data. This balance needs to be determined, and the safest procedure is to err on the side of too much data. It is not possible to go back in time and get lost information. Murphy's law applies here, in that the time when you do not record key information will be the time of poorest quality and highest yield loss. The converse is not true, unfortunately, in that taking the data will not prevent defects.

One of the best ways to avoid defects is to detect the potential problems before they occur. It is much more efficient to scrap a batch of contaminated coating solution than to scrap the finished coating. The focus should be on properties, information, and characteristics that bear directly on coating quality and on defects. A main incentive for following the procedure is to arrive at a cause of the defect, and then modify the coating process so that if the same circumstances are again present, they are detected before the product is coated. The earlier in the process that the conditions leading to a defect are eliminated, the better the process will be.

General Process Information

Both quantitative and qualitative data can be obtained from the process. Quantitative measurements are obtained from process instrumentation, from measurements with portable diagnostic instruments, and from the quality control and analytical measurements that are taken in each step of the process. These data can be stored in many formats. With today's computers, it is suggested that these measurements be obtained automatically and stored in digital format on magnetic tape or on disk for later recall and analysis. Then the subsequent analysis is considerably easier since the data can be retrieved from a data analysis program and analyzed. Also, the same computer program that records the data can also recall and plot the data for the time period desired. Of course, the data can also be recorded by hand on log sheets or log books and then entered manually into a data analysis program from which graphs can be prepared. Graphs make the identification of trends much easier. Table 1-5 is a summary of some of the quantitative measurements that can and should be made on a process, and which will be useful to analyze when defects occur.

Not all of the important process variables are recorded by instruments. Qualitative observations made by operating personnel are also essential for the troubleshooting process. The state of the shipping container needs to be checked to ensure that it is intact and has not been damaged in transit. The outside of the container should be checked to see if there is anything unusual on it. Dirt from a container top can inadvertently be added to a formulation kettle leading to defects. The appearance of all raw materials needs to be observed and anything unusual noted as part of the data collection process.

Table 1-5 Typical Quantitative Measurements

Raw materials
  Formulating and Compounding
    Mixing conditions: time–temperature profile, agitation rate, agitator motor amperage, filter type, pressure buildup, milling conditions, etc.
    Coating solutions: viscosity, surface tension, temperature, density, percent solids, percent active ingredients, viscoelastic properties
  Coating
    Premetered coating applicator: pump rpm, pump size, pressure if pressurized feed, flow rate, applicator temperature, gap between coating roll and applicator, bead vacuum level (if used)
    Roll applicator: roll speeds, gaps, roll speed ratios, roll diameter, roll surface
    Dried coating weight, dry coverage of active ingredients
Dryer
  For each drying zone: total air volume flow rate, velocity in representative nozzle slots or plenum pressures, air temperature, solvent level delivered to nozzles, pressure drop across filters, solvent removal unit conditions, return air temperature and solvent level, makeup air conditions, air balance
  Location of end of falling-rate period, and uniformity across web
Unwinding, web transport, and rewinding
  Coater line speed, web tension in each zone, drive roll conditions
Miscellaneous
  Air conditions: temperature, relative humidity, and balance in coating room and in the unwind and rewind area

The support to be coated on should be checked to ensure that it is the right material and it should be tested occasionally to ensure that the correct side is coated. It is our experience that almost all coating operators and coating engineers have inadvertently coated on the wrong side of the base or on the wrong support at one time or another. If a precoated support is used, the coating quality should be checked as the roll is being mounted on the unwind stand and any unusual conditions noted. The quality of the unwinding roll is all important. An unbalanced roll can send vibrations throughout the web and this can lead to coating chatter.

The overall condition of the coating complex is important. Is the coater/drying room unusually dirty? If so, that can lead to dirt contamination in the dryer and dirt can be wound up in the coated roll leading to pressure defects. Is the coater unusually warm or cold? A cold stream of air blowing on a roll can lead to condensation and defects. Are there drafts of air in the coating room that have not previously been present? This can change the air balance in coating applicator room and blow dirt on the coating instead of protecting it. Are there unusual vibrations felt in the building or coming from motors and bearings? This can lead to chatter. Has any maintenance been done in the building recently?

1.4.2 Analytical Data

Analytical data on the apparently defective material, on the raw materials used to manufacture them, and on the normal defect-free material is necessary and is a major part of the troubleshooting procedure. Once samples of the defects have been collected, a complete physical, and if warranted chemical analysis should be conducted to determine the exact nature of the defect and how it differs from normal material. The objective is to collect information on the defect in order to determine the cause. If spot defects are present, the chemical analysis is needed to determine if there are any foreign materials present and what they are. These tests should include both defective and standard material. Often, the cause of a defect may be differences in the level of a normal component of the coating. If the product is exiting the coater wet, the coverage should be measured to ensure that it is within limits. In the next chapter we discuss the varying analytical methods in more detail.

The testing of the raw materials used for the coated product is a key step in preventing and troubleshooting defects. The major materials are tested to ensure product performance. However, it is worthwhile to check that they are indeed the correct material at the time of use. If a surfactant typically is a clear liquid and today it has white strands, something has changed. This should be recorded and consideration given to using a new batch. There are many raw materials that go into the final product. All should be tested. Coating formulations include polymers, binders, surfactants, cross-linking agents, pigments, and solvents. They and the coating support are the obvious raw materials. The testing of the raw materials used for the coated product is a key step in preventing and troubleshooting defects. The materials need to be tested to ensure that they are, indeed, the material indicated on the label and that the chemical composition, activity, and impurity level are all within specifications.

1.4.3 Analyze the Data and Identify Potential Causes of Problem

The next step in the problem solving and troubleshooting process is the analysis of the data collected to determine the cause of the defect. One must integrate all the information that has been obtained. This is the most difficult step in the process. If the previous steps have been followed, the answer should be available. All the information suggested in Section 1.4.2 should be available for analysis. The difficulty is to find and isolate the relevant important information from the useless and false leads. It is very important at this step to ensure that the process used and the logic and the thinking encompasses all possible solutions and does not focus too rapidly on the most obvious solution, which may not be correct.

We recommend using a team approach to identify the causes, to formulate hypotheses, and to test these hypotheses. This will bring a wide perspective and will help to ensure that all possible causes are identified. There is a definite synergy from having several people participate. The team can be small or large depending on the size of the problem and the available resources. For a large, high-value problem, the data are collected by a number of people, all of whom should obviously participate in the analysis process. Representatives from manufacturing, engineering, quality control, and research and development should all participate. The team for a smaller problem can be two or three people discussing the problem over coffee.

For a large team of four or more people a structured process should be used and a brainstorming technique is recommended. Give the group the statement of the problem that had been developed earlier, and have the group develop a list of questions that need to be answered or are suggested by the problem statement. This list of questions will then be the basis for the analysis step. As an aid, Table 1-6 is a general-purpose list of questions to start the problem-solving process and to stimulate thinking.

Table 1-6 Questions on the Defect

When was the problem first observed?
Had it been seen previously?
Is it present for all products on the coater?
Is it seen more or less frequently on a particular shift?
Is it seen more or less frequently at any time period?
Is it seen more or less frequently at a particular season of the year?
Does the outside weather have an influence?
Is the defect seen more at a particular point in the process?
How does the problem change with time?
Is anything else going on in the facility at the same time as the problem?
Is there anything different in the coater or storage area?
Is the area cleaner or dirtier then usual?
Is the temperature and humidity in the coater different then usual?
Is the defect more prevalent with a particular applicator?
Is the defect more prevalent with a particular product?
Are all the defects really the same?
How are the defects distributed?
Have any new constraints been imposed on the coater?
Are the standards the same now as previously?
Is a different person interpreting the standards?
Are there any new sources of raw materials? What is the age of the raw materials?
Have any changes occurred in the coater or converting area?
Has there been a recent shutdown?

This list contains both general questions and specific questions for coating defects. The list of questions should be expanded to incorporate the specifics of the particular problem. The next step is to answer the questions with the data that have been accumulated.

After the questions have been answered, the possible mechanisms or causes of the problem are developed. One of the many brainstorming techniques should also be used for this as well as for developing the questions. A good way to record all the comments is to use Post-it notepads or index cards and affix them to a wall in the room. A basic rule is to be positive and not reject any idea. Record everything and encourage all to contribute anything that they think is reasonable. After every comment is recorded, they can be combined then and priorities assigned for testing the hypotheses.

A very useful methodology for the analysis phase of the procedure is the worksheet used by Kepner and Tregoe (1965). It provides for a sequential procedure to focus the information that has been collected and to help identify the changes that have led to the current problems.

There are four key elements to this worksheet, illustrated in Table 1-7. In the columns one puts down these four basic attributes of the problem or defect:

1. What is the deviation?
2. Where did it occur?

Table 1-7 Kepner and Tregoe Problem Analysis Worksheet

	Defect Is:	Defect Is Not:	What Is Distinctive in This
What	Bars across web Nonuniform-density chatter bars	Spot defects Streaks	Vibrations will induce hydrodynamic instabilities and cause chatter
Where	Across web	At edges only At center only	Uniform source of vibrations Uniform transmittal to web
When	At start of coating roll	During balance of roll End only	Vibrations only at coating start Vibrations disappear
Extent	Severe at start Decreasing severity to 200 feet	Not uniform Same intensity	Coating start hardware active in this time
Possible Suggested causes			
Coating start hardware malfunctioning			
Coating start hardware conditions set wrong			
Coating roll bearings defective			
Vacuum initiation source			

3. When was it seen?
4. To what extent is it present?

The top row gives the heading for the columns. The first column is for what is true about the attribute, and the second is for what is not true. This information is drawn from what has been collected in the initial part of the troubleshooting procedure. The third and last column, which is sometimes omitted, is to identify what is distinctive about the information, based on the first two columns. Finally, at the bottom of the worksheet, possible causes of the defect are listed.

For example, if there is a spot that occurs on the right side of the web, under *Where*, you indicate that it is on the right side and is not on the center or left side. This then leads to consideration of the process and to what is distinctive about the right side. It could be air ducts, or spargers, or web aligners, and then these have to be checked.

Variations of this form can be used. For a large group working on a complex problem, each point can occupy several pages. A simple one-page form, as in Table 1-7, can be used by the troubleshooter working alone. The format is flexible and can be modified as needed. Several examples are shown in Kepner and Tregoe (1965).

Table 1-7 is based on a chatter problem in a coated photographic film product. Chatter was found on an increasing number of rolls, and a small team was assigned to identify the cause and eliminate it. Table 1-7 is the worksheet as filled out after the data on the problem were assembled. It was used to organize the information that was obtained. The *What* was chatter or variable-density bars, which were uniform across the web. The *Where* was across the entire web, and not random, or only at the edges or at the center. *When* was only at the start of coating. The extent of the chatter was severe at the start and then fell off rapidly in intensity. The uniformity of the chatter suggested mechanical vibrations as the source, not hydrodynamic instabilities. Since it occurred only at the start, this suggests possible malfunctions of the coating start hardware. This hardware was checked and it was found that a part was misaligned when initially activated and caused the vibrations until it finally aligned itself into the correct position. Other causes were also suggested and are listed at the bottom of the worksheet, and these would have been checked if the first, most likely cause was shown not to be the problem.

Typically, the analysis will lead to several possible causes which have to be pursued either by obtaining more data or by running experiments in a laboratory coater, or even on the production facility. This can be time consuming and inefficient. It is easy to obtain an enormous amount of data very quickly and spend many hours in generating plots. You can end up more confused than before and can drown in the mass of data. The availability of personal computers and easy-to-use computer software programs can help the analysis, or, on the other hand, they can complicate the analysis and hinder the solution. Computers definitely make the job easier by providing more sophisticated analysis tools and reducing the time to prepare plots and calculate statistical parameters. Computer plots are faster to make and are more accurate than hand plots. However, if you are not careful, more plots can be prepared than can possibly be analyzed. This can and will slow down the troubleshooting process. All the computer does is simplify the mechanics; it does not analyze and formulate possible defect-causing mechanisms. It is not a replacement for creative thinking.

The analysis function can be simplified by recognizing that there are only two basic causes of problems or defects in a coater. The specific problems at hand can be based on two mechanisms: (1) something in the process has changed, and (2) an underlying physical principle is being violated. The authors believe that the majority of the problems that are typically encountered in production are in the first category. The product was being made without the defect and something has changed, causing the undesired effect. By the time a product is in normal production, all of the problems due to underlying science and physical principles have been eliminated. Thus the goal of the analysis procedure is to focus initially on what has changed in the process that could have led to the defect. This does not mean that the basic physical principles are not important to coating and can be ignored. An understanding of these principles is needed as a guide in looking for the changes and to determine if the changes identified can possibly have the effect that has been observed. Both of these mechanisms work together to guide you to a rapid solution.

This concept can be illustrated in several typical problems that are often encountered in a coating process. When coating chatter is observed, the possible changes that can lead to this are: (1) changes in vibrations at the coating applicator and web, (2) changes in the coating solution rheology, and (3) changes in the coating applicator conditions.

Thus data collection and analysis should focus on vibrations in the web, the building, at the coating applicator, and in the roll bearings and the feed systems. In addition, data on viscosity and coating weight data and on changes in the applicator conditions, such as vacuum level and die-to-coating web gap in slide coating, need to be obtained.

When bubbles are detected, the solution preparation kettles, pumps, and feed lines should be analyzed to determine if changes have occurred that lead to air leakage into the coating solution. If the film is leaving the dryer wet instead of dry, changes have occurred in the solvent load into the dryer or the solvent removal capability.

Correct identification of the defect or problem is a key step in this process, as described previously. It determines the possible changes to look for in the process and should be the first step in the analysis process. The defect should not be named until after it is characterized. In some cases it is possible to name the defect, such as chatter or bubbles, rapidly and correctly. For other defects, particularly spot defects, extensive analyses may be needed to identify them correctly. There are a wide variety of causes for spots, and many are due to contamination. Thus analytical identification of the contaminant is needed to determine possible sources and where to look for the changes.

Once the team has established the potential causes, laboratory or plant experiments are needed to verify the mechanism. This can be done by using different lots of raw materials, a different base, changing bearings, adjusting airflow to the dryer, adjusting solution viscosity, and so on. In all these cases the initial goal is either to prove the hypothesis correct or incorrect, that is, that the factor can lead to the defect or that it cannot lead to the defect. These experiments should be designed to gain a solid base of information about the defect. Several rounds of experiments may be needed. Statistically designed experiments are highly recommended to improve the quality of the results and the efficiency. The differences between good and bad product are often subtle, and the precision and accuracy of statistics are needed. Well-designed experiments are far more cost-effective than nonstatistical experiments.

A major error in tactics at this point is to try and make one good coating and resume production by combining all of the identified changes in one "Eureka" cure. Often, the defect is solved by this approach and everyone is satisfied that the task is apparently concluded successfully. Then there is no more effort. It could be that the changes solved the problem; it is also equally probable that they did not and the cure was a result of an unknown fortuitous change. Since the cause of the problem is unknown, it will usually reemerge and then the entire process must be repeated at considerable expense.

The two basic mechanisms for causing defects also apply to the laboratory coater. However, in the development of new products and new processes, it is important to define the basic physical principles involved and then operate without violating them. Once this has been done, any changes become the prime factor to be considered.

Computer simulations can also play a key role in analyzing problems and identifying solutions. Programs are available that can simulate coating flows for several coating processes. These can be used for problems such as chatter, ribbing, streaks, and unstable coating beads (Grald and Subbiah, 1993; Subbiah et al., 1993). It may be more effective to simulate the process change and gain an understanding of the underlying principles using a computer model as opposed to extensive coatings. The role of these programs is a prescreening tool. It will not replace some level of laboratory and plant experiments.

1.5 ELIMINATE THE PROBLEM

To summarize, at this point the problem or defect has been defined, the data collected, possible causes identified, and statistical laboratory experiments run to verify the hypotheses that have been developed. In addition, computer simulations may have been run to help define the mechanisms. The next step, then, is to select a preferred solution and take it into the production coater and implement it. Again, the use of a statistically based experiment is recommended to verify the laboratory results and to determine the optimum solution. When this experiment is run, extensive data should be obtained in the event that the problem is not completely solved. It may be necessary to go back to the preceding step and reanalyze the problem with the latest data.

1.6 DOCUMENT THE RESULTS

Even though the problem has been solved, it is important that a closing report to document the results be completed. Our experience in troubleshooting defects has shown that a significant percentage of defects will reoccur at some time in the future and their causes will be similar. In addition, we have observed that coaters and products on these coaters tend to have certain patterns of characteristic defects. Some products are prone to streaks and others are prone to agglomeration and spot defects. Some machines tend to give chatter while others tend to give similar types of winding problems. If complete information is easily available from previous problem-solving exercises at the start of the current problem, then when they do reoccur, this historical information can simplify the current procedure. Also, at this point in the problem-solving procedure, a great deal of time, effort, and money has been expended to obtain a wide variety of diagnostic information about the coater and the defect. It is cost-efficient to record the information so that it has ongoing value. This documentation will also help in the future operation of the coater and can also be a very useful training tool for new personnel.

There may also be preliminary progress reports presented as the problem-solving procedure is occurring. These reports can be in the form of weekly reports or of team reviews for management. It is a common practice in oral presentations to provide copies of the visual aids used, which then become a circulated record. Often preliminary data and possible mechanisms are presented in these preliminary reports, which turn out not to be true when the problem is solved. However, the early report may be found later and the reader will not be aware of the subsequent change. One final official report will prevent this and will be of ongoing value.

Documentation is often not considered as part of the problem-solving procedure for many reasons. The personnel involved have put in extra hours and effort in solving the problem and are expected to get back to their regular assignments or to solve the next problem. Also, many scientists and operating personnel are not comfortable with report writing and do not believe that it is as valuable as running a coater or doing experiments. Reports are often viewed as a research-type function and do not have a role in manufacturing operations. The mechanics of report writing can be a chore and very time consuming, particularly if the report is handwritten and then has to be retyped for legibility. There may not be the personnel to do the retyping, duplication, and distributing of the report. Typically, the work was done by several people, and writing a report in which many people contribute can lengthen the process to the point where the report never issues.

The final closing report is an important part of the problem-solving process and we recommend that it be made a part of the process. The report preparation does not have to be a major task and can be accomplished very easily by using the following guidelines:

1. Issue the report within 30 days after the problem is solved.
2. The preparation and issuing of the report are the responsibility of the team leader and are his or her last official functions on the team.
3. A report with 80% of the information issued on time is much better than a report with 95% of the information that never issues.
4. Prepare progress reports and summaries as the work is evolving. These can then be easily edited and modified for the final report.
5. The final report should be simple and easy to follow. A good format is to have an overall summary at the beginning. A couple of paragraphs on each problem-solving step will provide the reader with a good overview of the problem and its solution. Subsequent sections can be the details from each of the team members, analytical reports, defect samples, and so on.
6. The report can include copies of appropriate material and does not have to be all original writing.
7. Include the names of all team members and participants so that they can be contacted to discuss results.

REFERENCES

Branin, F. H., and K. Huseyin, *Problem Analysis in Science and Engineering*, Academic Press, New York, 1977.

Brown, S. I., and M. L. Walter, *The Art of Problem Posing*, Franklin Institute Press, Philadelphia, 1983.

Cohen, E. D., "Troubleshooting film defects: the initial name syndrome," presented at *IS&T 46th Annual Conference*, Cambridge, MA, May 1993.

Cohen, E. D., and R. Grotovsky, "Digital image analysis of photographic film defects," in *Symposium on Coating Technologies for Imaging Systems*, J. Troung and E. Cohen, eds., Society for Imaging Science and Technology, Springfield, VA, 1991.

Cohen, E. D., and R. Grotovsky, "Digital image analysis for coating defects," *XVIIIth International Conference in Organic Coatings Science and Technology*, Athens, 1992.

Cohen, E. D., and R. Grotovsky, "Application of digital image analysis techniques to photographic film defect test methods development," *J. Imag. Sci. Technol.* **37**, 133–148 (1993).

Grald, E. W., and S. Subbiah, "Analysis of the reverse roll coating process with NEKTON," presented at *IS&T 46th Annual Conference*, Cambridge, MA, May 1993.

Kane, V. E., *Defect Prevention: Use of Simple Statistical Tools*, ASQC Quality Press, Marcel Dekker, New York, 1989.

Kepner, C. H., and B. B. Tregoe, *The Rational Manager: A Systematic Approach to Problem Solving and Decision*, McGraw-Hill, New York, 1965.

Pokras, S., *Systematic Problem-Solving and Decison Making*, Crisp Publications, Los Altos, CA, 1989.

Rubinstein, M. F., *Patterns of Problem Solving*, Prentice-Hall, Englewood Cliffs, NJ, 1975.

Subbiah, S., J. Gephart, and B. Peppard, "Application of the spectral element method to modelling slide coating flows," presented at *IS&T 46th Annual Conference*, Cambridge, MA, May 1993.

CHAPTER II

Coater Diagnostic Tools

To be an effective problem solver, a working knowledge of a wide variety of diagnostic techniques is needed to provide the necessary information about the coater and the defects being studied. This information must be accurate and, when used in the problem-solving procedure, will help to eliminate the defect. In this chapter we review some of these specialized methods and show how they are used to provide information about the unique problems that arise in a coating operation. Each of the methods or tools to be described is well known and has been described extensively in the general literature and in formal courses. Our intent in this chapter is to describe the methods that are particularly useful use for coating problems and to focus on the specific type of information on coating defects that they will provide.

Problem solving is a discipline in the same way as chemical engineering, physical chemistry, physics, and so on, and requires its own tools to meet its objectives. To treat a sick patient, a physician requires specialized tools to obtain measurements on the patient's condition and to diagnose the illness. A stethoscope is used to listen to the heart, a sphygmomanometer to check the blood pressure, an ophthalmoscope to check the eyes, x-rays and ultrasound to see inside the patient's body, and a thermometer to check the temperature of the patient. These tools provide the physician with the measurements and data that are needed to diagnose and treat the disease correctly. The effective coater problem solver also needs specialized measurement tools to assess the status of the coater and to determine the cause of the problem. Over the years the authors have developed a working knowledge of a variety of diagnostic tools to diagnose and correct coating problems. In fact, one of the authors has a collection of portable instruments in his lab used to diagnose problems that are available for loan to other problem solvers. These are engraved with the author's name so that they can be returned when inadvertently left in the dryer or in a corner of the coating alley.

The three basic categories of diagnostic tools that are needed are:

1. Analytical methods to characterize the coating defects
2. Instruments to obtain more information about the overall coater behavior
3. Statistical techniques and computer technology to analyze the data and extract useful information from the measurements and coater operational data

Table 2-1 Basic Components of Problem-Solving Tool Kit

- Analytical techniques to characterize defects
  - Microscopy
  - Scanning electron microscopy and energy-dispersive x-ray
  - Electron spectroscopy for chemical analysis
- Coater characterization instruments
  - Rheometers
  - Portable data loggers
  - Air velocity meters
  - Temperature measurement devices
    - Noncontacting infrared thermometers
    - Resistance-temperature devices
    - Thermistor thermometers
    - Liquid-filled and bimetallic thermometers
    - Thermocouples
  - Tachometers
  - Tape recorder
  - Camcorder
- Data analysis techniques
  - Statistical methods
  - Computer analysis software
  - Experimental designs

Table 2-1 describes these tools in more detail and is also an outline for the material in this chapter. The basis for the selection of these tools is our own experience in problem solving and troubleshooting defects in film coaters. These are the tools we have used and have found to be consistently helpful over many years. There are a wide variety of analytical tools, computer programs, statistical methods, and instruments that are available from a variety of manufacturers. All of these can be used and will provide needed information. It is not our intent to rate these instruments as to effectiveness or to endorse specific vendors and equipment. We will cite specific instruments so that the reader is aware of how they function and of the information they provide to help the problem solver build his own collection of tools. Much of the selection is individual preference due to familiarity gained by using the tools over a period of years. If an analytical method or instrument is easier to use or gives better information, use it. A limited amount of data taken with a usable instrument is much better than no data at all. An inexpensive instrument that is used is better than an expensive, sophisticated instrument that is never used.

2.1 ANALYTICAL METHODS TO CHARACTERIZE DEFECTS

To eliminate a defect in a coating it is necessary to analyze the defect to determine its cause and source. The goal of this analysis is to rapidly determine the uniqueness of the defect compared to the main body of the coating. A complete and thorough analysis using a wide variety of techniques is the ideal state; however, there are practical factors that dictate a more selective approach to the analysis. Although there are many useful and expensive surface analytical methods available, it is neither possible nor economically feasible to run all these tests even when the equipment is available. There are many occasions when a rapid answer is needed to meet the needs of production and to get the coater running again. The test methods should be selected for the best overall value at the lowest cost. A partial analysis obtained quickly may be far more helpful than a detailed and elaborate analysis that takes several weeks. The analysis is only part of the overall procedure, and a partial analysis coupled with the other observations will often lead to a solution.

A practical example of this concept occurred in an adhesive coating problem that one of the authors was involved with early in his career. An adhesive layer was being coated on a metal support for a photopolymer printing plate. In the early stages of production, spot defects were found all over the web. Simultaneously, the filter pressure climbed to the point where the filters were plugged. Production was stopped, the filters were changed, a new batch of adhesive prepared, and production resumed. Again, the defect was seen after a short time. At that point technical assistance was requested by manufacturing and a small team was assembled. The problem was defined as agglomerates in the coating solution. Possible causes were incomplete dissolution of the adhesive, coagulation of the adhesive during pumping and holding, or contamination.

The first steps to gain information were to obtain the used filters to determine why the pressure was increasing, and to obtain samples of the coated defect for analysis. The filters were found to contain a white polymeric-appearing contaminant with the colored adhesive. The defects in the coating were also found to have the same contaminant in them. None of the adhesive normally contained any white components, which indicated a foreign contaminant. The major component of the adhesive was obtained from an outside vendor who had indicated that the equipment was used for many other syntheses. We had a list of these, and one that stood out as having similar properties to the white component was a cellulose nitrate–based lacquer. At that point more samples were in the process of being obtained and were to be taken back to the lab to be characterized by infrared spectroscopy. However, a quicker test was found. The contamination was collected and placed in an ashtray, and touched by the flame from a match. The material exploded. This was good evidence that it was indeed cellulose nitrate, a known explosive. The hypothesis was that the batch we were using had been contaminated in the manufacturing process. A call to the vendor confirmed that the batch previous to the one we were using was indeed made with cellulose nitrate. An earlier lot was used next for the solution preparation, and it coated without any problems. Subsequent spectral analysis did confirm that the material was indeed cellulose nitrate.

This is an example of the problem-solving procedure and how it can lead to a successful solution to a defect problem. Complex sophisticated analytical methods are not always needed. The microscope coupled with a simple analytical test can often lead to a successful solution in a very short time.

We recommend that the analysis be done in two stages: a preliminary analysis to gain some rapid characterization data followed by a more detailed analysis. The first step can be done in the quality control laboratory of the production facility or in the inspection area, with only limited instrumentation. It may give the operating personnel enough data to solve the problem. This first stage also helps determine the type of tests, if any, that are needed in the second stage. A more selective approach to the second stage will be far more efficient than running all the tests simultaneously.

For the first stage of analysis a good light microscope is the basic tool. The unaided eye is a poor analytical instrument, but the eye aided by the magnification of a microscope is excellent in determining the characteristics of a defect. Table 2-2 summarizes a wide variety of different types of microscopic methods and gives some of the characteristics of each. The advantages of the light microscope become very apparent. It is rapid and low in cost and has a good magnification range for the defects usually encountered in coating. The advent of enhanced video microscopy in the late 1980s has greatly advanced the utility of this technique. In video microscopy, the eyepiece of the microscope is replaced with a television camera and monitor (Maranda, 1987). Instead of viewing through an eyepiece, the image is viewed on a high-resolution monitor. This considerably reduces the eyestrain involved in looking at many samples. Magnifications up to several hundred times can be obtained from instruments available at reasonable prices. Microscopes with a variety of viewing modes (transmitted, reflected, or incident) should be used. These systems are relatively low in cost and are rugged enough to be installed in a coater and used by operating personnel. An additional advantage is that hard copies of these images can be made either from the monitor screen or by a special attachment on the microscope. The Polaroid photographic system is particularly useful for this application because the copy can be seen immediately and checked that there is a good record, before going on to the next sample.

A further enhancement of the microscopy technique is to integrate the microscope into a digital image analysis system. These systems provide a quantitative analysis capability and electronic storage for the images being studied. In these systems the image from the television camera is digitized, stored in a computer, and then quantitative characteristics of the defects are calculated using specialized software programs (Cohen and Grotovsky, 1993). There are a wide range of systems that can be assembled for this use, depending on the need and the funds available. Figures 2-1 and 2-2 show two typical systems that are easy to use and sufficiently rugged for a manufacturing operation. An additional advantage is that since these images are stored in computer-readable format, they can easily be recalled for study and comparison with previous samples. An advantage of all microscopic techniques is that they can be run rapidly and provide a fast result.

Microscopic techniques are very useful for discrete coating defects and those with moderately sharp boundaries, such as spots, streaks, ribbing, chatter, scratches, bubbles, contamination, and a wide variety of defects with poetic descriptive names such as "fleets." Microscopy is not very useful for subtle defects that have vague boundaries, such as mottle. However, microscopy is useful in selecting representative defects for the second-stage analysis. Often, when discrete defects are seen in a web, there are many of them and they may not all be the same type. It is more efficient to screen the defects with a conventional microscope and select those defects that are the most representative for further analysis. With conventional microscopy, an individual defect can be studied and characterized in 5 to 10 min. The second-stage techniques require about an hour per sample and are more expensive to run. This prescreening will ensure that the proper defects are characterized while reduc-

Table 2-2 Microscope Capabilities

Type	Acronym	Magnification Range	Resolution (nm)	Analysis Time[a] (s)	Pressure	Cost (U.S. Dollars)
Light microscope	LM	5–2,000	200	10	Ambient	2000–100,000
Scanning electron microscope	SEM	10,000–250,000	5	60	Vacuum	40,000–400,000
Scanning electron microscope field emitter	SEM/FE	10,000–40,000	1.0	90	Vacuum	200,000–600,000
Transmission electron microscope	TEM	1000–1,000,000	0.5	300	Vacuum	100,000–5,000,000
Scanning TEM with selective area diffraction	STEM/SAD	1000–1,000,000	0.15	30	Vacuum	100,000–5,000,000
Scanning laser acoustic microscope	SLAM	10–500	1000	60	Ambient	600,000–800,000
Confocal scanning microscope	CSM	100–10,000	0.1	20	Ambient	50,000–200,000
Scanning tunneling microscope	STM	5000–10,000,000	0.1	300	Vacuum	30,000–200,000
Atomic force microscope	AFM	5000–2,000,000	0.5	600	Ambient	75,000–200,000

Source: Based on Studt (1990)
[a]Time to acquire image.

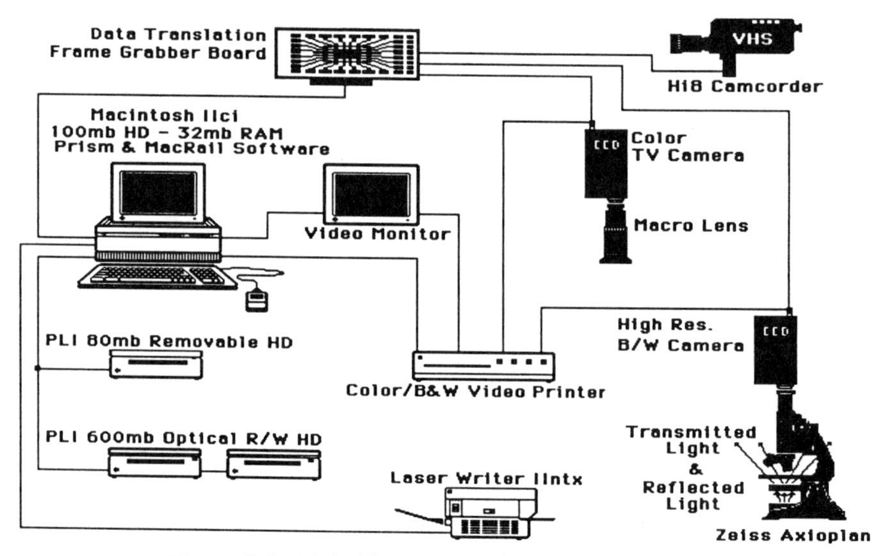

Figure 2-1. Digital image analysis microscopy system.

ing the costs and the time involved. Examples of defects that have been analyzed are shown in Figures 2-3 and 2-4, including some of the range of magnifications and the quantitative properties of the defects that can be calculated.

The second stage of the analysis uses a wide variety of surface characterization techniques to characterize the surface chemically and physically. Tables 2-2 and 2-3 show some of the more sophisticated analytical methods that are available for a detailed characterization of the surface. Of these, the scanning electron microscope (SEM) is the most useful for a wide variety of typical coating defects. SEM provides very high magnification and resolution and gives a three-dimensional view of

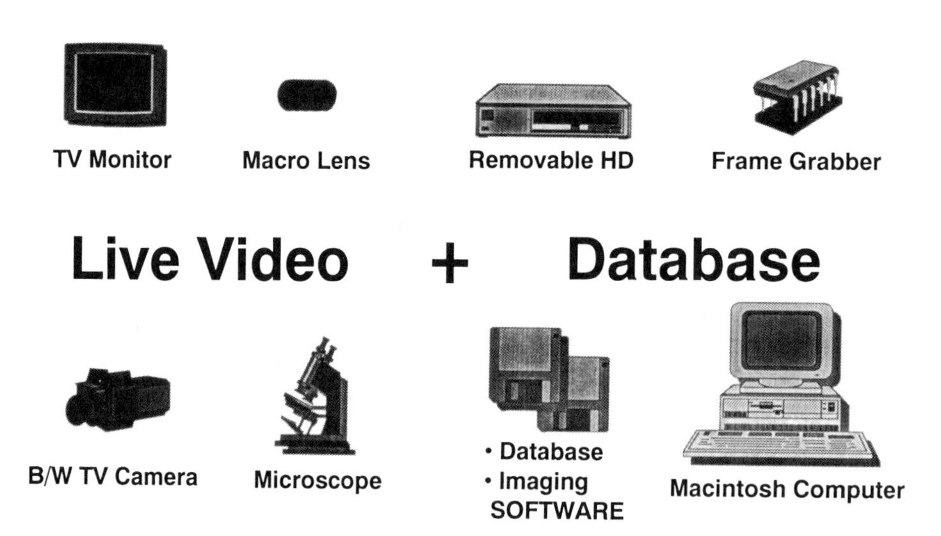

Figure 2-2. Digital image analysis microscopy components.

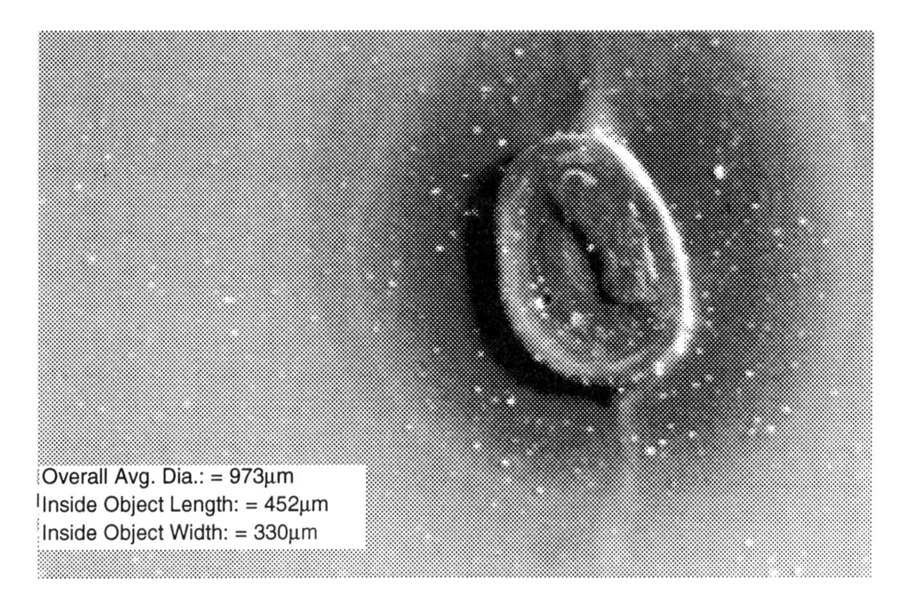

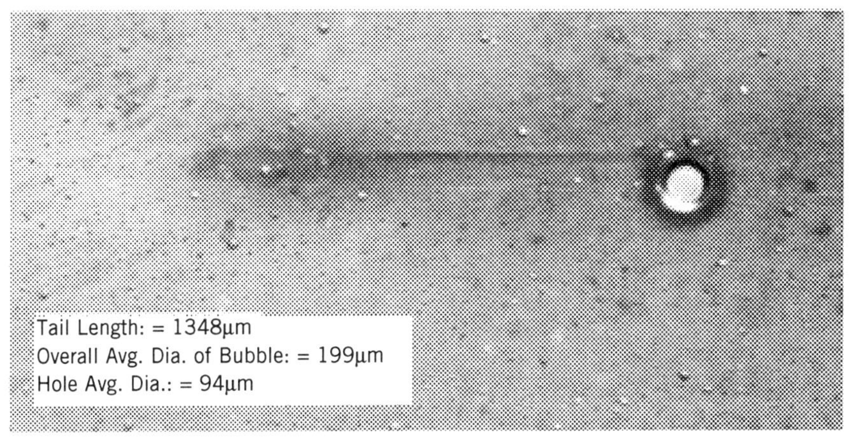

Figure 2-3. Defects as seen by digital image analysis: (*a*) particle contamination; (*b*) bubble with streak.

the image. All the needed detail can be seen. An additional advantage of this method is that an energy-dispersive x-ray analyzer (EDX) can easily be integrated into the system and used to measure the chemical composition, both qualitative and quantitative, of the defect and the normal sample area. The optical image in an SEM is developed by irradiating the surface to be studied with an electron beam in a vacuum, collecting the scattered electrons, and developing an image on a television monitor. These electrons also create x-ray emissions that can be analyzed to determine which elements are present along with an estimate of their quantity. The limitation is that only the elements heavier than sodium can be determined (Sawyer and Grubb, 1987). Organic compounds cannot be detected. However, there are other readily available methods to characterize organic compounds.

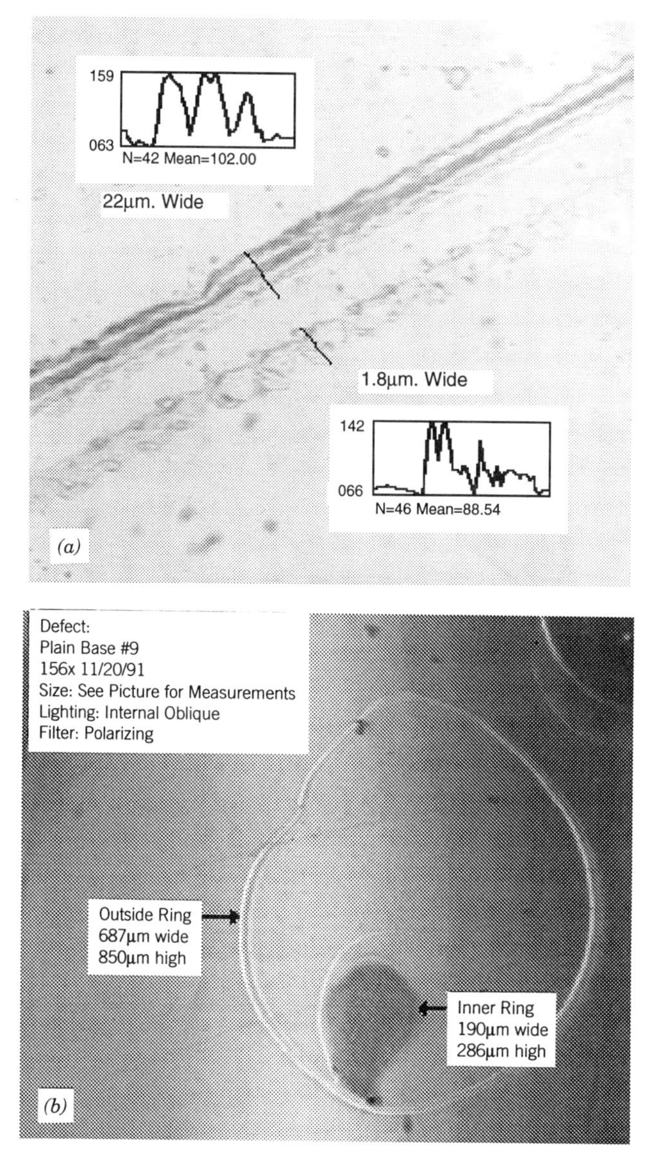

Figure 2-4. Defects as seen by digital image analysis: (*a*) scratch; (*b*) bacterial contamination on base.

In SEM/EDX the visual image and the x-ray images can be combined and presented in a variety of formats. An image of a particular element can be obtained and its composition displayed throughout the sample. Another format is to scan the beam in a single line both across the defect and through a normal sample. This will show how the element distribution varies from the standard material to the defective material. This combined image will show the physical appearance of the defect and which elements are associated with specific features of the defect. Also, if there are contaminants or particles present, the SEM and EDX can be combined into one

Table 2-3 Surface Chemical Analysis Instruments

Type	Acronym	Detectable Elements	Sample Size (cm)	Analysis Time[a] (s)	Depth of Resolution (Å)
Scanning auger microscopy spectroscopy	SAM	Lithium to uranium	20	10–100	50
Electron spectroscopy for chemical analysis	ESCA	Lithium to uranium	20	10–100	50
Secondary-ion mass spectrocopy	SIMS	Hydrogen to uranium	20	100	40
Energy dispersive x-ray	EDX	Sodium to uranium	2	100	50,000
Fourier transform infrared spectroscopy	FTIR	Organic functional groups	7	10–500	10,000
X-ray fluorescence spectroscopy	XRF	Magnesium to uranium	20	500	2×10^8
Laser ionization mass spectrocopy	LIMS	Hydrogen to uranium	20	10	20,000
Surface analysis by laser ionization	SALI	Hydrogen to uranium	2	50	20,000

Source: Based on Studt (1990), Karasek (1975), Karesek and Laub (1974), and Willard et al. (1974).
[a]Time to acquire image.

image. The images obtained from these techniques can be stored digitally and can be added into the defects catalog, which is described in Chapter 11.

The SEM/EDX technique is powerful since it combines both the physical and the chemical elemental analysis into one measurement. It is very useful for defects with discrete boundaries, such as spots and defects that are suspected to be from foreign contaminants. Once the chemical elemental composition of the defect is known, the source can be determined and eliminated. To help in the identification process, the composition of potential contaminants from the coater can be characterized and then compared with the defect to give a definite identification. In addition, samples can be made in the laboratory which have the suspected contaminants deliberately added to the coating composition. These can then be analyzed by SEM/EDX. There is also some utility in analyzing diffuse defects if the areas can be identified and are in the size range of the SEM. It can also be used to study surface effects, such as surface haze, orange peel, and reticulation.

A good example of the utility of this method is in the identification of metallic contaminants which can get into a coating. The metals used in the bearings, idler and coating rolls, pumps, and ducts are all different and have unique compositions of iron, nickel, chromium, carbon, molybdenum, and titanium. The following are some specific examples of the composition of stainless steels which are typically found in coaters (Perry, 1963):

Alloy	Composition, nonferrous (%)	(balance is iron)
Stainless, 17-7 PH	Cr 17 Ni 7 Al 1	C 0.09
Stainless, 17-4 PH	Cr 17 Ni 4 Cu 4 Co 0.354	C 0.07
Stainless, 316	Cr 18 Ni 11 Mo 2.5	C 0.10
Stainless, 304	Cr 19 Ni 9	C 0.08

If metal is found in a defect, its composition can be compared with known sources to aid in determining its source. The samples in SEM/EDX must be run under vacuum so that the high-voltage electron beam can be directed to the sample and then to the analyzer. At normal pressures the electron beam is scattered by the air and does not have enough energy to eject electrons from the sample. Also, the sample must be coated with a conductive coating, such as gold, so that current does not build up on the surface. This can limit the types of samples that can be used. The environmental scanning microscope, a recent development, can be used on samples without the conductive coating and this extends the range of samples that can be studied.

The other methods referred to in Table 2-3 are very powerful tools for surface analysis. They are really research tools and are not for routine plant tests. However, they should be considered if some of the specific characteristics are particularly needed. If there is a suspected support coatability problem, the atomic force microscope (AFM) or scanning tunneling microscope (STM) should be considered. These will give a good deal of information about the submicroscopic surface characteristics which cannot be seen with conventional microscopy. If there is a reason to suspect a difference in the properties in the depth of the sample, a confocal scanning microscope (CFM), which generates three-dimensional images, should be considered. This microscope can focus in varying planes in the depth of the defect and create a three-dimensional image.

In addition to the EDX, there are several surface characterization techniques that can be used to determine the chemical nature of the surface and to indicate the compounds present. Of these, electron spectroscopy for chemical analysis (ESCA) and Fourier transform infrared spectroscopy (FTIR) are the most widely used for defect analysis. ESCA will detect the elements heavier than lithium and also organic compounds, and it can indicate the type of bonding that is present. This can be used to identify specific compounds. A drawback is that it requires a larger sample size than EDX, and it is a surface measurement that penetrates only 50 Å. It is useful for detecting contaminants in large spot coating defects and for analyzing the base or support. Infrared spectroscopy is used widely to identify organic compounds. There are several sources of standard spectra that can be used for comparison and identification. FTIR is particularly useful for providing the infrared spectra of a surface. Both normal and suspect surfaces can be characterized to indicate the difference in organic composition. A limitation of this technique is the thickness of the layer. A thin layer may not give a strong enough signal for an analyzable spectrum. These instruments are available in most analytical laboratories and should be considered for use if organic compounds are suspected or if SEM/EDX does not give any chemical composition data.

2.2 COATER CHARACTERIZATION INSTRUMENTS

Specialized instruments are also needed to measure coater properties and provide information for the questions that have been posed in the initial part of the problem-solving procedure. Even though most modern coaters have extensive control and measurement systems, it is not economically feasible to measure all possible variables in a coater and to measure all properties in all locations. In addition, it is possible that instruments are not functioning and are out of calibration. Also, more frequent observations may be needed than obtained by the standard data collection procedures. As a result, additional measurements are often needed.

All of these specialized tools have been greatly improved by the electronics revolution of the last 10 to 15 years. The rapid advances in miniaturization and the advent of new electronic devices has resulted in a wide variety of powerful, low-cost devices that are readily available and easy to use. Also, powerful low-cost computers are now available that can process data orders of magnitude faster than the computers of the 1970s. Analog strip-chart data loggers have been replaced by multichannel devices that store the data in digital format for direct processing by the computers.

2.2.1 Rheological Measurements

The properties of the coating solution are obviously important to the ongoing performance of a coating process, and changes in these properties can be a source of problems in a coater. Optimization of these is also important in the development of new products. Measurement of the coating fluid properties is an important aspect of the coater troubleshooting process. Fluids are complex, and some of the properties are difficult to measure. Before the measurement techniques are discussed, some of the basics of fluid flow properties will be reviewed.

By definition a fluid is a substance that undergoes deformation when subject to shear or tensile stress. Viscosity is a key parameter to characterize this deformation and is a measure of the resistance of the solution to flow under mechanical stress. High-viscosity solutions require a high level of stress to flow rapidly, while a low-viscosity solution will deform and flow rapidly with a minimum of force. For example, water has a low viscosity and will flow readily out of a glass under normal gravity forces, whereas molasses, a high-viscosity material, will take a long time to flow from a glass.

Viscosity, often represented by the lowercase Greek letter eta (η) or mu (μ) is defined as the ratio of the shear stress to the shear rate. The shear rate, $\dot{\gamma}$ (lowercase Greek gamma with a dot over it), is the rate of change of the velocity with distance perpendicular to the direction of flow. In Figure 2-5 the shear rate is the velocity of the bottom plate divided by the gap, or V/G. As the units of velocity are distance per unit time, the units of shear rate are 1/time. As time is measured in seconds, shear rates are measured in 1/seconds, or s^{-1}. The shear stress, τ (lowercase Greek tau), is the force per unit area parallel to fluid flow. In Figure 2-5 it is the force to pull the bottom plate divided by the area of the upper block, or F/(LW). The units of shear

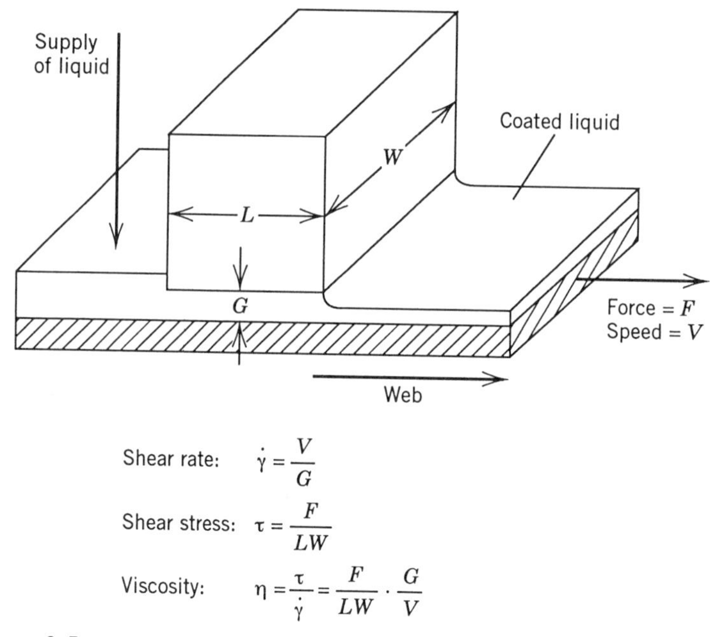

Figure 2-5. Terms in the definitions of shear stress, shear rate, and viscosity.

stress are force per unit area. In English units this is pounds per square inch. In SI units this is newtons per square meter, or pascals. In the cgs system this is dynes per square centimeter. As viscosity is shear stress divided by the shear rate,

$$\text{viscosity} = \frac{\text{shear stress}}{\text{shear rate}}$$

$$= \frac{\text{force/area}}{1/\text{time}} = \frac{\text{force} \times \text{time}}{\text{area}}$$

$$= \frac{\text{dynes} \cdot \text{s}}{\text{cm}^2} = \text{poise} \qquad \text{cgs units}$$

$$= \frac{\text{N} \cdot \text{s}}{\text{m}^2} = \text{Pa} \cdot \text{s} \qquad \text{SI units}$$

The usual units of viscosity in the cgs system are centipoise (cP). One cP is equal to 1 millipascal second (mPa · s). Typically, to measure viscosity the shear stress versus shear rate curves, at a given temperature or over a range of temperatures, are measured with one of a variety of viscometers, and the slope of the curve is the viscosity. When the curve is a straight line going through the origin, as in Figure 2-6, the slope is constant and the viscosity is independent of the shear rate.

The kinematic viscosity of solutions is measured with capillary, pipette, or cup type viscometers and is equal to the viscosity (shear viscosity) divided by the density. If one uses Newton's second law of motion (force equals mass times acceleration) to convert between force and mass, it is easy to show that the units of kinematic viscosity are distance squared divided by time. In the cgs system, 1 cm^2/s is called

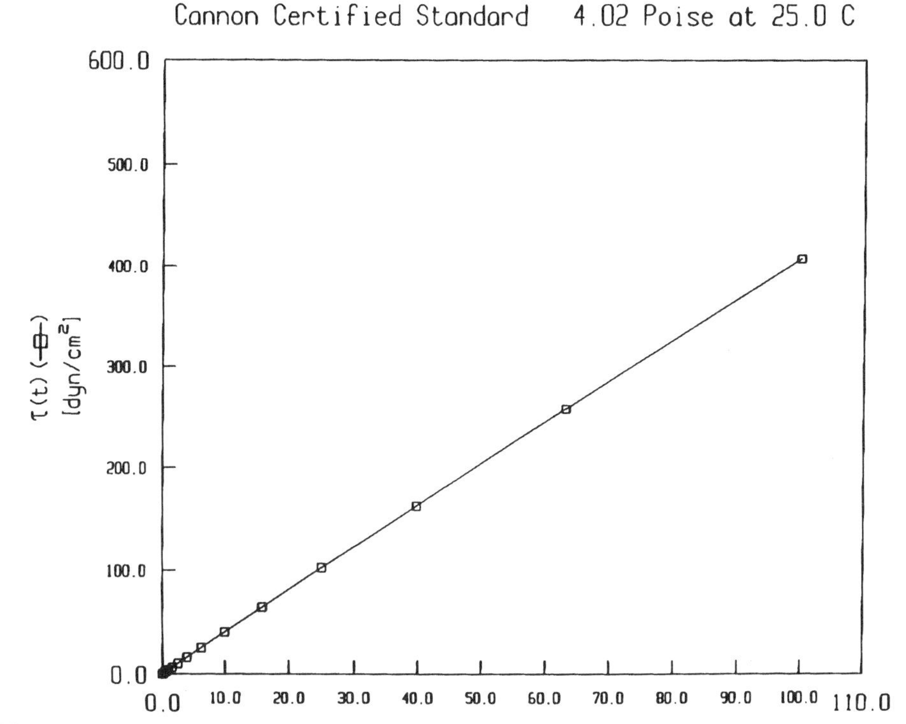

Figure 2-6. Shear stress versus rate of shear curve for a Newtonian liquid with a viscosity of 402 cP.

1 stoke (St). The more common unit is the centistoke (cSt). If the viscosity is 1 cP and the density is 1 g/cm^3, the kinematic viscosity is 1 cSt.

The viscosity can be independent of the shear rate or can be affected by it:

1. A Newtonian fluid has a constant viscosity independent of shear rate. The shear stress is a linear function of shear rate and the stress is zero at zero shear rate.
2. Pseudoplastic or shear-thinning behavior occurs when the viscosity decreases with increasing shear rate. This is typical of polymer solutions. The curve of shear stress versus shear rate bends toward the shear rate axis, as in Figure 2-7.
3. Dilatant or shear-thickening behavior occurs when the viscosity increases with increasing shear rate. In the shear stress–shear rate plot, the curve bends toward the shear stress axis, as in Figure 2-8.
4. A Bingham solid with yield-stress behavior occurs when the stress has to exceed some finite value before the fluid flows.

Figure 2-8 gives examples of these different flow behaviors. In some liquids the viscosity can also depend on how long the liquid has been flowing, giving rise to:

1. Thixotropic liquids, in which the viscosity decreases with time. This behavior is common in latex paints.
2. Rheopectic liquids, in which the viscosity increases with time.

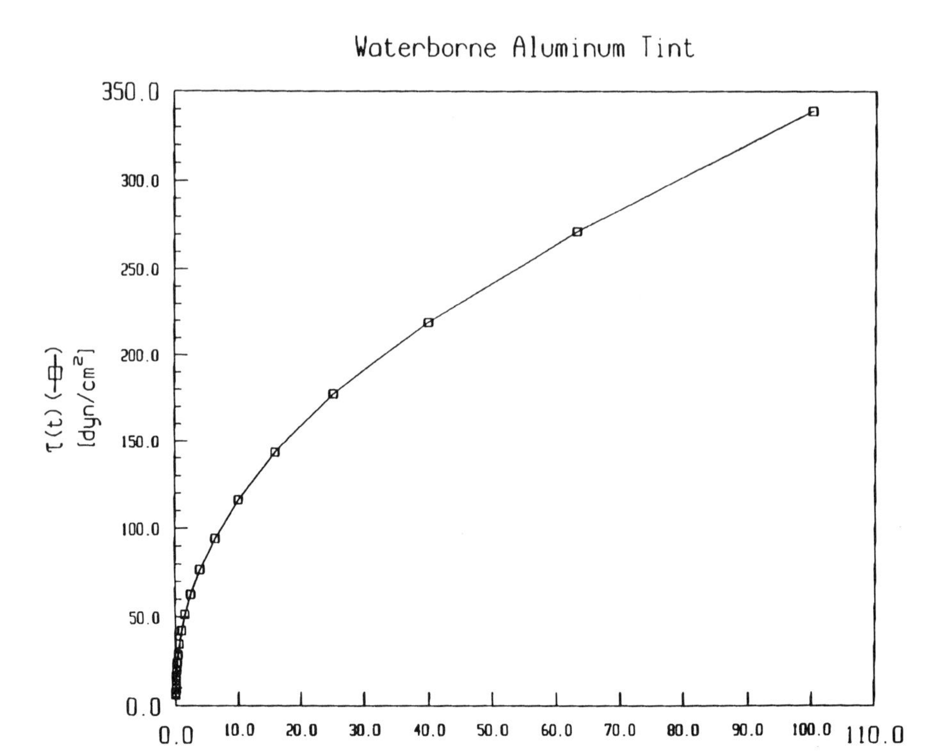

Figure 2-7. Shear stress versus rate of shear curve for a pigmented shear-thinning liquid.

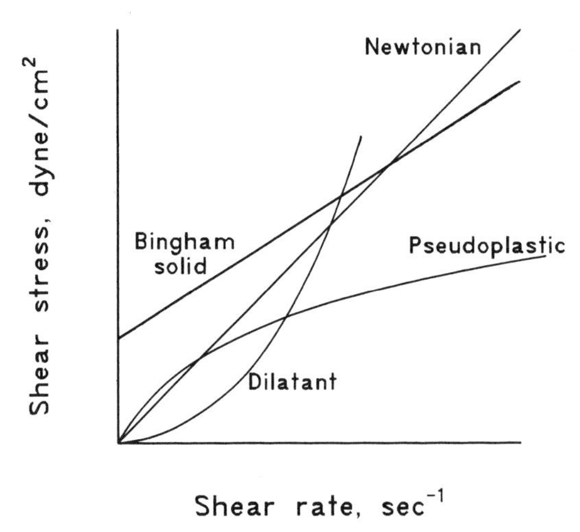

Figure 2-8. Shear stress versus rate of shear curves for a Newtonian liquid, a pseudoplastic or shear-thinning liquid, a dilatant or shear-thickening liquid, and a Bingham solid.

Routine measurements of viscosity prior to coating should be part of the normal quality control system for the coater. Often a problem is detected after the solutions are all consumed in the production process. If rheological data are not available, it may be very difficult to determine the source of the problem. It is possible to save samples and characterize them later. However, the viscosity of some liquids changes with time. If rheological data are not now being taken, they should be.

There is a great variety of commercial instruments available to measure all of these properties in a wide range of prices and complexity. For viscosity measurement there are two basic types: those with constant shear rate and those with variable shear rate. The constant-shear instruments provide a single-point viscosity measurement over a single or limited range of shear rates. They are convenient to use and are inexpensive and may be adequate for routine quality control. A disadvantage is that the inherent shear rate of the instrument rarely matches that of the coating process, and thus for shear-thinning liquids the data can be misleading. The variable-shear-rate viscometers can provide data over a wide range of shear rates. In some viscometers the shear rate is programmed to change automatically and a curve of viscosity versus shear rate is printed out; in others, the shear rates have to be set manually. The advantage of the variable shear rate is that the viscometers may be able to provide viscosities at the shear rates of interest in the coating applicator, and also, they may provide data on solution changes that are not seen in a single-point measurement. The disadvantages are the cost and the complexity of operation.

Examples of essentially constant-shear-rate viscometers are (1) glass capillary viscometers, (2) cup viscometers, (3) pipette viscometers, (4) falling-ball viscometers, and (5) electromagnetic viscometers. The constant-shear-rate viscometers are very useful for routine plant quality control measurements and are an excellent starting point if no routine measurements are now being taken. They are easy to use and inexpensive. When coupled with a quality control chart they can be very useful. The glass capillary viscometers, such as the Cannon–Fenske type, are an excellent research tool and are also good for routine measurements. They require only a small sample, on the order of 10 mL. They are used with a constant-temperature bath and give accurate and repeatable results. However, they are fragile and must be handled with care. With such care they can be recommended for use in a quality control laboratory. The cup viscometers measure the time for the solution to drain from a cup. This time can then be converted to a kinematic viscosity if the cup has been calibrated, and then, by multiplying by the density, to the normal shear viscosity. The individual cups have a limited viscosity range, and thus a series of these is needed. They are very easy to use in a plant situation and are widely used for spray applications because of their low cost and simplicity of use. Another advantage of the cups is that if used with a large volume of solution, the solution mass can provide a constant temperature and no additional temperature control is needed.

A very useful coating tool is the electromagnetic viscometer. This measures the viscosity and temperature simultaneously. It has the advantage that it can be used continuously in a flow line or even directly in a kettle, or off-line with samples (Wright and Gould, 1986; Wright and Gogolin, 1992). As with most modern instruments it can be interfaced directly to a data logging system.

The many types of rotational viscometers can give variable shear rates. A good basic unit is one in which a spindle rotates in a solution and the torque required to keep constant speed is measured and converted to viscosity. An example is the

Brookfield viscometer. One instrument with a variety of spindles can cover a very wide range of viscosities. It also has some ability to vary the rotational speed, and some simple shear rate–viscosity data can also be obtained.

The variable-shear-rate rotational viscometers, such as by Haake or by Rheometrics, give a wide range of shear rates, 1 to 20,000 s^{-1}, over a wide range of temperatures. Fluids can be extensively characterized to determine their viscosity–shear rate behavior and also the time-dependent behavior. These are particularly useful for optimizing a formulation in the preproduction process.

2.2.2 Surface Properties

The surface tension of coating liquids is important in relation to wetting the coating support. Because coating involves spreading a liquid out into a thin film with a very large surface area, surface forces play a large role. The terms *surface tension* and *surface energy* are sometimes used interchangeably. Surface tension is expressed as force per unit length, such as dyn/cm or mN/m (millinewtons per meter). Surface energy is expressed as energy per unit area, such as ergs/cm^2 or mJ/m^2 (millijoules per square meter). Numerically, all four terms are identical.

When we speak of liquids we usually refer to surface tension. We may find it difficult to visualize an undisturbed solid surface as being in a state of tension, and may find it easier to conceive of the surface of a solid as possessing surface energy. It is a matter of personal preference which term we use. In a freshly created liquid surface, the surface tension may change with time as dissolved materials diffuse to the surface to accumulate there, as surfactants do. This changing surface tension is called the *dynamic surface tension.*

When a drop of liquid is placed on a surface it may wet the surface and flow out or it may remain as a stationary drop. Figure 2-9 shows a drop on a surface when the forces are in equilibrium. The contact angle, θ, is a measure of the wettability of the surface by the liquid. With a liquid that spreads over the surface, the contact angle is zero. The contact angle formed on a support by a pure liquid, such as water, is a useful measure of the wettability of the support.

Surface tension is commonly found by the Wilhelmy plate method, where the force necessary to withdraw a thin platinum plate from the solution is measured, or the duNuoy ring method, where a ring made out of platinum wire is substituted. See, for example, Adamson (1982). These instruments are available from a variety of sources. These methods are static in that the surfaces have reached equilibrium and do not change with time. However, the coating process is dynamic in that the surface changes rapidly over a small time scale. Results from static surface tension may not be representative, but they are useful in ensuring that a solution has not changed between batches and has not changed with time. Dynamic surface tension measurements are considered to be a research tool only.

To study the surface wettablity of the support, optical comparators or goniometers that measure the contact angle of a drop of a solution on a support are available. A drop of liquid is placed on the surface and viewed through a microscope to measure the contact angle. It is also possible to use a series of calibrated and variable surface tension liquids for a qualitative test of the wettability of the surface. The drops are placed on the support and if they bead up, the surface tension of support is numerically below that of the liquids. If they flow out, the surface tension of the support

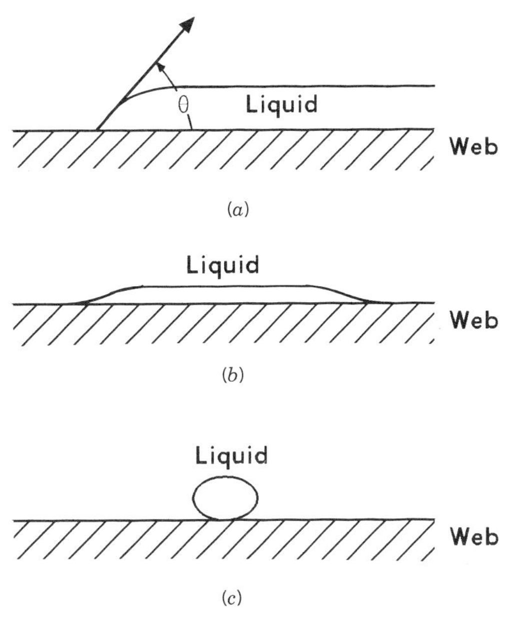

Figure 2-9. Contact angle between a liquid and a solid: (*a*) the contact angle is measured through the liquid; (*b*) when the contact angle is zero, a drop spreads out over the surface; (*c*) with complete nonwetting, the liquid balls up and the contact angle is zero.

is numerically higher than the liquids. With a series of liquids an estimate of the support surface energy or tension can be obtained.

2.2.3 Miscellaneous Tools

There are several additional characterization instruments that will improve the troubleshooting process and are described in this section. The selection of which instruments to use is up to the individual troubleshooter and the funds available to purchase them.

Temperature is an important parameter in coating solution preparation, coating application, drying, roll and raw material storage areas, and in the varying process rooms which contain the mixing kettles, the applicator, the unwinder, and the rewinder. In most coaters the key temperatures are recorded and controlled. However, there are many times when the problem solver needs to check locations that are not recorded, or to check whether the recorded readings are correct. As a result, temperature measurement should be considered a necessary diagnostic tool. There are several basic measurement techniques that are available to measure temperature:

1. Liquid-based thermometers, which use expansion of fluid as an indication of temperature
2. Bimetallic dial thermometers, in which a strip formed from two metals bends with temperature because of differences in thermal expansion
3. Thermocouples, in which an electrical potential arises in the junction of two dissimilar metals as a function of temperature

4. Resistance-temperature devices, in which the resistance of a platinum or other metal sensor is a function of temperature
5. Thermistors, which uses the change in the electrical resistance of a semiconductor as a function of temperature
6. Infrared thermometry, in which the energy emitted from the surface is an indication of the body temperature

Each of these has its advantages and fills different needs. Thermocouples are used when many measurements are needed at the same time. The thermocouples can be placed in a variety of locations and the data recorded in a central data logger. The wires that form the thermocouples are inexpensive and can be run over long distances without a loss of signal strength. Thermocouples can also be used to give an indication of the web temperature in a dryer. In single-sided dryers, where rolls are used to transport a web and drying air is not blown against the rolls, there is usually a large wrap of the web on at least some rolls. That roll surface will reach temperature equilibrium at or near the web temperature, and a thermocouple placed inside the roll, close to the surface, will then sense the temperature of the web. This is inexpensive and can be used to obtain data on the web temperatures in the different dryer zones.

Resistance-temperature devices are most accurate and are used when a few precision measurements are needed, such as in the basic dryer air control system. They are too expensive to use for many measurement points in a coater. They are ideal for a hand-held meter and can be obtained from laboratory supply houses.

If the temperature of a moving web, a coating bead, or any hard-to-reach location is needed, infrared thermometry is an excellent method to use, especially for remote measurements. Portable IR thermometers are available with a wide range of precision and accuracy. These are aimed at the remote object and the temperature measured. They are good for characterizing moving webs in the dryer and determining the temperature profile of a coating throughout the dryer. They can be more accurate than using a thermocouple in the roll. However, they are more expensive and may require calibration, as they depend on the emissivity of the surface. They are particularly useful in floater dryers, which have no rolls in the dryer. In addition to the single-point IR thermometers, there are also thermal imaging devices which can give an overall temperature scan of a wide area. The measured temperatures are converted to a color and then used to form the image of the part being measured. The image viewed on the monitor will be one uniform color if the temperature is constant over the viewing area, and will be many colored if there is a temperature variation. These devices are useful for locating leaks in heating and ventilating systems, heat losses in the dryer, and so on.

The SMART READER data logger discussed in the next section meets many of the needs for temperature measurements. They have their own sensors or they can be hooked up to thermocouples. The need for temperature and air velocity measurements can be met by having a hand-held meter that combines temperature and air velocity measurements in one long probe. These can be used to check the dryer airflow in plenums or the airflow from the dryer nozzles.

A tachometer is recommended to check the speed of the many rolls in a coater, the dryer fan speed, the coating pump speed, and by using a roll that rotates without slipping, the web speed. The commercially available models have accuracies and resolution on the order of 1% or better. They come in a variety of speed ranges, such as 5 to 10, 10 to 100, 100 to 1000, and 1000 to 10,000 rpm.

Tachometers are available in both mechanical or contacting models and strobe or noncontacting models. The contacting units require the measuring head to be held against the item for which the speed is needed. A variety of heads are available to get the best contact. Noncontacting stroboscopic devices should be used where the speed is very high or access is difficult. These devices need some form of indicating marker on the rotating wheel for the light to reflect from. Special reflective liquids and tapes are available from laboratory supply companies such as Cole-Parmer, Fisher, Edmund Scientific, and others. Care must be taken in using the stroboscopic tachometer not to be mislead by multiples of the true speed. Thus half the true speed will give the same stationary appearance.

The microcasette recorder is used to record both data read aloud from instruments and spoken observations, while one is in the coating area taking measurements or observing the operation. The tape recorder is preferred to writing because data can be recorded rapidly and recalled very easily. It provides flexibility in recording data while performing other duties. The coater is a high-speed operation with many functions going on simultaneously. It is easier to speak into a recorder than to write on a clipboard or a notebook. Also, it may be hard to read the notes at a later time, and written notes are more likely to be incomplete. A small voice-activated microcassette is preferred because it is light in weight and can easily be stored in a pocket. The voice-activated feature permits the recording of data hands-free while setting an applicator roll speed, adjusting a damper, and so on. These are relatively inexpensive and reliable machines and can be obtained for under $200 in any commercial electronics store.

Two-way radios are very useful for communicating during an experiment when several people need to coordinate actions or observations. Typically, a damper may be located far from the gauge that measures flow, or the tension control device is located away from the web where wrinkles are occurring. Two people are needed to set the conditions at the desired value, and the two-way radios will simplify the communications while reducing the time needed. It is always possible for one person to do this, but in a production unit with expensive raw materials, the time saved can be an appreciable cost factor. Careful coordination by radio can help to optimize a process condition without obtaining an adverse effect in the coater. When setting a tension in a coater, the observer can instantly see when the web goes unstable and communicate that fact to his colleague, who can adjust it. The time interval for one person to set the condition and walk to see the result could lead to web wrap and coater shutdown. Two-way radios are also relatively inexpensive and can be purchased at commercial electronic stores.

A visual recording device, from an inexpensive camcorder to a high-speed motion analyzer, can also be an important part of the tool kit. It can make visual permanent records of any point in the process where there is a problem. A recording of the coating bead or applicator roll can help define a quality problem. If there are creases in winding, a visual recording can help define the cause. The recording can then be played back for further analysis at a later time and can be played at lower speeds for slow-motion analysis. For additional analysis, key frames from the recording can be captured and analyzed in an image analysis system. This can be used to calculate properties such as the size of bubbles, the deflection of a web or the motion of a bead. A wide variety of equipment is available for this use. High-speed motion analyzers can record up to 4500 images per second. The low cost and the features of a high-quality home camcorder can meet most of the needs. Some additional equipment, such as lights and a tripod to hold the camera, should be used. The high-speed units are expensive, and consideration should be given to either renting equipment when needed, or using a service firm that specializes in making such recordings.

When taking measurements in the coater area there is another important factor that needs to be considered—personal safety. The taking of the temperature and pressure measurements on a catwalk, checking the process for problem areas, inspecting a moving web, installing thermocouples, and so on, all have the potential for causing an accident with injury to the troubleshooter and personnel in the vicinity. The troubleshooter may be in an unfamiliar situation because this is the first visit to the coater, or because modifications have been made since the last visit. While concentrating on the measurement and not on the surroundings, it is more likely to have an injury—by inadvertently tripping and falling over a new wire, by getting a hand caught between a roll and the moving web, and so on. To minimize the dangers one should be aware of the area safety rules. One should wear the proper protective equipment and should consider all actions before doing them. Make sure that the coater operating crew is aware of your presence and knows what you are doing and where you will be. Consider the safety aspects when you run wires or set up specialized equipment. If there are many thermocouple couple wires, bundle them with tape and use a route to get them to the recording device that is not a walkway. If need be, tape them to the floor or wall with bright tape and make a sign to mark them. Be aware of the equipment and how it operates, and consider the possible interactions with the coater environment. In a solvent environment explosion-proof equipment is needed.

2.3 DATA LOGGERS

The sensors described earlier in the chapter can generate a significant amount of process data and measurements, all of which need to be recorded for further analysis. In this section we describe some of the various data logging devices that are available to record the data from these sensors. Most modern coater control systems have some form of data logging systems that preserve the measurements in electronic format for further analysis. These will give the troubleshooter much of the needed data. However, we have found many occasions where additional measurements must be taken. Data loggers will provide the tool to record these measurements automatically without requiring a person to be present.

Both analog and digital devices are available. Analog strip charts provide a continuous record of the process variable versus time and have been in widespread use for many decades. They have the advantage of easily obtaining qualitative information by visual examination, time trends stand out clearly, and they are inexpensive. However, quantitative information is harder to obtain for detailed statistical calculations because the troubleshooter must transcribe each point to a digital format for detailed statistical analysis. In addition, practical experience has shown that the pen may clog or run out of ink at the most critical point in the data set, and much data may be lost because of this.

The preferred devices are digital data logging devices that can record the desired sensor measurements in digital format, will add the time of the measurement to the record, and will store both in some magnetic format, such as on magnetic tape, a floppy disk, or a hard disk. These units have several advantages. Since the data are in digital format, they can be transferred directly to a computer for analysis and plotting. The time for analysis will be greatly reduced, leading to a faster solution to the problem. Costs are reasonable, under $10,000. Flexibility is enhanced. The same equipment can monitor pressure, temperature, or viscosity, depending on the sensor to which it is attached. Most of the hardware is light and portable and can easily be transported into the coater. The functions these data loggers perform is to take the input sensor signal, which is usually in either milliamperes or millivolts, convert this signal to the desired units through the use of an appropriate algorithm, and then record the results. A central program is required to convert the signal, provide the time base, and control the recording function. These devices are all basically digital computers and vary in the range of functions that they can perform. Within this category there are many types of devices to record a variety of signals and to store data on a wide variety of recording media.

Personal computers, of both the IBM PC and Macintosh types, are excellent for these data logging functions. Special boards are made to be placed in the expansion slot of the computers, which will attach the sensors in the field to the computer. Typically, these boards also come with the necessary computer software programs. These programs are easy to use and allow the user to specify the data storage, the conversion of signals to desired units, the time sequence, and any other calculations that are needed. A wide variety of boards are available from manufacturers, such as those listed in Table 2-14. Once the data are recorded through these boards and programs they can be analyzed in the same computer, and this saves time. It is much easier to prepare reports and analyze data by using this method than with analog devices.

The portable data loggers are the most useful of these devices. They can record all of the sensor information described previously, along with the time and date of the reading. Temperature data are frequently needed in the coater, and specialized temperature measurement data loggers are available. They can record up to 1000 channels per second and can be used with either thermocouples or resistance-temperature devices. All of the conversion algorithms are provided and the reading is directly in °F or °C. These cost from $1000 to $3000, depending on the sophistication desired.

An example of a very useful portable data logger is the SMART READER unit from ACR. This is a combination of sensor and data logger and is very useful for troubleshooting. Temperature, relative humidity, pressure, and the time and date of the readings can be recorded. The data logging function, recording function, battery, and the sensors are all contained in a small self-contained unit, 4.2 × 2.9 × 0.9 in. (107

× 74 × 23 mm), which weighs only 5 ounces (140 g). To aid in the positioning of the devices, magnets are attached to the outside case so that they can easily be positioned in ducts, dryers, or almost any location where a reading is needed. The memory can contain up to 32,000 readings, and measurements can be recorded over an extended period, depending on the frequency of the measurements. Units record continuously and the recording memory is of the wraparound type. Thus it will record over the first readings when the memory is full. With this feature, the logger can be left in place and removed when a problem occurs. Table 2-4 gives the types of measurements that can be made. Because of the small size and low cost, \$400–\$800, these units can be carried in the tool kit and used whenever needed. Just record time in and time out of the process so that the correct data are identified for analysis.

The SMART READER functions with a personal computer, which is used to set the timing of the desired measurements and to move the data from the data logger to the computer for subsequent analysis. A companion program plots the data and formats it for analysis by statistical software packages. It is also easy to isolate particular time periods. An example of data obtained from this unit is shown in Figure 2-10. In Chapter 9 we give an example of how they were used to diagnose a drying problem.

Table 2-4 SMART READER Data Logger

Temperature and relative humidity	
Temperature	Various ranges from −30 to 490°F, using thermistors
Relative humidity	10–90%
	Accuracy of ±4%
Single point	
Pressure and temperature	
Pressure	0 to 0.5 psi
	0 to 5 psi
	0 to 30 psi
	0 to 100 psi
Temperature	Various ranges from −30 to 490°F, using thermistors
Can also be equipped for relative humidity 10–90%	
Single point	
Thermocouples	
Three channels using type J, K, or T thermocouples	

Type	Range (°F)	Resolution (°F)
J	0 to 370	2.5
	−22 to 1100	9
K	−10 to 440	3.0
	−145 to 1650	12.0
T	−30 to 390	3.0
	−825 to 930	12.0

Eight-channel logger	
Eight thermistor channels	

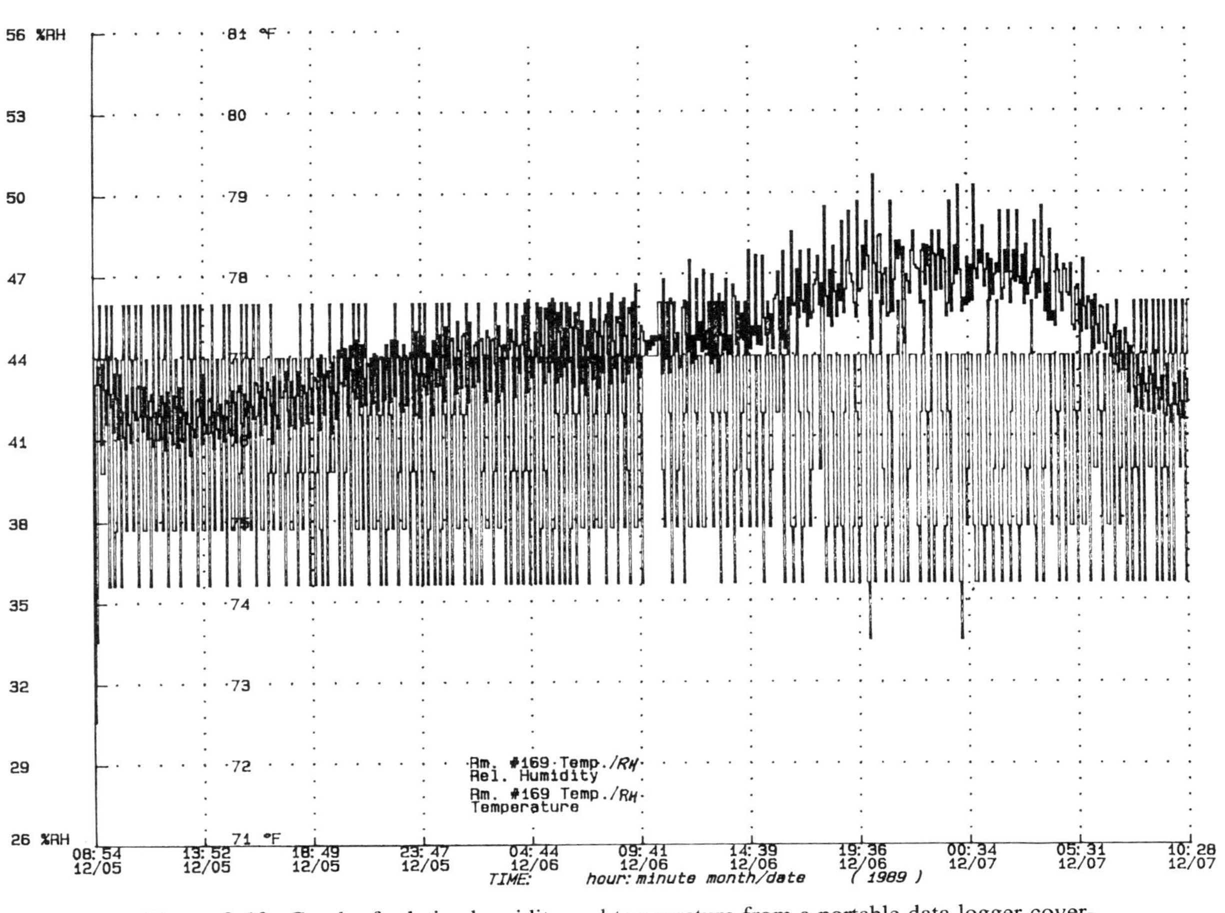

Figure 2-10. Graph of relative humidity and temperature from a portable data logger covering 2 days.

2.4 DATA ANALYSIS TECHNIQUES

Modern analytical and characterization methods in conjunction with computer-based data logging have the ability to create a vast quantity of data. These data then need to be analyzed and used to support or disprove the various hypotheses that have been developed. It is often easier to generate the data than it is to analyze them. Most organizations have statisticians to help the troubleshooter with the analysis and to guide him or her in the formulation of statistically designed experiments for the action step. However, we believe that it is essential that the troubleshooters have a basic working knowledge of data analysis techniques, so they are able to do a portion of the statistical analysis themselves. There are many occasions when a preliminary analysis of the data by operating personnel is needed to help solve a problem. A good rapid analysis of a small data set is better than no analysis of a larger data set. Also, there may not always be a statistician available to help, and then there is no other choice than for the troubleshooter to do it. In addition, operating personnel can bring a valuable essential reality check to the conclusions that are reached from the data. It is easy in statistics and data analysis to develop correlations that are mathematically valid but have no relation to the reality of the process and to the science behind it.

Once the data have been collected and assembled, there are basically three parts to the subsequent statistical analysis: (1) selecting the appropriate statistical parameters to be calculated, (2) calculating the selected statistical parameters, and (3) interpreting the parameters to decide which relations are valid and how they answer the questions that were originally posed.

The bottleneck or rate-limiting step in the statistical analysis is the calculation of the selected parameters. The appropriate parameters for an analysis could be determined quickly through consultation with a statistician, discussion with a colleague who has background in statistics, or by referring to one of the many references in statistics or statistical process control (Frost and Gutoff, 1991; Davies, 1958; Natrella, 1966; Brooks et al., 1979). It is relatively easy in most problem-solving situations to get recommendations on what to calculate or to plot, and to get assistance in interpreting the data once the calculations are performed. It is, however, much more difficult to get someone to do the actual calculation, particularly if there is a large data set to be processed.

In the last decade the widespread availability of personal computers has revolutionized the calculation procedure and has eliminated this as the bottleneck in the statistical analysis procedure. In the next section we focus on the calculation steps and discuss some of the basic parameters and computer tools that we have found helpful in our own troubleshooting efforts. These computer-based tools are readily available, inexpensive, and can very quickly generate the parameters needed to characterize and analyze the data. A basic assumption is that everyone has access to personal computers either in the coater or in the office. For a more detailed discussion of the statistical parameters to calculate, refer to the references on statistics cited previously.

Typically, at this point there is a collection of data from measurements of coater process parameters, such as dryer temperature, web tension, coating solution viscosity, support thickness variations, support wettability, and so on. In addition, there can be measurements on the defect problem, such as streak frequency in rolls, streak location in web, chatter frequency and location, spot count versus web location, adhesion level in a series of rolls, and so on. All of these measurements typically cover a variety of time periods. Both perceived normal operation as well as problem periods should be covered. These time intervals can either represent the current production or cover a range of productions, and the time can be measured in regular time units, such as hours and minutes, or may be in terms of roll identification numbers.

The overall objective of the analysis at this stage is to determine which of the process variables have changed and how this change influences the defects or problem under analysis. To do this analysis, a series of statistical properties need to be calculated:

1. Statistical properties that describe the parameters of the specific data set, such as the average or the mean and the range or standard deviation
2. Statistical properties that determine if there are significant differences between specific data sets, such as the t-statistic, the F-statistic, and so on
3. Statistical tests that determine if there is a valid cause-and-effect relationship between coater variables and the defect being studied, such as a regression analysis or a correlation matrix

There is a wide variety of readily available, easy-to-use computer software tools for this function, all of which are effective. It is possible to build up an expensive library of specific programs for each of the specific analyses that can sit on a shelf and look impressive. However, the vast majority of the basic calculations can be done with just one program, and that program is a commercial spreadsheet. These all have several built-in statistical and data plotting routines that can perform the needed calculations. Spreadsheets also have a wide range of other uses. A spreadsheet is recommended for all problem-solver tool kits. If only one program is to be purchased, make it a spreadsheet. Table 2-5 lists some of the widely available

Table 2-5 Statistical Analysis Programs

Program	Vendor
Spreadsheets	
Excel	Microsoft
Works	Microsoft
Lotus 1-2-3	Lotus
Resolve	Claris
Quattro Pro	Borland
Graphing programs	
Cricket Graph	Computer Associates
Delta Graph	DeltaPoint
Kaleidagraph	Synerfy Software
PC-WAVE	Visual Numerics
Statistical programs	
Costat	Cohort
DataDesk	Data Description
JMP	SAS Institute
Minitab	Minitab
SPSS	SPSS
Statistica	StatSoft
StatView	Abacus Concepts
Systat	SSTAT
SigmaStat	Jandel Scientific

spreadsheet programs that run on the Macintosh and the PC. To use a spreadsheet for statistical analysis the data are imported into the spreadsheet in column and row format. If the data logging devices referred to earlier in this chapter have been used, the data are already in this form electronically and can easily be imported into the spreadsheet. It may already be in Excel or Lotus format. Similarly, data from any computer generated historical record can be arranged in a spreadsheet readable format and then copied onto a disk for use in the spreadsheet. If local area networks are available, the data can be transmitted between the computers. For small amounts of data, manual entry into the spreadsheet is possible.

An example of a typical analysis using a basic spreadsheet such as Works 3.0 or Excel 4.0 is shown in Table 2-6. The data of viscosity, temperature, time, and date were obtained for a coating solution using an in-line viscometer and the SMART READER data logger. The data were extracted in spreadsheet format with one row (horizontal data) giving a complete reading. The complete data set is not shown, only the first 3 hours of the data. For this data set, the average, standard deviation, and range were felt to be adequate to characterize the measurements. To do the calculations, a summary table was set up at the top of the spreadsheet and the text describing the property entered in the first column. Then the equations are entered in the appropriate cell to calculate the desired property. This is facilitated by the spreadsheet with its preprogrammed functions to calculate statistical properties. All that needs to be entered is the function name and where the data are. The following commands are used for the temperature data in the Excel 4.0 spreadsheet shown in Table 2-6:

Cell	Equation	Calculates:
C6	= AVERAGE(c16:c324)	Average
C7	= STDEV(c16:c324)	Standard deviation
C8	= MAX(c16:c324)	Highest value
C9	= MIN(c16:c324)	Minimum value
C10	= COUNT(c16:c324)	Total number of points
C11	= 100*c7/c6	Coefficient of variability as %

Each spreadsheet has its own set of functions and syntax, which must be followed exactly. These are spelled out clearly in the manuals. Another advantage of a spreadsheet is that it is easy to edit the data and to remove suspect data. The data set in Table 2-6 was taken over several days. There were times when the coater was not coating with no solutions flowing, giving low temperatures and viscosities. Data were still being recorded for these periods. Using such data in the calculations would give completely erroneous results. These inactive periods were detected by the wide range of values reflected in the minimum readings. These data were then deleted from the spreadsheet and the recalculated values obtained instantly. Another spreadsheet advantage is that new properties can easily be calculated from the recorded data and added to the analyses. In this example, the elapsed time is calculated from the time data, using the first point as the start. The elapsed time is in the last column. A statistical function, the coefficient of variability (COV), is not available as a function in Excel but is a good way to express the overall variability of the data set. The formula for the COV is the standard deviation divided by the average expressed as percent and can be calculated and included in the summary, as shown above.

Table 2-6 Example of Statistical Analysis with Spreadsheet

	Temperature (°F)	Viscosity (CP)
Average	97.77	9.67
Standard deviation	1.24	0.61
High	100.98	18.50
Low	91.18	6.86
Number of Points	320	320
COV, Standard deviation/Average	1.27%	6.28%

Data Obtained from In-Line Viscometer in Plant Coating Using SMART READER

Time	Date	Temperature (°F)	Viscosity (cp)	Elapsed Time (h) Calculated
0:36	11-Feb	91.18	18.5	0.00
0:41	11-Feb	93.63	10.36	0.08
0:46	11-Feb	96.08	10.36	0.17
0:51	11-Feb	96.08	9.77	0.25
0:56	11-Feb	96.08	9.77	0.33
1:01	11-Feb	96.08	9.77	0.42
1:06	11-Feb	98.53	9.77	0.50
1:11	11-Feb	96.08	9.77	0.58
1:16	11-Feb	96.08	9.77	0.67
1:21	11-Feb	96.08	9.77	0.75
1:26	11-Feb	96.08	9.77	0.83
1:30	11-Feb	96.08	9.77	0.92
1:35	11-Feb	96.08	9.77	1.00
1:40	11-Feb	96.08	9.77	1.08
1:45	11-Feb	96.08	9.77	1.17
1:50	11-Feb	96.08	10.36	1.25
1:55	11-Feb	96.08	10.36	1.33
2:00	11-Feb	96.08	10.36	1.42
2:05	11-Feb	98.53	10.36	1.50
2:10	11-Feb	96.08	10.36	1.58
2:15	11-Feb	98.53	10.36	1.67
2:20	11-Feb	96.08	10.36	1.75
2:25	11-Feb	96.08	9.77	1.83
2:30	11-Feb	96.08	9.77	1.92
2:35	11-Feb	96.08	9.77	2.00
2:40	11-Feb	96.08	9.77	2.08
2:44	11-Feb	98.53	9.77	2.17
2:49	11-Feb	98.53	9.77	2.25
2:54	11-Feb	98.53	9.19	2.33
2:59	11-Feb	98.53	9.77	2.42
3:04	11-Feb	98.53	9.77	2.50

A simplifying step in the use of spreadsheets is to write a standard blank template, in which the commands are written for a wide range for the data in several columns, and with text descriptions given. This template can be called up, the data pasted in, and the spreadsheet will calculate the results automatically. The template can also be customized to include plotting. Analysis can be done in under 10 minutes.

The next step is to plot the data for a visual analysis. Graphing of data is a very powerful analytical tool and should be used when possible. The eye and the human brain are an excellent combination for spotting trends and determining relations. Often, they reveal clues and general answers to some questions. A plot will rapidly determine if there is a justification for pursuing an analysis. The spreadsheets can plot data in many formats and are an excellent tool for this. However, it is our experience that graphing programs, such as Cricket graph, are easier to use for this function. They can all prepare a wide variety of graphs. They will plot the data and do regression analysis correlations that are excellent for preliminary analysis and for presenting the data to a group. It is also easy to prepare overlaying plots and expand or contract time periods as needed. These graphing programs also work very well with the spreadsheets, in that the data can be transferred directly from spreadsheets without further manipulation and then graphed. A variety of colors can be used to illustrate specific points. Figure 2-12 is the plot of data from a similar spreadsheet.

The graphing programs can also be used for stand-alone data analysis. An example of this is shown in Figure 2-11, where the data of a spot defect frequency from Frost and Gutoff (1991) are plotted in quality control chart format with upper and lower control limits. The data are saved in tabular format and can easily be updated and a new plot made. Another example of the analytical power of the graphing programs is shown in Figure 2-12. This shows the coating weight or coverage of a coating solution in the transverse or cross direction of the web. In this figure, two different equations, a simple linear equation and a polynomial equation, are both calculated from the data using a least-squares method. The equation describing the data along with the square of the correlation coefficient (r^2). The square of the correlation coefficient indicates how much of the variability in the data is explained by the equation. A perfect fit gives a correlation coefficient of 1.0. An r^2 of 0.828, as with the linear equation in

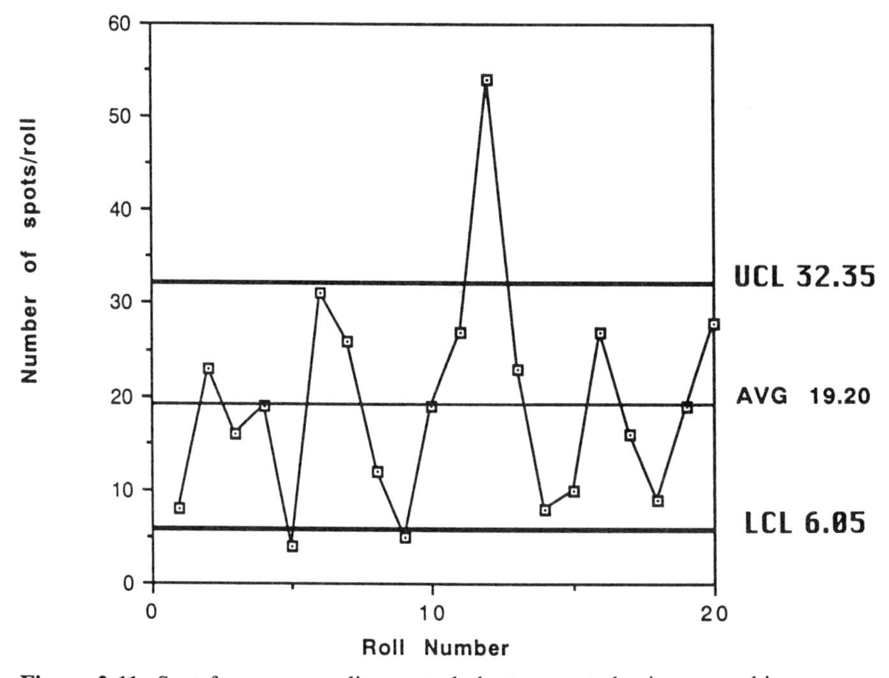

Figure 2-11. Spot frequency quality control chart generated using a graphing program.

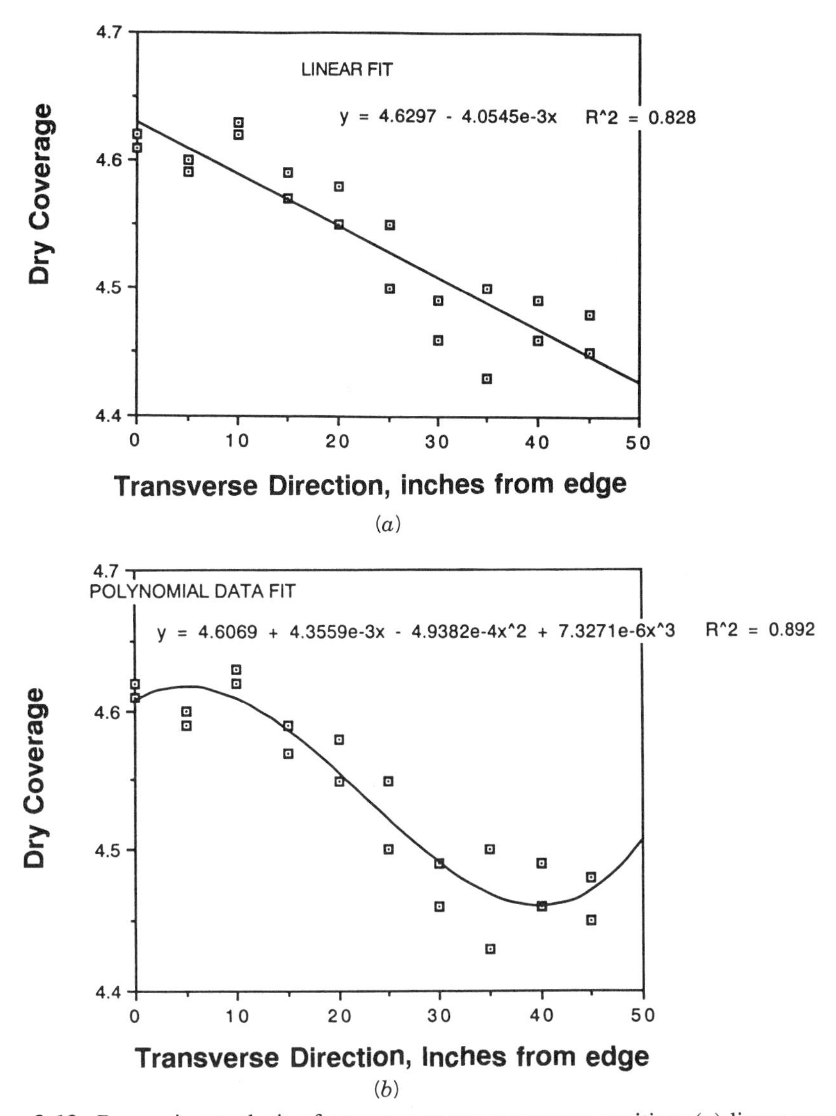

Figure 2-12. Regression analysis of coverage versus transverse position: (*a*) linear equation; (*b*) cubic equation.

Figure 2-11, indicates that 82.8% of the variability in the data is explained by the regression line that is shown, and 17.2% is not. There is always a concern whether the regression line is meaningful or whether it may lead to erroneous conclusions. This is particularly true for small data sets. To determine if the regression equation is significant, Table 2-7 from Davies (1958) may be used. In this particular data set there are 21 data points. As 2 degrees of freedom are used for the constants in the linear regression equation, 19 degrees of freedom are left for the correlation coefficient. Table 2-7 indicates that for 19 degrees of freedom and a 95% confidence limits ($p = 0.05$), an r value greater then 0.433, or an r^2 greater than 0.19, is significant. Therefore, the regression line is meaningful.

Table 2-7 Values of Correlation Coefficient for Different Levels of Significance

Degrees of Freedom, ϕ	P					Degrees of Freedom, ϕ	P				
	0.10	0.05	0.02	0.01	0.001		0.10	0.05	0.02	0.01	0.001
1	0.988	0.997	1.00	1.00	1.00	16	0.400	0.468	0.542	0.590	0.708
2	0.900	0.950	0.980	0.990	0.999	17	0.389	0.455	0.528	0.575	0.693
3	0.805	0.878	0.934	0.959	0.991	18	0.378	0.444	0.515	0.561	0.679
4	0.729	0.811	0.882	0.917	0.974	19	0.369	0.433	0.503	0.549	0.665
5	0.669	0.754	0.833	0.874	0.951	20	0.360	0.423	0.492	0.537	0.652
6	0.621	0.707	0.789	0.834	0.925	25	0.323	0.381	0.445	0.487	0.597
7	0.582	0.666	0.750	0.798	0.898	30	0.296	0.349	0.409	0.449	0.534
8	0.549	0.632	0.715	0.765	0.872	35	0.275	0.325	0.381	0.418	0.519
9	0.521	0.602	0.685	0.735	0.847	40	0.257	0.304	0.358	0.393	0.490
10	0.497	0.576	0.638	0.708	0.823	45	0.243	0.287	0.338	0.372	0.465
11	0.476	0.553	0.634	0.683	0.801	50	0.231	0.273	0.322	0.354	0.443
12	0.457	0.532	0.612	0.661	0.780	60	0.211	0.250	0.295	0.325	0.408
13	0.441	0.514	0.592	0.641	0.760	70	0.195	0.232	0.274	0.302	0.380
14	0.426	0.497	0.574	0.623	0.742	80	0.183	0.217	0.256	0.283	0.357
15	0.412	0.482	0.558	0.605	0.725	90	0.173	0.205	0.242	0.267	0.337
						100	0.164	0.195	0.230	0.254	0.321

Source: Reprinted with the permission of Simon & Schuster, from *Statistical Methods in Research and Production*, by O. L. Davies, Macmillan, New York, 1958.

In addition to calculating the basic statistical parameters described above, using functions that calculate a specific parameter, advanced spreadsheets such as Excel 4.0 can perform a wide variety of multiparameter advanced statistical analyses. In addition to the 70 statistical analysis functions shown in Table 2-8, there are over 20 preprogrammed statistical analyses tools. These tools differ from the functions commands described above in that a series of steps are combined into one macro

Table 2-8 Examples of Statistical Analysis Functions in an Advanced Spreadsheet from Excel 4.0

Function	Returns
AVEDEV	Average of absolute deviations of data points from their mean
AVERAGE	Average of arguments
BETADIST	Cumulative beta probability density function
BETAINV	Inverse of the cumulative beta probability density function
BINOMDIST	Individual term binomial distribution
CHIDIST	One-tailed probability of the chi-squared (c2) distribution
CHIINV	Inverse of the chi-squared (c2) distribution
CHITEST	Test for independence
CONFIDENCE	Confidence interval for a population
CORREL	Correlation coefficient between two data sets
COUNT	Tally of the arguments that are numbers
COUNTA	Tally of the nonblank values in the list of arguments
COVAR	Covariance, the average of the products of paired deviations

Table 2-8 (*Continued*)

Function	Returns
CRITBINOM	Smallest value for which the cumulative binomial distribution is less than or equal to a criterion value
DEVSQ	Sum of squares of deviations
EXPONDIST	Exponential distribution
FDIST	F probability distribution
FINV	Inverse of the F probability distribution
FISHER	Fisher transformation
FISHERINV	Inverse of the Fisher transformation
FORECAST	Returns values along a linear trend
FREQUENCY	Frequency distribution as a vertical array
FTEST	Result of an F-test
GAMMADIST	Gamma distribution
GAMMAINV	Inverse of the gamma cumulative distribution
GAMMALN	Natural logarithm of the gamma function, $G(x)$
GEOMEAN	Geometric mean
GROWTH	Values along an exponential trend
HARMEAN	Harmonic mean
HYPGEOMDIST	Hypergeometric distribution
INTERCEPT	Intercept of the linear regression line
KURT	Kurtosis of a data set
LARGE	kth largest value in a data set
LINEST	Parameters of a linear trend
LOGEST	Parameters of an exponential trend
LOGINV	Inverse of the lognormal distribution
LOGNORMDIST	Lognormal distribution
MAX	Maximum value in a data set
MEDIAN	Middle value in a data set
MIN	Minimum value in a data set
MODE	Most common value in a data set
NEGBINOMDIST	Negative binomial distribution
NORMDIST	Normal cumulative distribution
NORMINV	Inverse of the normal cumulative distribution
NORMSDIST	Standard normal cumulative distribution
NORMSINV	Inverse of the standard normal cumulative distribution
PEARSON	Pearson product moment correlation coefficient
PERCENTILE	Value from a range at the kth percentile
PERCENTRANK	Percentage rank of x among the values in a data set
PERMUT	Number of permutations for a given number objects
POISSON	Poisson probability distribution
PROB	Probability that values in a range are between two limits
QUARTILE	Quartile from a data set
RANK	Rank of a number in a list of numbers
RSQ	r^2 value of the linear regression line
SKEW	Skewness of a distribution
SLOPE	Slope of the linear regression line
SMALL	kth smallest value in a data set
STANDARDIZE	Normalized value

Table 2-8 (*Continued*)

Function	Returns
STDEV	Estimate of standard deviation based on a sample
STDEVP	Standard deviation for a population based on the entire population
STEYX	Standard error of the predicted y-value for each x in the regression
TDIST	Student's t-distribution
TINV	Inverse of the Student's t-distribution
TREND	Values along a linear trend
TRIMMEAN	Mean of the interior of a data set
TTEST	Probability associated with a Student's t-test
VAR	Estimates variance based on a sample
VARP	Calculates variance based on the entire population
WEIBULL	Weibull distribution
ZTEST	Two-tailed P-value of a z-test

Source: User's Guide Book 2, Microsoft Excel 4.0, Document Ab26297-0492, 1992.

procedure that performs a series of steps with the one command. The data are analyzed and placed into a table with only one command. Table 2-9 lists these macro procedures. As an example, Table 2-10 is an analysis of the same data as Table 2-6. However, it was done with only one command by using the descriptive statistics command, and then entering the range of data. As can be seen in

Table 2-9 Statistical Analysis Tools Procedures in Excel 4.0

ANOVA: Single-factor
  Two-factor with replication
  Two-factor without replication
Correlation
Covariance
Descriptive Statistics
Exponential Smoothing
Fourier Analysis
F-test: Two-sample for variances
Histogram
Moving average
Random number Generation
Rank and percentile
Regression
Sampling
t-Test
  Paired two-sample for means
  Two-sample assuming equal variances
  Two-sample assuming unequal variances
z-Test: Two-sample for means

Source: "Analyzing and Calculating Data," in Chapter 1, *User's Guide Book 2*, Microsoft Excel 4.0, Document Ab26297-0492, 1992.

Table 2-10 Statistical Analysis Using Spreadsheet Statistical Tools

	Temperature (°F)	Viscosity (cp)
Mean	97.77203125	9.67
Standard error	0.069596152	0.03
Median	98.53	9.77
Mode	98.53	9.77
Standard deviation	1.244973819	0.61
Variance	1.549959811	0.37
Kurtosis	1.668231394	141.62
Skewness	−1.179757163	9.30
Range	9.8	11.64
Minimum	91.18	6.86
Maximum	100.98	18.50
Sum	31,287.05	3095.28
Count	320	320

Data Obtained from In-Line Viscometer in Plant Coating Using SMART READER[a]

Time	Date	Temperature (°F)	Viscosity (cp)	Elapsed Time (h) Calculated
0:36	11-Feb	91.18	18.5	0.00
0:41	11-Feb	93.63	10.36	0.08
0:46	11-Feb	96.08	10.36	0.17
0:51	11-Feb	96.08	9.77	0.25
0:56	11-Feb	96.08	9.77	0.33
1:01	11-Feb	98.08	9.77	0.42
1:06	11-Feb	96.53	9.77	0.50
1:11	11-Feb	96.08	9.77	0.58
1:16	11-Feb	96.08	9.77	0.67
1:21	11-Feb	96.08	9.77	0.75
1:26	11-Feb	96.08	9.77	0.83
1:30	11-Feb	96.08	9.77	0.92
1:35	11-Feb	96.08	9.77	1.00
1:40	11-Feb	96.08	9.77	1.08
1:45	11-Feb	96.08	9.79	1.17
1:50	11-Feb	96.08	10.36	1.25
1:55	11-Feb	96.08	10.36	1.33
2:00	11-Feb	96.08	10.36	1.42
2:05	11-Feb	98.53	10.36	1.50
2:10	11-Feb	96.08	10.36	1.58
2:15	11-Feb	96.53	10.36	1.67

[a]Not all data shown.

Table 2-10, a much wider range of parameters is calculated than in the previous analysis.

Another example of statistical analysis using the tools macro procedures is shown in Tables 2-11 and 2-12. Table 2-11 is a set of data from Frost and Gutoff (1991) which contains the optical density of negative film in several transverse direction lanes across the web and for several rolls. The objective of the analysis is to determine

Table 2-11 Data for Statistical Analysis: Negative Film Density Versus Lane and Roll

Roll No.	Lane 1	Lane 2	Lane 3	Lane 4	Lane 5	Roll Average
1	1.66	1.71	1.70	1.73	1.67	1.694
2	1.64	1.70	1.71	1.70	1.69	1.688
3	1.67	1.69	1.72	1.68	1.68	1.688
4	1.65	1.72	1.74	1.73	1.67	1.702
5	1.66	1.70	1.72	1.71	1.65	1.688
6	1.65	1.71	1.71	1.70	1.68	1.690
7	1.68	1.70	1.72	1.72	1.67	1.698
8	1.66	1.68	1.74	1.68	1.66	1.684
9	1.67	1.69	1.73	1.71	1.65	1.690
10	1.62	1.67	1.68	1.67	1.63	1.654
11	1.67	1.69	1.74	1.70	1.66	1.692
12	1.65	1.71	1.75	1.72	1.69	1.704
13	1.68	1.72	1.74	1.71	1.68	1.706
14	1.66	1.68	1.72	1.73	1.67	1.692
15	1.64	1.70	1.70	1.70	1.66	1.680
16	1.68	1.74	1.72	1.73	1.69	1.712
17	1.66	1.73	1.74	1.72	1.70	1.710
18	1.69	1.73	1.75	1.70	1.70	1.714
19	1.69	1.71	1.74	1.73	1.69	1.712
20	1.68	1.74	1.75	1.72	1.67	1.712
Mean	1.66	1.71	1.73	1.71	1.67	1.69

Source: From P. J. Frost and E. B. Gutoff, *The Application of Statistical Process Control to Roll Products*, 2nd Ed., Table 4-C2-1, PJ Associates, Quincy, MA, 1991, by permission of the publisher.

whether there are differences in the various lanes. To do this the descriptive statistics, analysis of variance (ANOVA), and t-test functions were used to generate the analysis in Table 2-12. The descriptive statistics summarized the basic statistical parameters for the data and suggest that lanes 1 and 5 were different from lanes 2, 3, and 4. The analysis of variance with the large value of the F-statistic indicated that there was differences among the lanes. This was then confirmed by using the t-test and comparing pairs of lanes. The t-test determines if there is a difference in sample means by testing the hypothesis that there is no difference in the means (this is the null hypothesis). The calculated t-statistic is compared with a value from a standard table to determine whether there is any significant difference. For 19 degrees of freedom and a confidence limit of 0.05 (95% probability), a t-statistic greater than 1.7 indicates a significant difference. The calculation of t for lanes 1 and 2 gives a value of 10, and therefore the difference between them is very significant. Even the difference between lanes 1 and 5, the two lanes with the closest values, is significant, as the t-statistic of 2.3 is greater than 1.7. Therefore all the lanes differ significantly from each other.

There are also computer software programs specifically designed for statistical analysis and plotting. These perform a wide range of analysis, such as regression analysis, correlation matrices, nonparametric statistics, and so on. They are all reasonably priced. Some widely used programs are listed in Table 2-5, and Table 2-13

Table 2-12 Analysis of Data in Table 2-11 Using an Advanced Spreadhseet, Excel 4.0

Using Descriptive Statistics Tool

	Lane 1	Lane 2	Land 3	Lane 4	Lane 5	Roll Average
Mean	1.663	1.70575	1.72725	1.7085	1.673	1.6955
Standard error						0.0032245
Median	1.66	1.7025	1.7275	1.71	1.67	1.693
Mode	1.66	1.7	1.74	1.7	1.67	1.688
Standard deviation	0.01809333	0.01981991	0.01831558	0.01755443	0.01809333	0.01442038
Variance	0.00032737	0.00039283	0.00033546	0.00030816	0.00032737	0.00020795
Kurtosis	0.14706136	−0.62324752	0.75462072	−0.14079389	0.14706136	2.23411378
Skewness	−0.49878347	0.16954709	−0.8578079	−0.65495137	−0.49878347	−1.04309049
Range	0.07	0.07	0.07	0.06	0.07	0.06
Minimum	1.62	1.67	1.68	1.67	1.63	1.654
Maximum	1.69	1.74	1.75	1.73	1.7	1.714
Sum	33.26	34.115	34.545	34.17	33.46	33.91
Count	20	20	20	20	20	20
Confidence level (95%)	0.00792959	0.00868629	0.008027	0.00769341	0.00792959	0.00631988

Single-Factor ANOVA: Summary

Group	Count	Sum	Average	Variance
Lane 1	20	33.26	1.663	0.00032737
Lane 2	20	34.12	1.706	0.00039368
Lane 3	20	34.52	1.726	0.00037263
Lane 4	20	34.19	1.7095	0.00033132
Lane 5	20	33.46	1.673	0.00032737
Roll average	20	33.91	1.6955	0.00020795

Table 2.12 (*Continued*)

ANOVA: Source of Variation

	SS	df	MS	*F*	*P*-value	*F* crit.
Between groups	0.05598	5	0.011196	34.2679482	3.072E-21	2.29390906
Within groups	0.037246	114	0.00032672			
Total	0.093226	119				

t-Test: Paired Two-Sample for Means

	Lane 1	Lane 2	Lane 3	Lane 4	Lane 1	Lane 5
Mean	1.663	1.706	1.726	1.7095	1.663	1.673
Variance	0.00032737	0.00039368	0.00037263	0.00033132	0.00032737	0.00032737
Observations	20	20	20	20	20	20
Pearson correlation	0.48966659		0.3235484		0.45337621	
Pooled variance	0.00017579		0.00011368		0.00014842	
Hypothesized mean difference	0		0		0	
df	19		19		19	
t	−10.004415		3.3801153		−2.3639448	
$P(T \leq t)$ one-tail	2.6129E-09		0.00157131		0.01444346	
t Critical one-tail	1.72913133		1.72913133		1.72913133	
$P(T \leq t)$ two-tail	5.2257E-09		0.00314263		0.02888691	
t Critical two-tail	2.0930247		2.0930247		2.0930247	

Table 2-13 Statistical Program Comparison: Statistically Significant Differences[a]

	DataDesk 4.1	JMP 2.0	Minitab 8.2	SPSS 4.0.4	Statistica 3.1	StatView 4.01	SYSTAT 5.2.1
General							
Company	Data Description	SAS Institute	Minitab	SPSS	StatSoft	Abacus Concepts	SYSTAT
Phone	607/257-1000	919/677-8000	814/238-3280	312/329-3500	918/583-4149	510/540-1949	708/864-5670
Toll-free phone	○	○	800/448-3555	800/543-6609	○	800/666-7828	○
Price	$595	$695	$995 ($595 academic users)	$498	$595	$595	$895
Regression/ANOVA							
Linear/multiple linear	●/●	●/●	●/●	●/[b]	●/●	●/●	●/●
General nonlinear/polynomial	○/●	●/●	○/●	●/●	●/●	○/●	●/●
One-way ANOVA/n-way ANOVA	●/●	●/●	●/●	●/●	●/●	●/●	●/●
Repeated measures ANOVA	●	●	●		●	●	●
ANCOVA	●	●	●	●	●	●[c]	●
Factor analysis/principal components	○/●	●/●	●/●	●/●	●/●	●/●	●/●
Logistic regression	○	●	●		●	○	●
Cluster analysis							
Hierarchical	●	○	○	●	●	○	●
Single/complete/centroid linkage	●/●/○	○/○/○	○/○/○	●/●/●	●/●/●	○/○/○	●/●/●
Time series							
Linear/nonlinear smoothing	●/●	●[d]/●[d]	●/●	[e]/[e]	●/●	○/○	●/●
Moving average	●	●[d]	●	[e]	●	●	●
Autocorrelation/ARIMA	●/○	○/○	●/●	[e]/[e]	○/○	○/○	●/●
Fourier analysis	○	○	○	[e]	○	○	●

Table 2-13 (*Continued*)

	DataDesk 4.1	JMP 2.0	Minitab 8.2	SPSS 4.0.4	Statistica 3.1	StatView 4.01	SYSTAT 5.2.1
Nonparametric							
Kendall rank/Pearson	●/●	○/●	●/○	●/●	●/●	●/●	●/●
Spearman rank-order	●	●[d]	●	●	●	●	●
Wald–Wolfowitz/Mann–Whitney *U*	○/●	○/○	●/●	●/●	●/●	●/●	●/●
Kolmogorov–Smimov/Wilcoxon	○/●	○/●	○/●	●/●	●/●	●/●	●/●
Kruskal–Wallis/Lilliefors	○/●	●/○	●/○	●/●	●/●	●/●	●/●
Goodman–Kruskal	○	○	○	●	○	●	●
Friedman/Cochran *Q*	○/○	○/○	●/○	●/●	●/●	●/●	●/○
Graphics							
Pie/bar/line	●/●/●	●/●/●	○/●/●	f/●/f	●/●/●	●/●/●	●/●/●
X-Y/regression	●/●	●/●	●/●	●/f	●/●	●/●	●/●
Density/probability	●/●	●/●	○/○	○/○	●/●	○/○	●/●
Boxplots/stem-leaf	●/○	●/○	●/●	○/f	●/●	●/●	●/●
Scatterplot matrix	●	●	○	○	●	○	●
3-D scatter (spin)/surface	●/●	●/○	○/●	○/○	●/●	g/○	●/●
Pareto/control	○/○	●/●	●/●	○/○	○/○	h/h	○/○
Color draw tools	●	○	○	f	●	●	●
Color titles/title fonts	●/●	●/●	●/●	f/f	●/●	●/●	●/●
Label overlays	●	●	○	f	●	●	●

Source: Seiter (1993).
[a] ●, Yes; ○, no
[b] With $495 SPSS Advanced module.
[c] With $100 SuperANOVA option.
[d] Two-step process.
[e] With $395 SPSS Trends module.
[f] With $195 CricketGraph module.
[g] With $100 MacSpin module.
[h] With $100 Quality Control module.

compares the functions of these programs for Macintosh computers. These tend to be more sophisticated and they require more knowledge of statistics to use than do the spreadsheets. A compensating factor is that the manuals for these are often an excellent reference for statistical methods. They teach the statistics as well as how to use the program. The Minitab program evolved from a course in statistics at Pennsylvania State University. Obtaining a program and using the manual may be a good way to gain an introductory knowledge of statistics (Ryan et al., 1985).

2.5 ON-LINE INSPECTION SYSTEMS

The last diagnostic technique that will be discussed in this chapter is on-line inspection systems for defect detection. These are sophisticated image analysis systems that are capable of detecting film nonuniformities, such as spots, streaks, dirt, and so on, in a moving web. They can be located either on the coating line or in the converting hardware and will give an instantaneous indication of a possible defect in the web. Their function is to inspect the web for defects and to indicate their location so that areas of the web that are unacceptable for the customer can be identified. These systems originated in late 1960s and came into widespread use in the 1980s as they became economically competitive due to progress in electronics and in laser technology. They were originally designed to replace the human operators who inspected the web visually. As coating speeds increased and coating uniformity demands became more stringent, it was essential to automate this function to improve its reliability. They are particularly useful in the coating of precision products such as photographic film.

These systems could not be useful until the development of reliable lasers, which provide a coherent spot light source, and of the computer analysis capability. The laser beams are focused on a rotating mirror that effectively scans the beam across the web. A collecting device such as light pipe then collects the light and analyzes for distortions that are a result of the defects. These distortions are then processed by the computer to give quantitative information about the defect and its location. These can be used either in the reflected mode or in transmission. Nonlaser sources can be used, but they are not as sensitive as laser sources. Other approaches to this are the use of camera systems that focus on the web and detect the difference in optical density that results from the defect. Table 2-14 lists some of the suppliers of these inspection systems.

These systems do not currently fit into the problem solver's tool kit as a routine tool, since they are relatively sophisticated to use, very expensive to purchase, and complex to operate. However, they are a technique that will grow in importance, will gain more widespread use, and will be a mainstay of the problem-solving process in the future. At present, while they do indicate the presence of a defect, they do not give a detailed characterization of the defect, and additional microscopic analysis is needed. There is also a limitation on the size of the defect to be detected, which depends on the type of the instrument, the type of defect, the optical properties of the coating, and the speed of the web. However, the technology is advancing at a rapid rate, resulting in system improvements and cost reductions. As a result, it is now becoming practical to consider using these as a on-line troubleshooting tool. As defects are detected, the coater can instantly be shut down for the troubleshooting

Table 2-14 Equipment Sources

General instrumentation	
Cole-Parmer	Chicago, IL
Edmund Scientific	Barrington, NJ
Fisher Scientific	Various
Temperature measurements	
Omega Engineering	Stamford, CT
Rheological measurements	
Automation Products	Houston, TX
Brabender	Hackensack, NJ
Cambridge Scientific	Cambridge, MA
Haake	Paramus, NJ
Mettler	Hightstown, NJ
Paul Gardber	Pompano Beach, FL
Kruss USA	Charlotte, NC
Nametre	Metuchen, NJ
Rheometrics	Piscataway, NJ
Data logging	
ACR Systems, Inc.	Canada
ADAC	Woburn, MA
DATAQ Instruments	Akron, OH
Industrial Computer Source	San Diego, CA
Keithley Metrabyte	Taunton, MA
National Laboratories	Austin, TX
Omega Engineering	Stamford, CT
Strawberry Tree	Sunnyvale, CA
Film inspection systems	
SICK Optic-Electronic, Inc.	Eden Prairie, MN
Flow Vision, Inc.	Charlotte, NC
Sira, Inc.	Darien, CT
Intec Corporation	Trumbull, CT
Systronics, Inc.	Norcross, GA
Dalsa, Inc.	Waterloo, Ontario
Motion Vision Systems	Waterloo, Ontario

process to begin. In addition, as experiments are being run on the coating applicator or dryer, the inspection system can provide rapid feedback and greatly speed up the troubleshooting process.

2.6 KEEPING CURRENT

All of the tools covered in this section are rapidly changing. Both new tools and new improved versions of old tools are being introduced all the time. The troubleshooter should be aware of these because the new tools can help do the job more effectively and the old tools do need to be replaced when they wear out or get broken. A convenient method of staying current and having a convenient source of information in this area is to get on the subscriber list for monthly magazines specializing in equipment. These are usually free. Once you are on one of these mailing lists, more subscriptions will follow. Manufacturers of these tools advertise in these journals and will send you information. Some useful references in this category are *Scientific Computing and Automation*, *Laboratory Equipment*, and *Omega Encyclopedia*.

Another source of information is the multivolume *Thomas Register of American Manufacturers*, which is issued yearly. It lists manufacturers of a wide range of commercially available products and service, has information on the companies, and has extracts from published catalogs. Most laboratory equipment houses will gladly send you their annual catalog, which contains descriptive material on many instruments. Table 2-14 lists some sources of information for the tools described in this chapter. Another way of keeping current is by attending trade shows or technical society meeting which have expositions, such as those of the American Chemical Society, the American Institute of Chemical Engineers, and the Technical Association of the Pulp and Paper Industry. Manufacturers of all types of instruments, computers, and software programs tend to exhibit their new equipment at these shows.

REFERENCES

Adamson, A. W., *Physical Chemistry of Surfaces*, 4th ed., Wiley, New York, 1982.

Brooks, C. J., I. G. Betteley, and S. M. Loxston, *Fundamentals of Mathematics and Statistics for Students of Chemistry*, Wiley, New York, 1979.

Cohen, E. D., and R. Grotovsky, "The application of digital image analysis techniques to photographic film defect test method development," *J. Imag. Sci. Technol.* **37**, 133–148 (1993).

Davies, O. L., ed., *Statistical Methods in Research and Production*, Hafner, New York, 1958.

Frost, P. J., and E. B. Gutoff, *The Application of Statistical Process Control to Roll Products*, 2nd ed., P.J. Associates, Quincy, MA, 1991.

Karasek, F. W., "Detection limits in instrumental analysis," *Res. Dev.*, pp. 20–24 (July 1975).

Karasek, F. W., and R. J. Laub, "Detection limits in instrumental analysis," *Res. Dev.*, pp. 36–38 (June 1974).

Maranda, B., "An economical, powerful system for video-enhanced microscopy," *Am. Lab.* (Apr. 1987).

Natrella, M. G., *Experimental Statistics*, NBS Handbook 91, U.S. Dept. Commerce, Washington, DC, 1966.

Perry, J. H., *Chemical Engineers' Handbook*, 4th ed., Table 23-5, pp. 23-31 to 23-47, McGraw-Hill, New York, 1963.

Ryan, B. F., B. L. Joiner, and T. A. Ryan, *Minitab Handbook*, 2nd ed., Duxbury Press, Boston, 1985.

Sawyer, L. C., and D. T. Grubb, *Polymer Microscopy*, Chapman & Hall, New York, 1987.

Seiter, C., "The statistical difference," *Macworld*, pp. 116–121 (Oct. 1993).

Studt, T., "Surface science pushes theoretical limits," *Res. Dev.*, pp. 64–72 (Aug. 1990).

Willard, H., L. Merritt, Jr., and J. A. Dean, *Instrumental Methods of Analysis*, Van Nostrand, New York, 1974.

Wright, H., and L. E. Gogolin, "In-line viscosity measuring improves control in coating process," presented at *AICHE Spring Meeting*, New Orleans, LA, 1992.

Wright, H., and L. Gould, "Using electromagnetics to measure viscosity," *Sensors* (Oct. 1986).

CHAPTER III

Problems Associated with Feed Preparation

The coating solution preparation procedure as well as the system that delivers the coating solution to the apparatus can also be a source of defects and problems. In addition, poor or improper preparation and poor control of flow to the coating stand can lead to a number of coating defects and problems, such as the presence of particulates, agglomerates, or bubbles in the coating; poor cross-web uniformity due to concentration and viscosity variations across the coating stand; poor down-web uniformity due to viscosity fluctuations; plugging of filters due to colloidal instability or to incomplete dispersion in the coating liquid; and down-web coverage fluctuations due to poor flow control to the coating die. If the coating liquid is not suitable, a high-quality coating cannot be achieved, and if a good feed system to the coating die is not available, uniform coverage cannot be achieved.

3.1 DIRT AND OTHER PARTICULATES

Filtration is needed to remove dirt and other particulates from the coating liquid. Particulates can come from almost anywhere. Dirt can be carried in with the raw materials. Particles can come from dust in the air. Corrosion in the pipes leads to dirt in the coating. Pumps can chew up gaskets to form particulates. Colonies of bacteria can slough off a pipe into the flow stream. Some polymer or resin may not dissolve, but only swell, and these swollen, undissolved polymer globs in the liquid, called fisheyes, then become spot defects when coated on a web. Some particles of active ingredients may be too large in size and can show up in the coating as defects, and others, small in size, may agglomerate to form large clumps. However they arise, all these particulates must be removed before the coating liquid is coated. Filters must be used to remove them. Filters should be used on all coating lines, even when the coating liquid, as is often the case, contains dispersed active ingredients. The filters must permit the active ingredient, which is of small particle size, to pass through while preventing the passage of larger, unwanted particles.

Particulates may be classified into one of two types: hard and soft. Hard particles do not deform and are are relatively easy to filter out. They comprise dirt, agglomerates, corrosion products, pieces of gaskets, and so on. Dried coating liquid may be hard or soft. Soft particles deform, and can elongate to squeeze through pores in a filter under a high-enough pressure drop, to re-form their spherical shape after passing through. If soft particles are present, the pressure drop across the filter must be limited to a low value, perhaps 3 to 5 psi (20 to 35 kPa). This means that a large area of filtration (more filters) may be necessary to maintain the flow rate with a low pressure drop. Swollen but undissolved polymer and bacterial colonies are the main soft particulates. It is desirable to prevent the formation of soft particles completely. Sterilization by heat or with bactericides should be used to prevent bacteria from growing. The polymers and resins should be completely soluble; if not, a better source of supply should be found. The solution preparation method should be such that the polymer does dissolve completely.

The dissolution of polymers and resin requires special care. If powdered polymer is dumped into a solvent, the outside of the mass will soak up solvent and will start to swell. However, this outer layer of swollen polymer will act as a skin to greatly slow down the flow of solvent to the interior. The dissolution time may then go from minutes to many hours. It is generally recommended that polymer be added slowly to the eye of a vortex in the agitated solvent. With particular polymers and solvents other procedures, and special temperature cycles, may be recommended.

3.1.1 Filtration

Filters come in two types: surface filters and depth filters. Surface filters can be represented as a membrane with pores, as in Figure 3-1. Particles larger than the largest pore cannot pass through (assuming that the particles are hard). Thus these filters can be considered absolute filters when the pore size is carefully controlled and fairly uniform. The active area for filtration is just the surface area. Filter clothes used in filter presses are surface filters, as are the membranes used to remove yeast cells from beer in room-temperature sterilization. Surface filters are used when it is imperative to remove every single large particle. When a heavy solids loading is present, a filter cloth may just act as a support for the filter cake, which does the actual filtration.

Depth filters function differently. They generally consist of a filament wound around a central core in a special pattern, to give a cylindrical element perhaps 1

Figure 3-1. Surface filter. Hard particles larger than the largest pore cannot pass through.

in. (2.5 cm) thick (Figure 3-2). The openings are relatively large. The particles are removed by impaction. A particle entering a pore may strike a filament before changing direction, and then stick to the filament. If it passes through the surface openings, it still has many more opportunities to hit a filament and stick to it in the interior. Thus some small particles may be trapped and some large particles may pass through. For example, if the initial material had a particle-size distribution as shown in Figure 3-3, the filtrate—the filtered coating liquid—may still have particles with a broad, but narrower, particle-size distribution such as shown in the same figure. Essentially all the particles larger than a certain size are removed, while most of the small particles remain. The fraction of particles retained in the filter increases with increasing size.

Depth filters do not have a standard rating. The filter shown in Figure 3-3 can be rated anywhere from the largest size that passes through to the smallest size that is retained. Different manufacturers use different rating systems. Thus you cannot compare ratings from one manufacturer to another. A 10-μm filter from one manufacturer may be larger than a 20-μm filter from another. All you know is that a 10-μm filter from one manufacturer is smaller than a 20-μm filter from the same manufacturer.

Depth filters are used where the liquids have relatively low loading of solids to be removed. They are commonly used in coating lines. For dirty liquids it is sometimes desirable to use a coarse filter followed by a fine filter. Filters must be installed properly. It is possible to mount a depth filter with the gasket improperly seated. Such a filter will not filter. All the liquid will bypass the filter to flow through the opening formed by the improperly seated gasket.

Depth filters are made from filaments and fibers. It is not uncommon for fibers to slough off in first use. Also, the filters are often sealed on their ends with a potting compound, which can leach out low-molecular-weight species on first use. It

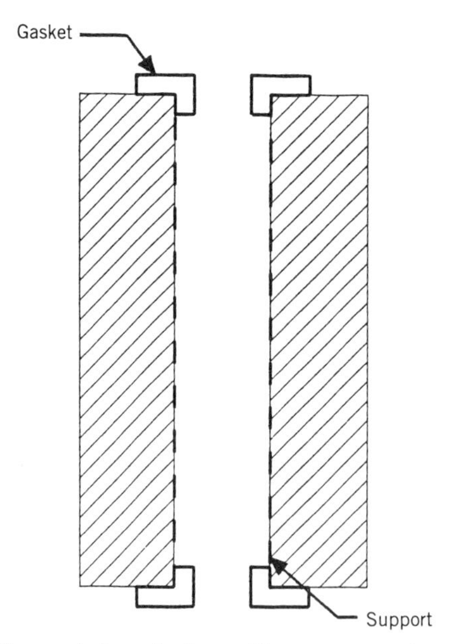

Figure 3-2. Cartridge filter made by winding a filament around a support is a typical depth filter.

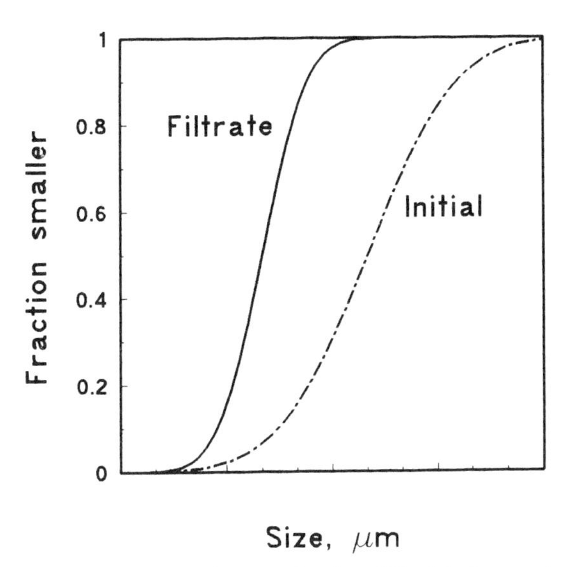

Figure 3-3. Particle-size distribution of contamination, "dirt," before and after filtration with a depth filter.

is advisable to flush the filters before using. If material leaching out of the filter is not a concern, recirculating coating liquid can be used to flush the filter, as this will remove all the loose fibers. Otherwise, solvent should be used.

Depth filters are made of many different fibers, including cotton and all the synthetics. The proper material must be chosen for chemical compatibility. The surface properties would also affect how well particles stick to the filament, and this should be checked by experiment.

Soft particles such as swollen polymers or gels can be forced through pores much smaller in size than the particles. Therefore, when these soft particles are present and must be removed, the filters have to operated under low differential pressures, which means having extra filtration area for a given flow. Both surface and depth filters are used for removing these soft particles. Bag filters, a type of surface filter, have been recommended for removing these soft particles.

Two sets of filters, one in operation and one a spare, are recommended, so that when the operational filter plugs, it can be changed quickly. Also, as particulates can break off in the lines to the coating station, it is recommended that an additional, small, relatively coarse filter, such as of wire mesh, be installed just before the coating stand.

3.1.2 Agglomerates

Agglomerates exist in a coating liquid because of initial poor dispersion or because of agglomeration of dispersed particles due to poor colloidal stability. The initial dispersion may be simple or difficult to achieve depending on the material. With some solids consisting of weak agglomerates, gentle agitation is adequate for good dispersion; for others, one of many mills must be used. The *Chemical Engineers' Handbook* (Perry and Chilton, 1984) gives a good summary of the tumbling mills, stirrer and sand mills, vibratory mills, roller mills, disk attrition mills, colloid mills, and so on, that are available.

Once a solid has been dispersed, it must be kept dispersed. Surfactants and polymers are used for this purpose. The surfactants used here are called *dispersing agents*. Most often the dispersing agents are anionic surfactants. The surfactants adsorb onto the surface and impart a charge to the particles. With anionic surfactants the charge is negative. As like charges repel each other, the particles gain stability. High concentrations of ionizable salts, especially salts with polyvalent ions, decrease this charge stability and cause agglomeration. In some cases, use of two different types of surfactants, such as an anionic with a nonionic surfactant, may give increased stability.

Polymers adsorb onto the surface with chains dangling out around the particle, keeping similarly coated particles away. This protection against agglomeration by adsorbed polymers is called *steric stabilization*. On the other hand, if the polymer concentration is very low, a polymer chain dangling free from one particle can attach itself to an uncoated area on another particle, causing agglomeration. The polymer sensitizes the dispersion to agglomeration. Thus a polymer at very low concentrations can be an agglomerating or coagulating agent, while at higher concentrations it can be a dispersing agent. This is illustrated in Figure 3-4.

With some dispersions the pH can be an important factor in stability, as the pH can strongly affect the ionic charge on some particles. If latices have carboxylic acid groups on the surface, they will become ionized as the pH is raised, thus developing a negative charge and increasing in stability.

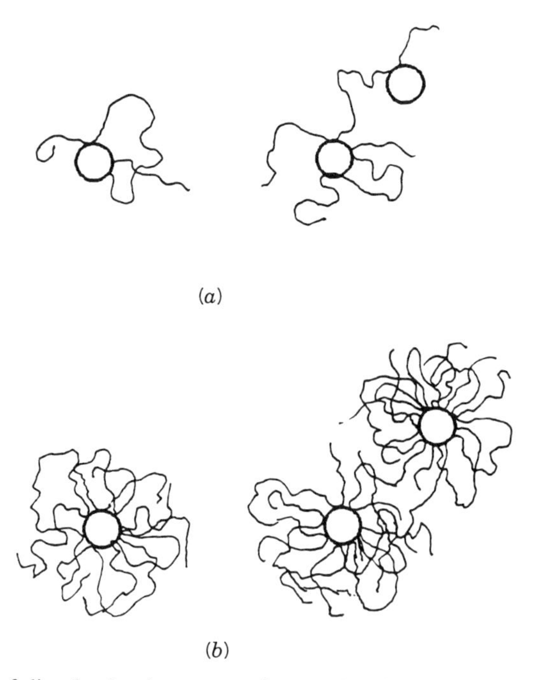

Figure 3-4. Effect of dissolved polymers on dispersed solids: (*a*) sensitization to agglomeration at low concentrations, where polymer bridges tie particle together; (*b*) protection against agglomeration at higher concentrations.

Agglomeration can be caused by the pumps used to feed the coating applicator from the solution preparation vessels. The high shear rates in centrifugal pumps, and the highly localized high shear in gear pumps where the gear teeth wipe the wall, can cause agglomeration of dispersed materials. This is of particular concern in coating processes where there is a recycle stream from the coating stand back to the feed vessels, thus subjecting particles to many cycles of this destabilizing force. In such a system the defects, if not removed by filtration, would increase with time. If agglomeration due to pumping is suspected, set up a recirculating system of a feed vessel and a pump, and monitor the solution for agglomeration with time. An in-line filter in such a system should show little increase in pressure drop once the initial particulates are removed, unless new agglomerates are formed. Agglomeration from whatever cause often shows up in rapid plugging of the filters.

3.2 BUBBLES

Bubbles form when air is introduced into a liquid, or is dissolved in the liquid and comes out of solution. Bubbles can form in the coating process and in the dryer, but in most cases they arise from air introduced in the feed preparation and handling. Care should be taken to ensure that air is not introduced into the system. It is usually much easier to prevent air from entering the system than it is to remove bubbles once they are formed. However, as one never knows when air will be introduced, it is well worthwhile to have debubbling devices in place to remove bubbles in case any do form.

Air can be introduced in a number of ways:

1. During mixing of liquids if the agitation is too strong.
2. When liquid is added to another liquid and splashes on the surface.
3. When liquid is pumped into another vessel, and the line, initially filled with air, enters below the surface.
4. At a pump suction, if the seal leaks and the suction pressure is below atmospheric.
5. From air pockets in the lines. To avoid air pockets, all lines should slope upward.

As this is usually impossible, it is desirable to have all lines slope, with, if possible, no more than one high point, and have a bleed valve at that high point.

Bubbles can also form from air that is dissolved in the coating liquid. The air can come out of solution when the solution temperature is raised, or where the pressure is low. Low pressure occurs:

1. At the ends of lines, such as at the coating stand.
2. Where the velocity is increased. This causes the pressure to be reduced and occurs in valves and other constrictions.

It is therefore desirable to remove bubbles and to reduce the amount of dissolved gas, to reduce the likelihood of bubbles forming in the lines. It should be noted that if one fills a glass of clear water and lets it stand for several hours, bubbles will appear attached to the glass walls.

To reduce the amount of dissolved gas, one has to reduce the solubility of the gas. The solubility of gas is less at lower pressures and higher temperatures. Normally, there is a temperature limitation. Liquids are usually degassed by pulling a vacuum. One should also allow enough time for the bubbles to form and to rise to the surface. The nucleation of the gas to form bubbles can be aided by ultrasonic energy. A shallow depth is advisable, for less time is required to rise to the surface.

Degassing is frequently done on the coating liquid before it is pumped into the system. The vacuum should be high but insufficient to cause boiling. Mild agitation is helpful but is often not used. The length of time that should be allowed varies with the viscosity of the liquid and its depth. It can vary from minutes to hours. The amount of solvent that evaporates during this process is normally negligible.

Debubbling, the removal of bubbles that already exist, is normally done in the line from the coating liquid to the coating stand. Often, the liquid is pumped onto a spherical dome, to flow down the sides while a vacuum is pulled. The vacuum should be high but not enough to cause boiling. The vacuum causes the bubbles to expand and thus rise faster. This unit has to be elevated to a height sufficient to support that vacuum (perhaps 32 ft for water), or else a pump with a level control must be used to remove the liquid.

Another way to remove bubbles without the use of vacuum is to use a simple trap, with liquid entering below the surface near the top, and leaving at the bottom. There should be enough residence allowed for any bubbles to rise to the surface. Ultrasonic energy is sometimes added to nucleate air coming out of solution. Centrifugal devices are also used, as are units similar to liquid cyclones that use the fluid energy to generate the centrifugal force.

Some filters will filter out bubbles, much as entrainment separators remove drops of liquid from a gas stream. The wettability of the fibers in the filter is important. Here the liquid should flow downward at a velocity slow enough that the bubbles caught by the filter can agglomerate to a larger size and rise up against the flowing stream. Other types, such as the use of sonic energy to drive the bubbles out of the flow stream, are under development but are not yet on the market (Hohlfeld and Cohen, 1993).

3.3 POOR CROSS-WEB UNIFORMITY

The cross-web coverage uniformity is mainly dependent on the construction of the coating system; however, nonuniformities in the concentration in the coating liquid from the center to the wall of the line may cause nonuniformities in coverage. The liquid in the center of the line may go to one part of the coating, and the liquid near the wall may go to another part. And nonuniformities in the liquid can arise just from the fact that it is flowing; a dispersion flowing in a line will segregate to some extent, with the maximum concentration somewhere between the center and the wall. As the flow in coating lines is usually laminar, there is little or no mixing in the lines. Therefore, an in-line mixer should be installed just before the coating stand.

The temperature may also vary from the center to the wall of the line. This will happen if the liquid is heated by flowing through a jacketed line, or if the liquid temperature is other than ambient such that the wall of the line will heat or cool the liquid. This can cause problems because the viscosity is a strong function of temperature. If liquid of one temperature goes to one part of the coating stand and liquid of a different temperature to another part, the viscosity will vary across the coating stand causing the flow onto the web to vary, and thus the coverage will vary across the web. Both in roll coating and in a coating die, the flow is dependent on the viscosity. (In a coating die the average flow just depends on the pumping rate, but the coverage at any cross-web position will vary as the viscosity varies.) For this reason, too, an in-line mixer should be installed just ahead of the coating stand.

3.3.1 In-Line Mixers

There are both dynamic and static in-line mixers. Dynamic mixers are essentially an agitated vessel in the line. They provide excellent mixing and are high shear units. They have a moderate holdup; the lowest holdup unit we are aware of is 600 mL. Because of the rotating mixing blades, they do introduce high-frequency pulsations. These can, however, be damped out just by a short length of plastic tubing. Many companies that make mixers and agitators also make these dynamic mixers.

Static mixers are small units with no moving parts, which are the diameter of the line and contain internal elements that essentially cut and recombine the liquid. In one type an element cuts the fluid into two slices. With 21 elements, which is a common number, there would then be 2×10^6 slices. These units give fair-to-good mixing, have very low holdup, and are low shear devices. Bubbles may be trapped in them unless there is upflow, depending on the internal construction. Kenics, Koch, and Sulzer are three suppliers of static mixers.

3.3.2 Temperature Control

In roll coating the coverage is strongly dependent on the liquid viscosity, and this in turn is a strong function of temperature. In a coating die, if the viscosity, and therefore the temperature, is not uniform across the full width, the flow will not be uniform. The coating liquid and the coating die should be at the same temperature. Thus unless one coats at room temperature and room temperature does not fluctuate appreciably, and the coating solution is at the same room temperature, temperature control is necessary.

The container for the coating liquid can be jacketed to heat or cool the coating liquid, a heat exchanger in the line to the coating stand can be used, or jacketed lines to the coating stand can be used to adjust the temperature. If jacketed lines are used to heat the liquid, the heat transfer liquid—usually water—should be hotter than the desired temperature, with the jacket flow controlled to give the desired temperature of the coating solution. Jacketed in-line mixers act as heat exchangers and can be used as such. If the jacket water is just at the desired coating liquid temperature, the coating liquid will only approach that temperature, but not reach it. Jacketed lines with the jacket water at the desired temperature should be used only to maintain the coating liquid at the desired temperature, when it had already been brought to that temperature.

3.4 FLOW CONTROL FOR DOWN-WEB UNIFORMITY

In roll coating the wet coverage depends on the geometry of the system, the liquid viscosity, and the speeds of the various rolls at the coating station. In slot, extrusion, slide, and curtain coating, the wet coverage depends on the pumping rate of liquid to the coating die, as well as the web speed and the coating width. It is therefore necessary to have good flow control from the feed vessel to the coating die in these systems. Good control of the web speed is also important for good down-web uniformity.

3.4.1 Feeding the Coating Liquid

In general, there are three methods of feeding the coating liquid to a coating die, as shown in Figure 3-5. Gravity feed can be used, where excess liquid is pumped to an elevated head tank, with the overflow returning to the feed tank. The head tank provides a completely constant head or pressure in the lines, making flow control, indicated symbolically by a flowmeter connected by dashed lines to a control valve, relatively easy. The system requires little maintenance, the tank is easy to refill, and

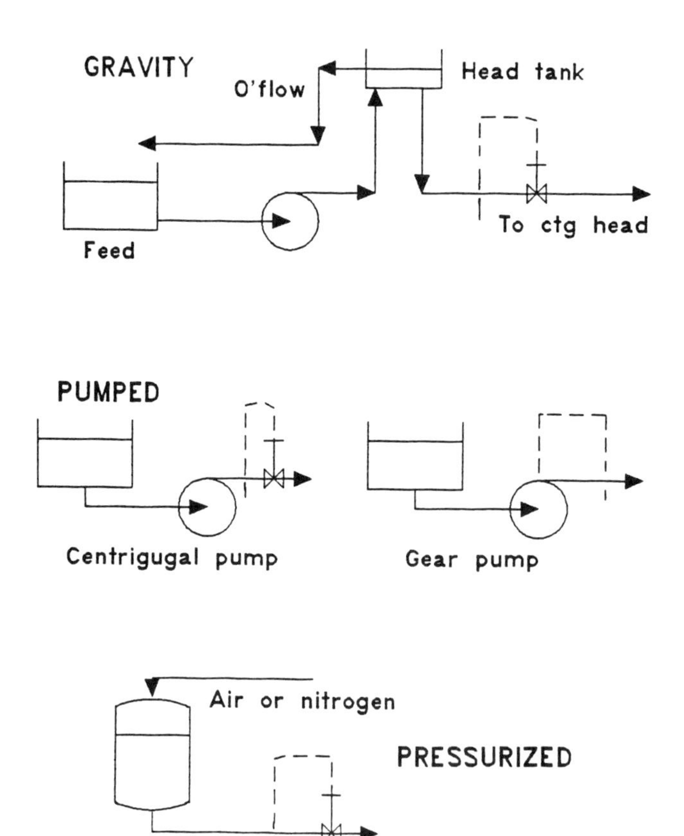

Figure 3-5. Methods of feed coating liquid to the coating die.

the system is pulsation-free. However, the elevated tank requires long lines and the head obtainable is still relatively low. The holdup in these long lines is large, and the system is open. Several decades ago this was the recommended system; now it no longer is.

Pressurized systems, shown in the bottom of Figure 3-5, are also pulsation free. A pressure vessel is required for this system, which is completely closed. There is the likelihood of dissolving gas in the liquid, which could lead to bubble formation later. If the liquid is not agitated, not very much gas will dissolve, and if the coating liquid is withdrawn from the bottom of the vessel, this should not be a problem. It is difficult to refill the vessel while it is pressurized, so two vessels would probably be required for long runs. It may be difficult to achieve bumpless transfer from one vessel to the other, so a single chatter bar would probably result when switching vessels. Also, it is difficult to empty a vessel completely. If one attempts to, and if air (or nitrogen) gets into the flow line, the pressure drop in the line will decrease because the viscosity of air is very low, the velocity in the line will therefore increase, and a slug of liquid and gas will come shooting out of the coating die. This has to be avoided at all costs. Pressurized systems are used successfully, but are not recommended for new installations except in pilot areas.

The pumped systems are the ones we recommend. The vessels are easy to refill and can be pumped dry if desired. Pulsations may arise from both centrifugal and from gear pumps, but these pulsations are easily damped out. Usually, plastic tubing (such as polyethylene tubing) after the pump provides enough damping to eliminate any pulsations at the coating die. Pumps do require regular maintenance, but this is a minor item. With a gear pump the flow meter would control the speed of the gear pump rather than adjust a control valve.

For feeding a very low flow rate stream, or low flow rates of an additive into the main stream, ahead of an in-line mixer, a syringe pump is usually used. For longer runs two pumps can be used, one filling while the other is emptying, but it is difficult to achieve bumpless transfer from one to the other.

3.4.2 Flow Control

The flow to a coating die should be metered and controlled. One should not depend on a calibrated valve or gear pump to control the flow unless only very crude coverage control is needed. One should measure the flow and have feedback control to a valve or to the speed of a gear pump. There are many different types of flowmeters available, but we recommend three types: magnetic flowmeters, ultrasonic meters, and Coriolis meters. All three can achieve flow accuracies of 0.5% of full scale.

Magnetic Flowmeters

A conductor passing through a magnetic field generates a voltage. In the magnetic flowmeter, illustrated in Figure 3-6, the fluid itself is the conductor, and the magnetic field is external to the nonconductive section of the pipe. The electrodes measure the voltage generated. The voltage to the coils of the magnet can be either direct current or alternating current, giving two classes of meters. These meters work very well, but the liquid in the pipe must have some conductivity, essentially limiting its use to

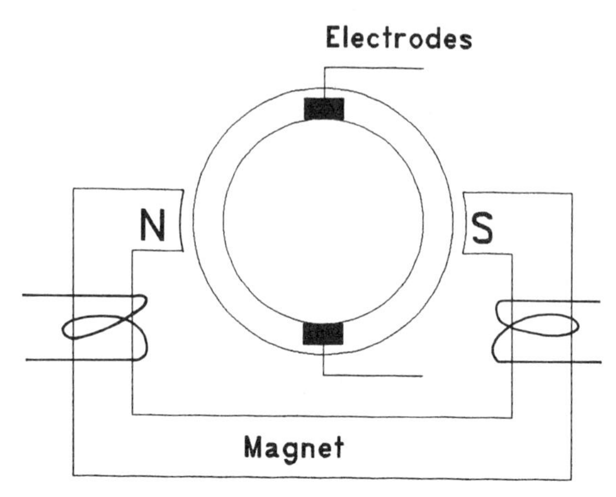

Figure 3-6. Principle of the magnetic flowmeter. A conductor, the liquid, passing through a magnetic field, generates a voltage that is a function of the flow rate.

aqueous coating liquids. In some situations the electrodes can foul and give erratic readings, but it is not difficult to clean the electrodes.

Ultrasonic Meters

In an ultrasonic meter (Figure 3-7) the transit time for an ultrasonic pulse to travel upstream is compared with the transit time to travel downstream. The transit time downstream with the flow is less than against the flow, and the difference is a measure of the flow rate of the liquid.

Coriolis Meters

Coriolis meters are true mass flow meters, compared to the others which measure the volumetric flow rate. They are based on Coriolis forces, which are the same forces that cause a swirl in the water draining from the bathtub. One type is shown in Figure 3-8. The liquid is pumped through a stainless steel U-tube, which is vibrated at its natural frequency. As the liquid moves through the tube it take on the tube's motion up or down. When the tube is moving up, liquid moving into the tube resists this

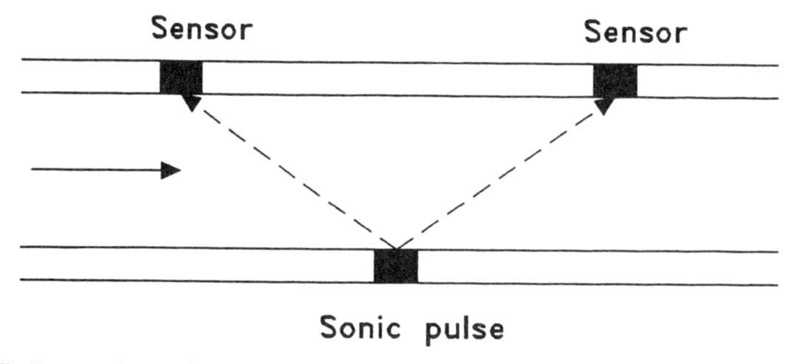

Figure 3-7. In an ultrasonic meter, the difference in transit time for an ultrasonic pulse to travel upstream or downstream is a measure of the flow rate.

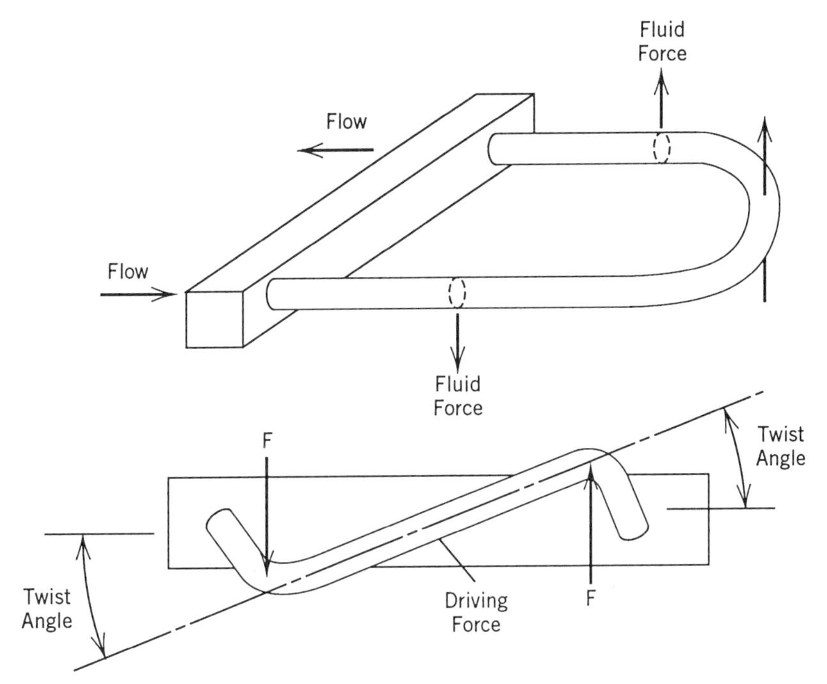

Figure 3-8. In a Coriolis meter the twist in the tube is a measure of the mass flow rate. [From Micro Motion, Inc., Product Bulletin, Boulder, CO 80301].

upward motion, effectively pushing downward on the tube. On the other hand, liquid flowing out of the tube tends to keep moving upward, and pushes upward against the walls of the tube. This combination of forces imparts a twist to the tube. This twist is measured and is directly proportional to the mass flow rate of the liquid.

REFERENCES

Hohlfeld, R. G., and E. D. Cohen, "Ultrasonic debubbling using acoustic radiation pressure," paper presented at *Annual Meeting of the Society for Imaging Science and Technology*, Cambridge, MA, May 1993.

Perry, R. H., and C. H. Chilton, *Chemical Engineers' Handbook*, 6th ed., McGraw-Hlll, New York, 1984.

CHAPTER IV

Problems Associated with Roll Coating and Related Processes

4.1 COATING METHODS

In this chapter we briefly cover coating problems in roll coating and related processes, that is, in all coating methods other than those where the wet coverage is determined solely by the flow rate to a coating die and the coating speed. Coyle (1992) gives a working definition of roll coating as *a coating method whereby the fluid flow in a nip between a pair of rotating rolls controls both the thickness and the uniformity of the coated film.* Extending this definition to include a roll of infinite radius covers flat sheets, and considering a roll of zero velocity, covers knives and blades. The term *roll coating* is generally understood, and we will now illustrate some specific configurations.

In forward-roll coating (Figure 4-1), the web travels over a backup roll while an applicator roll, turning in the same direction as the web and with a small gap separating it from the web, applies the coating. As the web and the applicator roll are turning in the same direction, the coating liquid carried into the nip on the applicator roll splits at the nip exit with some remaining on the applicator roll and the rest leaving as the coating on the web. The web and the applicator roll may be moving at the same speed, or the speeds may differ. The coating liquid can be supplied to the applicator roll by many different methods. Figure 4-1 shows a pan-fed applicator roll where the rotating applicator roll picks up liquid from the pan in which it sits. The liquid can also be supplied by a slot die aimed upward to give a fountain-fed system. A fountain roll can pick up the liquid from the pan and transfer it to the applicator roll to give a three-roll pan-fed forward-roll coating system. In another system a third roll butts against the applicator roll, with the liquid supplied to the nip between them and contained between end dams, to give a three-roll, nip-fed forward-roll coating system. Any number of rolls can be used, and the rolls can be smooth chrome-plated steel or they can be rubber covered. Ceramic rolls are coming into use.

In reverse-roll coating the backup roll is usually rubber covered, and the web travels in the opposite direction to the applicator roll, presses against it, and wipes off all the coating from the applicator roll, as in Figure 4-2. The amount of liquid on

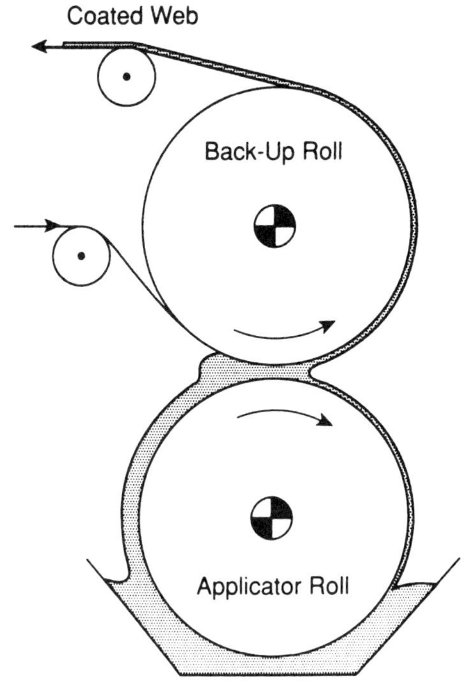

Figure 4-1. Two-roll, pan-fed, forward-roll coater. [From Coyle (1992). Reprinted with permission by VCH Publishers © (1992).]

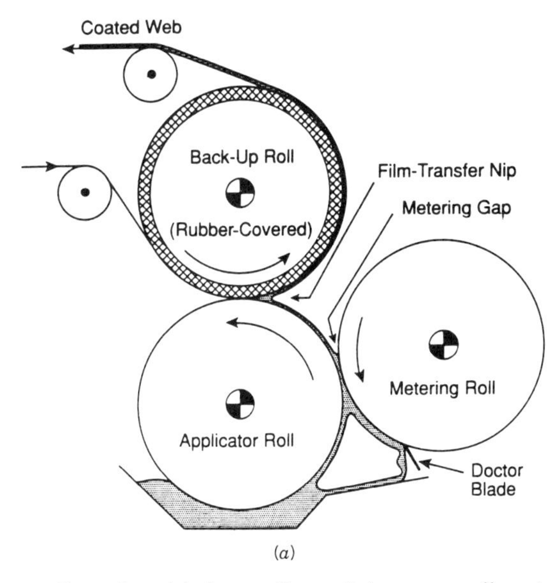

Figure 4-2. Reverse-roll coating: (*a*) three-roll pan-fed reverse roll coater; (*b*) four-roll pan-fed reverse roll coater; (*c*) three-roll nip-fed reverse-roll coater. [From Coyle (1992). Reprinted with permission by VCH Publishers © (1992).]

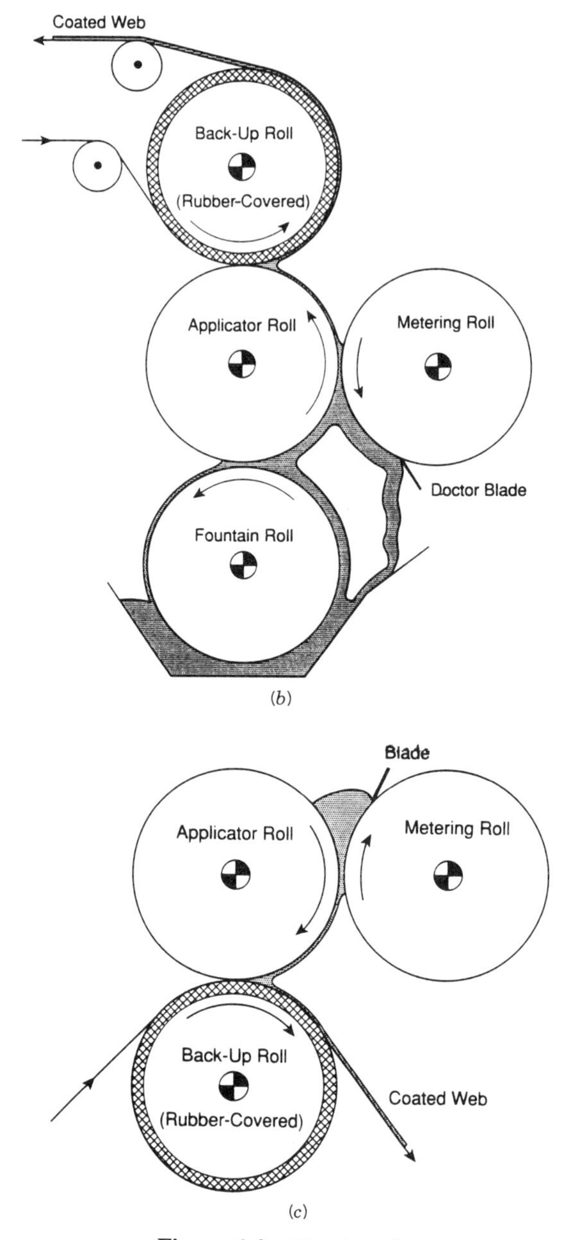

Figure 4-2. (*Continued*)

the applicator roll is controlled by a metering roll that rotates in the opposite sense. Figure 4-2 shows several different reverse-roll configurations. Note that in every case a doctor blade is used to wipe off all the coating liquid from the metering roll shortly after it does its job of metering the fluid on the applicator roll.

In gravure coating the gravure roll has indentations or cells that hold the liquid after the surface is wiped off by a doctor blade. The liquid is then transferred to the web moving on a rubber-covered backup roll, as in Figure 4-3*a*. This a direct

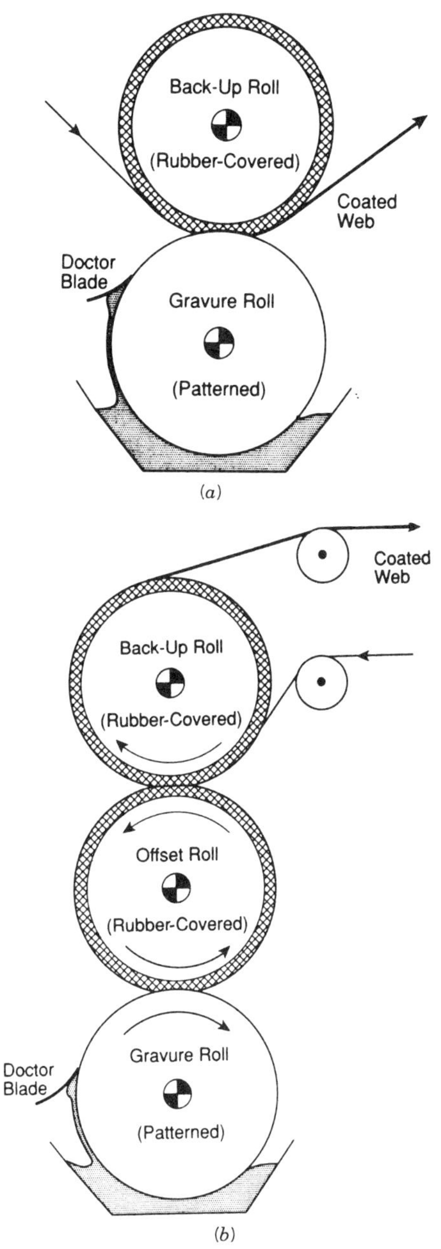

Figure 4-3. Gravure coating: (*a*) direct gravure; (*b*) offset gravure. [From Coyle (1992). Reprinted with permission by VCH Publishers © (1992).]

gravure. If the web travels in the direction opposite to the gravure roll, we have reverse gravure. If the liquid is first transferred to an intermediate rubber-covered roll, called an *offset roll*, we have offset gravure, as in Figure 4-3*b*. The amount of liquid transferred, and thus the coating weight, is controlled by the volume of liquid in the cells.

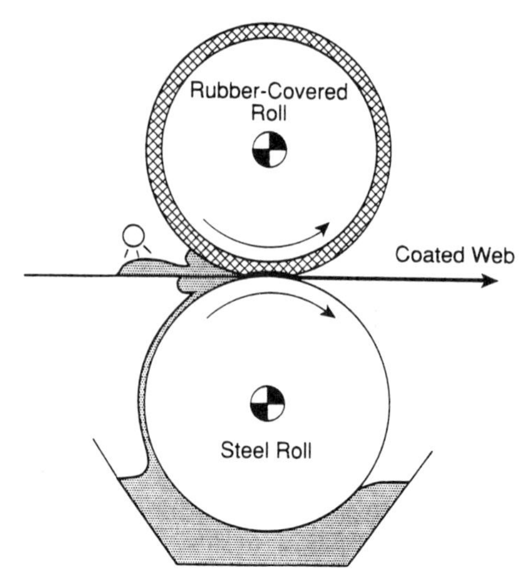

Figure 4-4. Two-roll squeeze roll coater. [From Coyle (1992). Reprinted with permission by VCH Publishers © (1992).]

In forward-roll coating, if a rigid roll presses the web against a rubber-covered roll, we have squeeze coating (Figure 4-4). Again, in forward-roll coating, if the liquid thickness on the applicator roll is less than the gap between the roll and the web, we have meniscus coating, as in Figure 4-5. The lower roll is not necessary; in meniscus coating the web and the backup roll can be just above the surface of a pool of liquid. Surface tension keeps the liquid against the web. In kiss coating (Figure 4-6), the

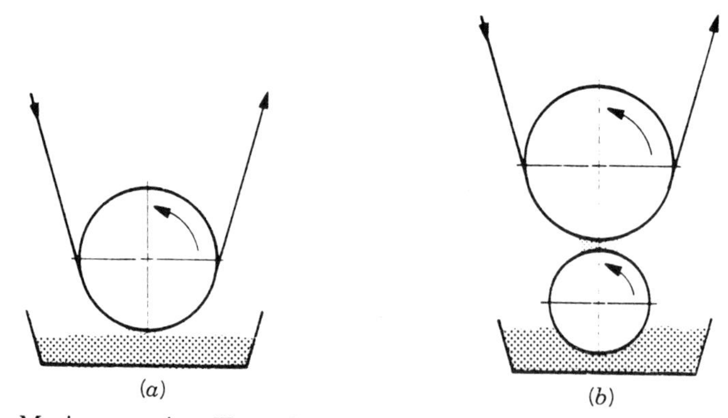

Figure 4-5. Meniscus coating. [From Grant and Satas (1984). Reprinted from *Web Processing and Converting Technology and Equipment*, D. Satas, ed., Van Nostrand Reinhold, New York].

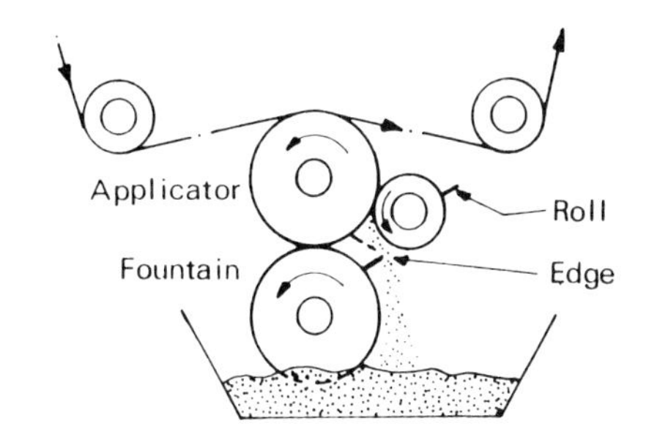

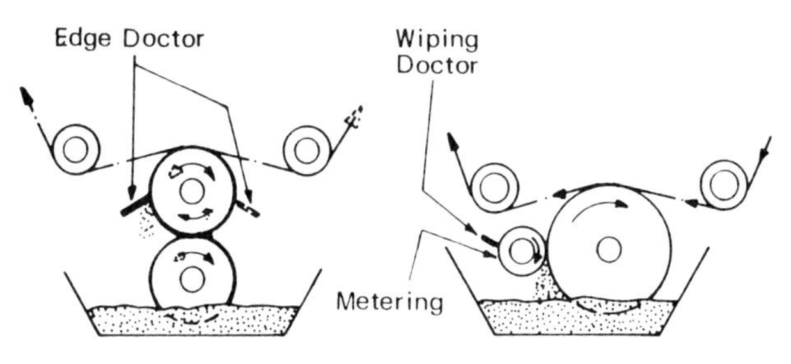

Figure 4-6. Kiss coating. [From Weiss (1977).]

web, traveling in the same or opposite direction to the applicator roll, is held against the applicator roll only by the web tension.

In rod coating a wire-wound rod, a Mayer rod, doctors off excess coating liquid (Figure 4-7). Excess coating liquid is applied to the web ahead of the coating rod by any convenient method. The amount of liquid remaining in the coating is related to the free space between the high points of the wires. The rod is usually slowly rotated so as to wear evenly and to free any particles caught against the rod that would create streaks. The direction of rotation should be such as to tighten the wire.

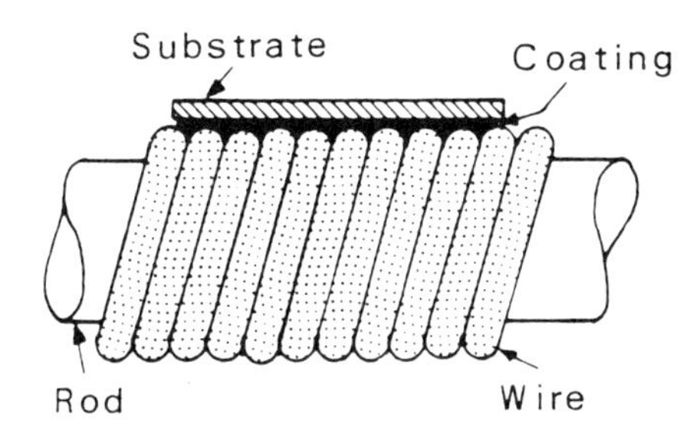

Figure 4-7. Rod coating with a wire-wound rod. [From Weiss (1977).]

In knife coating a rigid blade that is held perpendicular to the web and close to or against the web doctors off excess liquid. In Figure 4-8*a* a knife-over-roll coater is shown. The roll can be steel, in which case a fixed gap should be maintained; or the roll can be rubber-covered, with the knife pressing against the roll. The knife can also be held against an unsupported web as in the floating knife coater (Figure 4-8*b*). Here the control of web tension is very important.

In blade coating a thin blade is pressed against the web at an angle, as in Figure 4-9*a*, to doctor off the excess liquid. In beveled blade coating the blade is rigid (Figure 4-9*b*). If not originally parallel to the web, the bevel would soon become parallel due to wear. In flexible blade coating the blade bends under the pressure, and the flat of the blade, rather than the edge, often does the doctoring, as in Figure 4-9*c*.

In air-knife coating a blast of air (Figure 4-10) doctors off the excess liquid. Air-knife coating can operate in one of two possible modes. At lower air pressures the plane jet of air acts just like a doctor blade to control coating weight, rejecting the excess fluid which cannot pass and flows back to the coating pan. In higher-pressure

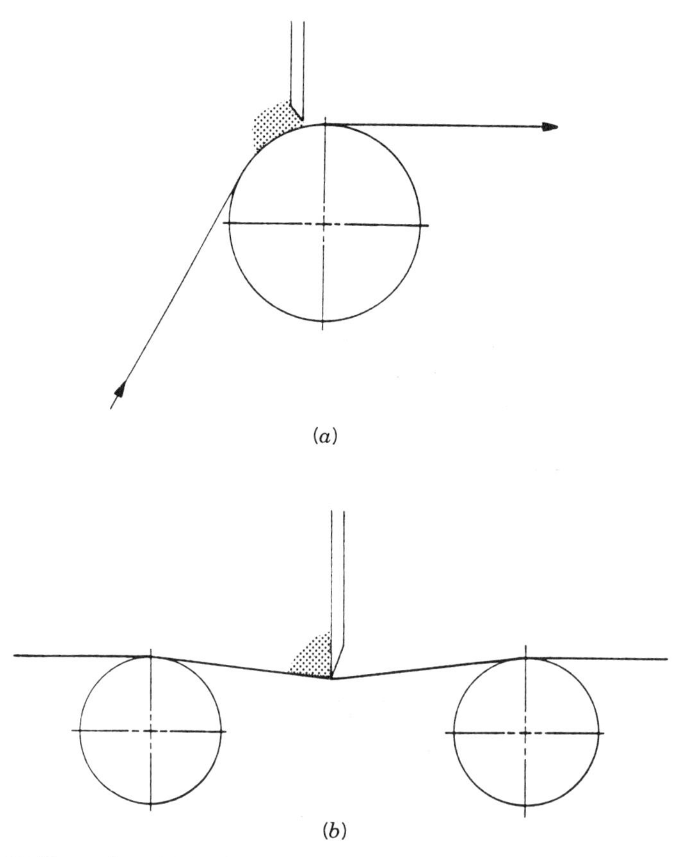

Figure 4-8. Knife coating: (*a*) knife-over-roll coating; (*b*) floating knife coating. [From Grant and Satas (1984). Reprinted from *Web Processing and Converting Technology and Equipment*, D. Satas, ed., Van Nostrand Reinhold, New York].

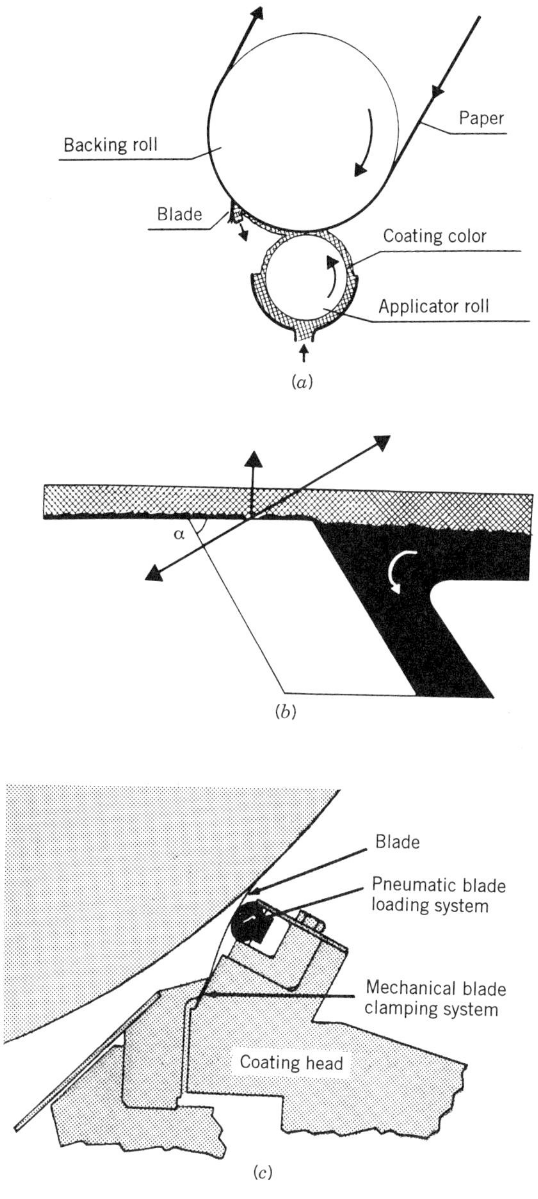

Figure 4-9. Blade coating: (*a*) general arrangement; (*b*) the beveled blade in blade coating; (*c*) flexible blade coating. [From Eklund (1984). Reprinted from *Web Processing and Converting Technology and Equipment*, D. Satas, ed., Van Nostrand Reinhold, New York].

systems the air acts as a squeegee and actually removes excess fluid from the web. At these higher pressures the resulting spray has to be contained. Air-knife coating works very well, but containing the spray in the high-pressure systems can be a problem. Air-knife coating consumes much energy in compressing the large volumes of air required.

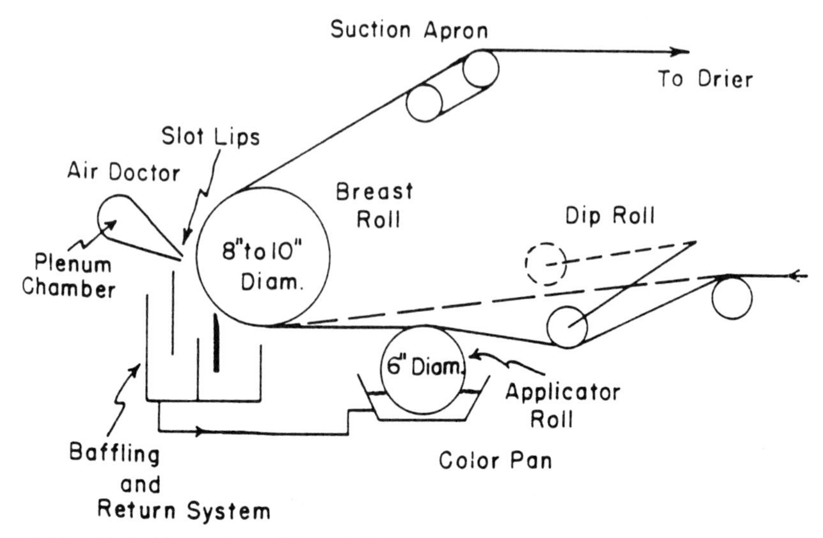

Figure 4-10. Air-knife coating. [From Eklund (1984). Reprinted from *Web Processing and Converting Technology and Equipment*, D. Satas, ed., Van Nostrand Reinhold, New York].

In dip coating the web is carried into and then out of a pool of coating liquid (Figure 4-11). The web carries off whatever liquid does not drain back into the bath. The web can travel around several rolls in the tank, and one or both sides can be coated. A doctor blade may be used to reduce the coated thickness. This is one of the oldest methods of coating and is still used for coating webs, fibers, and discrete objects.

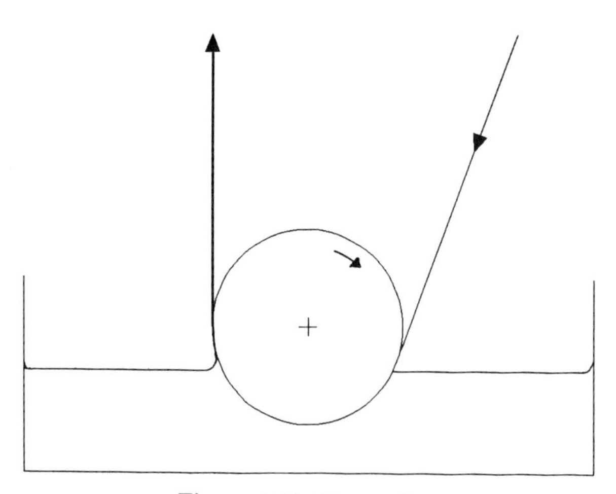

Figure 4-11. Dip coating.

4.2 CONTROL OF COATING WEIGHT

The scientists who design a coating and the customers who buy the coating are interested primarily in the dry coating weight. But to the coating operators and engineers who manufacture it, the wet coating weight is most important. The wet coating is what is applied, and the wet coating coverage controls coatability. When we discuss coating weight here we refer to the wet coating weight. As long as the coating liquid is well mixed and homogeneous, the dry coating weight will be proportional to the wet coating weight.

Control of wet coating weight involves three factors: the control of the average coating weight, the control of coating uniformity in the machine or down-web direction, and the control of coating uniformity in the cross-web direction. In the machine direction coating weight often varies cyclically, as illustrated in Figure 4-12*a*. Across the web the coating weight is controlled by the construction of the system. A typical cross-web coating weight profile is shown in Figure 4-12*b*.

In most roll-coating operations the average coating weight is controlled by the geometry of the system, the speeds of the web and of the rolls, the forces applied (as web tensions, or to knives and blades, or against rubber-covered rolls), and the viscosity or rheology of the coating liquid (Coyle, 1992). In air-knife coating the air pressure is also an important variable (Steinberg, 1992). In gravure coating the gravure pattern on the cylinder is the major factor in coating weight control.

Good control of coating weight can only be obtained by measuring the coverage and making adjustments. Rolls are difficult and expensive to change, so this is rarely done. Changing the rheology or concentration of a coating liquid involves making a new batch, and that takes time and the ingredients may be expensive. The operator will adjust those items that are easy to adjust and which do not interfere with production. Thus the operator will rarely reduce the coating speed. But he will adjust roll speeds and coating gaps or applied forces until the coating weight is correct. An automatic system with an on-line sensor, such as, perhaps, a beta gauge, and feedback control is desirable. This can adjust a roll speed or a coating gap to get the correct coating weight automatically.

In blade coating the coating weight is controlled by the force of the blade against the web. When the angle between the blade and the web is low, under about 20°, as in bent-blade coating, the coating fluid develops a pressure that opposes the force on the blade and tends to lift it. This fluid pressure depends on the geometry, with a gentle angle causing a greater pressure, and is proportional to the coating speed and the fluid viscosity.

Blade coating is often used to coat clay and other heavily loaded liquids. Those liquids that contain a high percentage of suspended solids, such as a clay, can behave strangely. Most polymer solutions will show a drop in viscosity as they are sheared or stirred faster and faster (described in Chapter 5). However, liquids with high loadings of suspended solids show the opposite behavior; their viscosity increases at high shear or stirring rates. Liquids that exhibit this shear-thickening behavior are called *dilatant*. A plot of shear thinning and shear thickening behavior was shown in Figure 2-8.

In blade coating clay suspensions the clay loadings may be so high that the coating liquid is too dilatant, which could cause the viscosity to increase greatly in coating. This could increase the fluid pressure against the blade sufficiently to lift it away

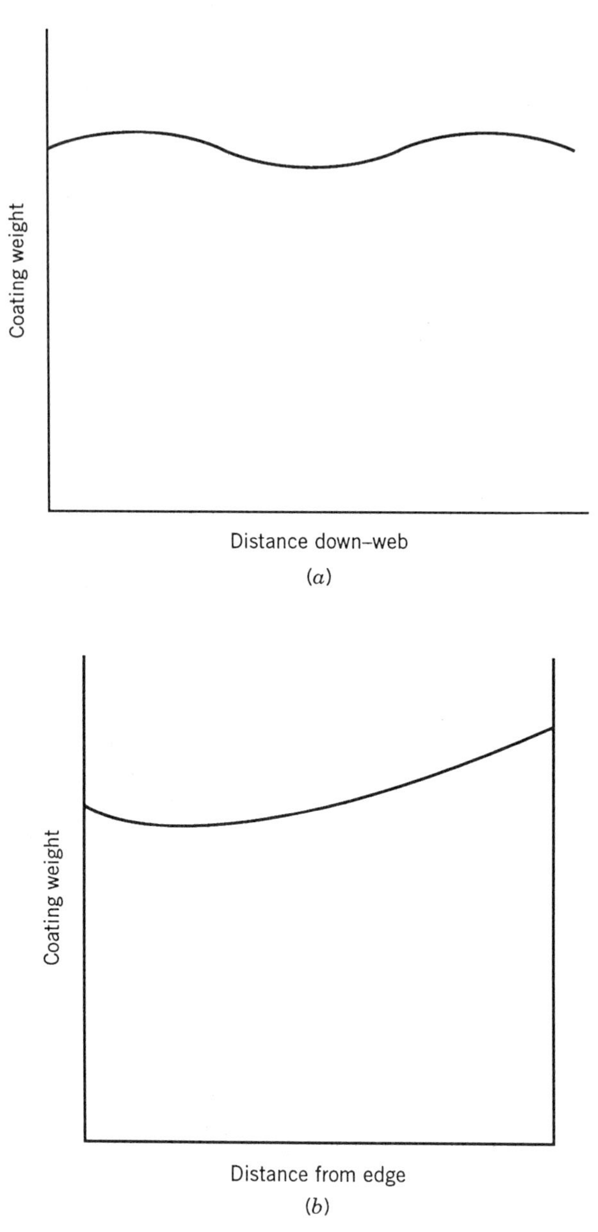

Figure 4-12. Typical coating weight profiles: (*a*) in the machine direction; (*b*) cross-web profile.

from the web, allowing excessive liquid to be coated. If the external force on the blade cannot be increased enough to bring the coated thickness down to the desired level, the blade angle should, if possible, be increased. Otherwise, one would have to thin down the coating liquid to reduce its dilatantcy.

The coating uniformity is limited by the inherent properties of the system. Most roll-coating systems cannot do better than a 5 to 10% variation in coverage. However, precision reverse-roll systems, with very precise bearing, are capable of keeping the variation within 2%.

Like precision reverse-roll coating, gravure coating can give very uniform coverages. However, on a very small scale the uniformity may not be that good, for the coating liquid from the cells in the gravure cylinder, after transfer, has to flow together and level. Large land areas—these are the flats on the gravure cylinder between the cells—tend to promote printing, or the retention of the cell pattern on the web. This is undesirable for uniform coating. Therefore, for uniform coverage and for good leveling the land areas should be as small as possible for a given cell volume. Leveling is also promoted by reverse gravure rather than forward gravure, as this tends to smear the cells together. Differential gravure, where the gravure cylinder runs at a different speed than the web or the offset cylinder, should also help in leveling.

Sometimes in gravure and other roll-coating operations, a smoothing bar—a stationary or slowly rotating smooth rod or bar—is used to aid in leveling. As it is an additional mechanical device, it is preferable not to use a smoothing bar unless absolutely necessary.

In gravure coating the volume factor, that is, the total cell volume per unit area of the gravure cylinder, is the main determinant of the coating coverage. A volume divided by an area is a length. Thus the cell volume per unit area of the cylinder is the height of coating liquid that would be on the gravure cylinder if the cylinder were smooth. If all this liquid were transferred to the web at the same speed, this would be the wet coverage. The volume factor, if expressed as cm^3/m^2, is exactly equal to what the wet thickness, in micrometers, would be if all the cell volume transferred. But not all of this liquid transfers to the web. Pulkrabek and Munter (1983) found that with a number of different knurl rolls, the amount of coating liquid transferred is only about 59% of the cell volume. The wet thickness is then about 59% of the volume factor. Coyle (1992) suggests that for all gravure rolls the amount transferred is 58 to 59% of the cell volume. In part, this could be due to the web or the offset roll being pressed into the cells and forcing some liquid out ahead of the nip. This liquid, plus excess liquid that may not have been doctored off, will form a rolling bank or pool that does not transfer to the web. This rolling bank, however, aids in leveling by helping merge the transferred liquid from adjacent cells.

If, as is most common, the gravure cylinder turns at the same speed as the web, the wet coverage should be just the 59% of the volume factor. However, the predictions of the amount of liquid in the cells that transfers to the web are not always correct. It is common to buy a test cylinder to check this. If the coverage is not correct, this test cylinder can be reworked until the correct coverage is obtained.

At times the coverage in gravure coating is much less than this 59% of the volume factor. The problem is usually due to too low a force between the impression roll and the gravure cylinder, or, in offset gravure, the force between the gravure and offset cylinders. The minimum force needed depends on the viscosity of the coating liquid, the type of web, and the cell type and size. Obviously, in direct reverse gravure, the force between the gravure cylinder and the impression roll has to be much less than in direct gravure, or else the web and the gravure cylinder would not slide past each other. This lower force may then result in incomplete pickout, and this may be the reason that reverse gravure is not often used.

4.3 RIBBING

Ribbing is the formation of lines running down the web and very uniformly spaced across the web, as if a comb were drawn down the wet coating. Ribbing has also been called comb lines, phonographing, rake lines, and corduroy. It can occur in just about every type of coating. In fact, the brush marks or lines formed in painting a wall with a bristle brush are most often ribs from the hydrodynamics of the coating process, and not, as generally thought, caused by the bristles of the brush. And in coating with a wire-wound rod, where one would expect ridges or ribs to be spaced at the wire spacing (the wire diameter), Hull (1991) has found that the rib spacing is often different from the wire spacing and is related to interactions between the wire spacing and the hydrodynamic ribbing frequency. Here, too, it is the coating hydrodynamics that causes ribs.

In rigid knife coating the geometry of the gap between the knife and the web determines whether or not ribs form, and also the appearance of the ribs. When the gap is purely converging with no diverging zones beyond the narrowest clearance, as illustrated in Figure 4-13, no ribs were observed by Hanumanthu et al. (1992). Diverging zones, which would be present if a round rod were used in Figure 4-13, seem to promote rib formation.

When ribs form, they will persist into the dried coating unless the liquid levels well, that is, unless liquid flows from the ridges into the valleys. This leveling is more likely to occur with low-viscosity liquids and will be aided by a longer time in the liquid state. High-viscosity liquids are unlikely to level, as are liquids that are dried very quickly or cured very quickly by ultraviolet light or electron beam radiation. High surface tensions aid leveling. As described in Chapter 6, when a curved interface—as between a liquid and air—exists, there will be a pressure differential across this interface. This pressure difference is called capillary pressure and is proportional to the surface tension. The pressure will be higher on the inwardly curving, or concave, side of an interface. Thus when ribs exist as shown in Figure 4-14, under the high points of the ribs the pressure in the liquid will be above atmo-

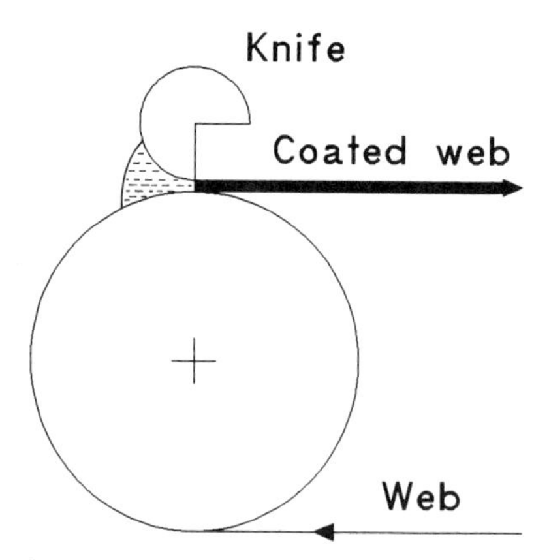

Figure 4-13. In knife coating diverging zones, as would exist if the missing quadrant were present, should be avoided to prevent ribbing.

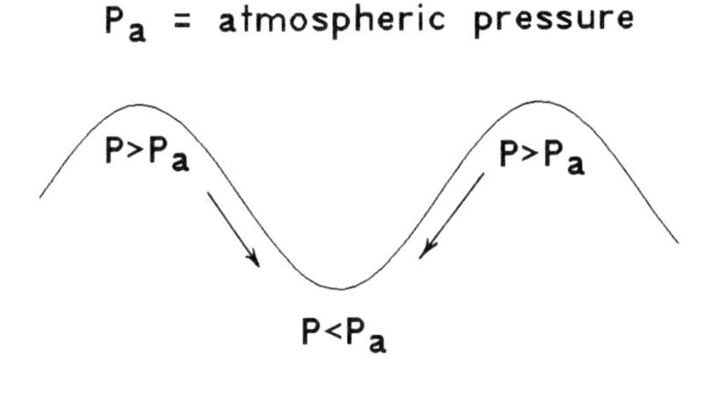

Figure 4-14. High surface tension aids leveling. Due to surface tension the pressure on the concave, or inwardly curving, side of an interface is higher than on the convex, or outwardly curving, side. This pressure difference promotes flow from the peaks to the valleys.

spheric. At the low points, however, the inwardly curving side is above the interface, and the air pressure is higher than the pressure in the liquid. As liquid will flow from regions of higher pressure to regions of lower pressure, liquid will flow from the peaks of the ribs to the valleys. This will aid in leveling. Gravity also aids in leveling.

When ribs are present in knife or blade coating, the geometry should be altered to ensure that there are no diverging zones for the liquid flow. If this is not possible, the surface tension should be as high as possible while still permitting good wetting on the support, and the viscosity should be as low as possible, to aid in leveling. Adding an additional leveling rod with no diverging zones is sometimes used, but it is better practice to avoid the initial rib formation if at all possible.

4.3.1 Ribs in Forward-Roll Coating

Rib formation almost always occurs in forward-roll coating. The diverging geometry of the flow exiting the nip between the rolls and also the viscous forces that are always present favor rib formation. As we have just discussed, surface tension opposes rib formation. As a result, the ratio of viscous forces to surface forces is an important variable in determining rib formation. The ratio of viscous to surface forces, called the capillary number, is equal to $\mu \overline{U}/\sigma$, the product of the viscosity and the coating speed divided by the surface tension. Here μ is the viscosity of the liquid, σ is the surface tension, and $\overline{U}$ is the average surface speed of the two rolls—the applicator roll and the backup roll—if they are different.

The ratio of the gap between the web and the applicator roll to the roll diameter is also important in rib formation. If the applicator and the backing rolls are of different size, then in the gap-to-diameter ratio a special average diameter should be used, such that

$$\frac{1}{D} = \frac{1}{2}\left(\frac{1}{D_1} + \frac{1}{D_2}\right)$$

Figure 4-15 shows the operating conditions under which ribs form as a function of these two ratios: the capillary number or ratio of viscous forces to surface forces, and the gap-to-diameter ratio.

In Figure 4-15 we see that as the capillary number increases there is a greater tendency to form ribs. This means that ribs are more likely to form at higher coating speeds, higher viscosities, and lower surface tensions. Also, ribs are more likely to form at lower gap-to-diameter ratios. This means that tight gaps and large-diameter rolls favor rib formation. Wide gaps, which would reduce the tendency to form ribs, give thicker coatings, so the gap cannot be changed arbitrarily.

If we take a typical coating speed of 200 ft/min or 1 m/s, a typical surface ten-

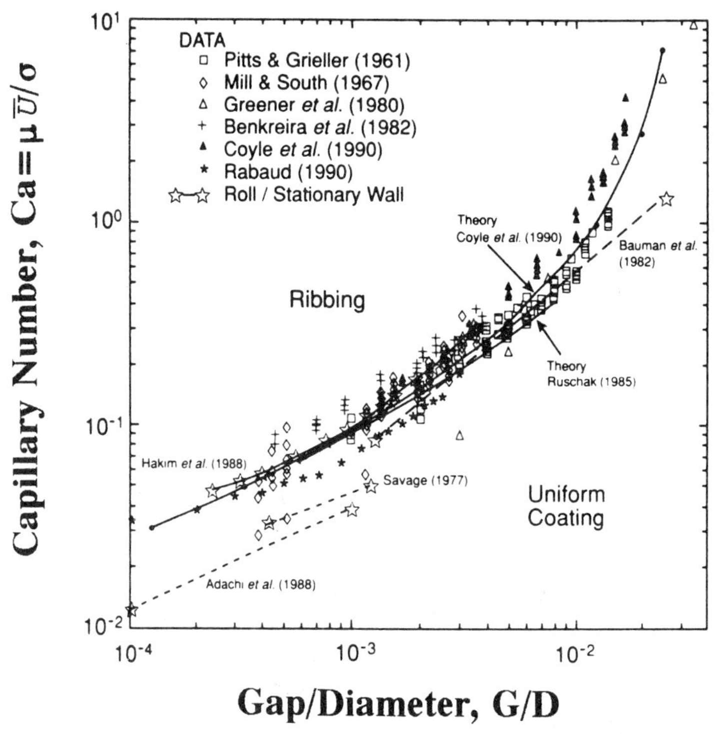

Figure 4-15. Ribbing in forward-roll coating. [From Coyle (1992). Reprinted with permission by VCH Publishers © (1992).]

sion of 40 dyn/cm or 0.040 N/m, and a typical viscosity of 100 cP or 0.1 N · s/m^2, the capillary number would be (0.1 N · s/m^2)(1 m/s)/(0.040 N/m), or 2.5. At this high capillary number ribs would form except at extremely large gap-to-diameter ratios.

This demonstrates that in forward-roll coating, rib formation is the usual occurrence. Forward-roll coating should not be used unless rib formation is not objectionable, the liquid levels well, or ribs in the product do not affect its usability. Because of this great tendency to form ribs, forward-roll coating is used much less now than in the past.

Viscoelastic liquids are liquids that, while they flow like ordinary liquids, show some elastic effects. Egg white is a common viscoelastic liquid. In fact, all polymer solutions and polymer melts are viscoelastic to a greater or lesser extent. Viscoelastic liquids show ribbing at much lower critical capillary numbers than do the simple Newtonian liquids used for the data in Figure 4-15. Viscoelastic liquids also can have a greater rib height than do Newtonian liquids. We may note that most coating liquids contain dissolved polymers and would therefore show viscoelastic behavior under the conditions that exist in most coating operations. Thus they would form ribs even under conditions where Figure 4-15 predicts no rib formation.

We may conclude that it is almost impossible to avoid rib formation in forward-roll coating. However, rib formation is minimized by operating at low speeds (we normally want to run at as high speeds as possible), and with low viscosities and wide gaps (which would give thick coatings).

As mentioned earlier, the use of a leveling bar with no diverging zones is sometimes helpful. A better solution appears to be the use of a string or wire stretched across the machine and touching the liquid in the film splitting region between the applicator roll and the web on the back-up roll, as suggested by Hasegawa and Sorimachi (1993).

4.3.2 Ribbing and Cascade (or Herringbone or Seashore) in Reverse-Roll Coating

In reverse-roll coating the most critical conditions exist in the gap between the metering roll and the applicator roll, and not where the applicator roll meets the web. After the metering operation all of the coating on the applicator roll is transferred to the web, and any defects formed prior to that point will also transfer.

Cascade or Herringbone or Seashore

Cascade, also called herringbone or seashore, is a distinct defect that occurs at high capillary numbers (based on the applicator roll speed) and at high ratios of the metering roll speed to the applicator roll speed. As Coyle (1992) points out, cascade is a periodic cross-web disturbance with a sawtooth pattern. It is caused by air being entrained in the gap between the metering roll and the applicator roll. Thus this defect shows up as areas of low coverage. Let us see how this can occur.

Figure 4-16 illustrates the process. At a given capillary number, meaning at a constant applicator roll speed, as the metering roll turns faster, the meniscus retreats into the gap. This is as one would expect. With the higher speeds, as the meniscus

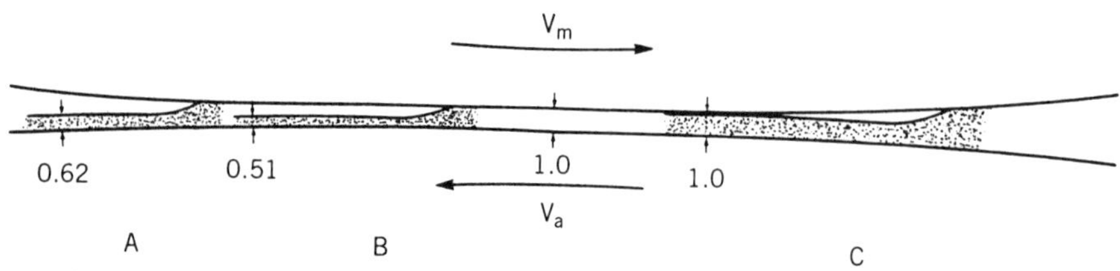

Figure 4-16. Formation of entrained air–cascade–in-reverse-roll coating. [From D. J. Coyle et al., *AIChE J.*, **36**, 161–174 (1990). Reproduced with permission of the American Institute of Chemical Engineers. © 1990 AIChE. All rights reserved.]

approaches the center of the gap the metered film thickness gets smaller. However, once the meniscus passes through the center of the gap, the separation of the roll surface increases and the metered film gets thicker again. When the metered film thickness becomes greater than the clearance at the center of the gap, air is trapped. But this condition can no longer exist at a steady state. When the liquid reattaches at the minimum gap, the film again becomes thinner, then the meniscus retreats back through the gap, with the film becoming thicker and again entrapping air, with the cycle repeating itself. This gives the periodic cross-web disturbance called cascade.

Figure 4-17 illustrates the defect-free operating regions in reverse-roll coating. At very low capillary numbers (low speeds and low viscosities) the coating is defect-free. At low ratios of the metering roll speed to the applicator roll speed, with higher values of the capillary number (higher speeds), ribbing occurs. At higher ratios of the metering roll speed to applicator roll speed, cascade occurs. But note that at high capillary numbers there is a range of speed ratios between the ribbing and the cascade regions where defect-free coatings can be made, which is not possible at lower capillary numbers. We see in Figure 4-17 that at some speed ratios one can go from ribbing to a defect-free coating by increasing the speed. This is one case where increasing the coating speed makes coating easier. This ability to get defect-free coatings at high coating speeds is one reason why reverse-roll coating is so popular.

In the discussion on ribbing in forward-roll coating we used an example where a typical capillary number might be 2.5. Figure 4-17 shows that there should be no trouble getting a rib-free coating in reverse-roll coating at this capillary number if the ratio of the metering roll speed to the applicator roll speed is about 0.3. It does not matter whether the gap is wide (1000 μm or 40 mils) or tight (125 μm or 5 mils).

We also see in Figure 4-17 that at tighter gaps the defect-free operating region is narrower. In some cases it can even disappear. There is even a suggestion (Coyle et al., 1990) that ribs are always present, but in the "stable" region they are just of extremely low amplitude and almost undetectable by the naked eye, and unobjectionable.

As in the case of forward-roll coating, viscoelastic liquids such as polymer solutions appear to be more prone to form ribs than simple or Newtonian liquids. Thus in some cases the defect-free operating region may disappear completely. In such cases one could add a leveling bar with no divergent zones, as mentioned earlier.

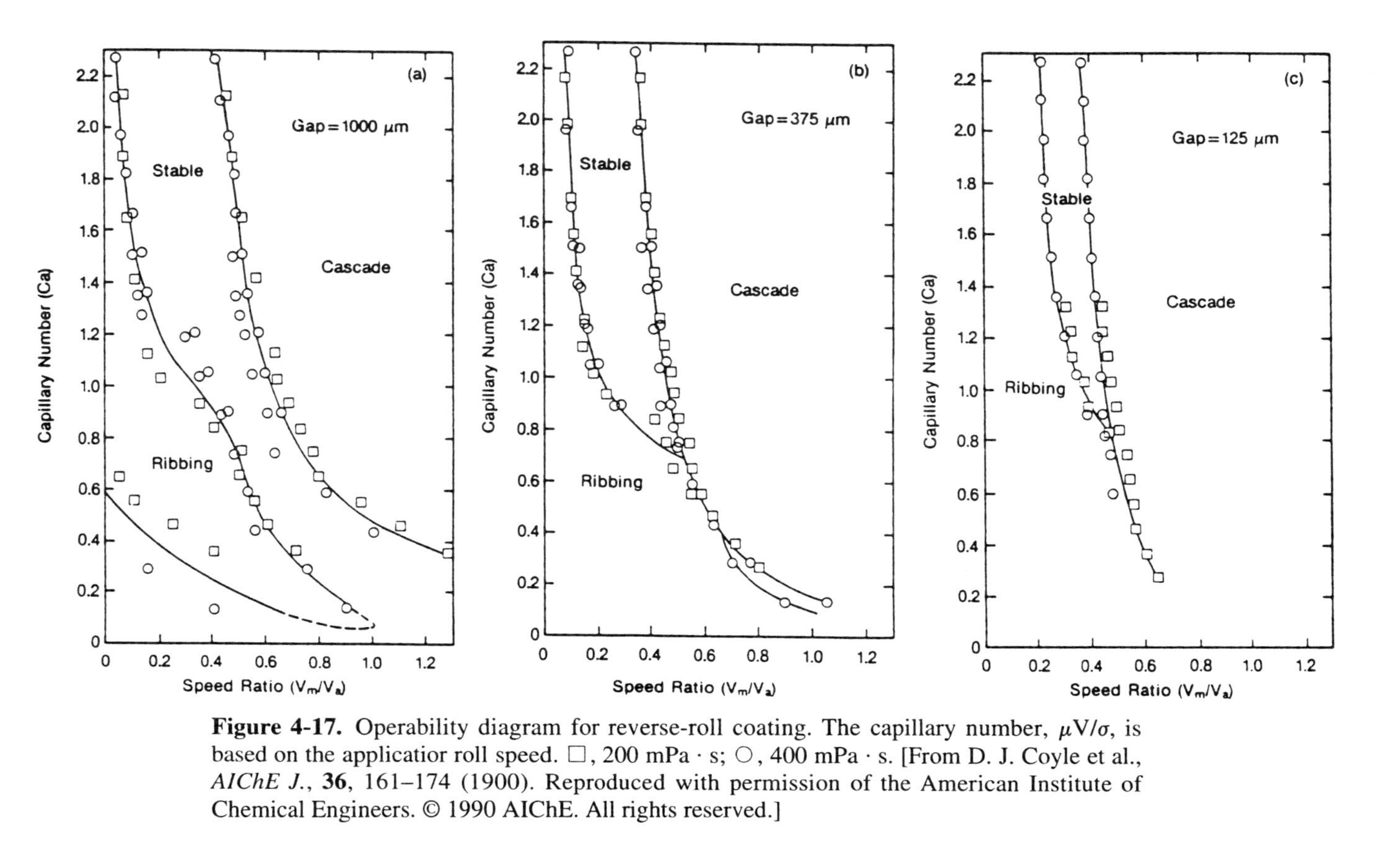

Figure 4-17. Operability diagram for reverse-roll coating. The capillary number, $\mu V/\sigma$, is based on the applicatior roll speed. □, 200 mPa · s; ○, 400 mPa · s. [From D. J. Coyle et al., *AIChE J.*, **36**, 161–174 (1900). Reproduced with permission of the American Institute of Chemical Engineers. © 1990 AIChE. All rights reserved.]

4.4 GRAVURE COATING

Defects in gravure coating mainly concern pickout—the transfer of the coating liquid from the gravure cylinder to the web or to the offset roll. The three major problems that can occur are flashing, combined or multiple pickout, and misting. Flashing occurs when the liquid in one or more grooves or rows of cells transfers only partially, leaving uncoated lines on the web. In combined or multiple pickout one or more adjacent grooves or cells are picked out to form a single line or a large, combined cell. This large, wide line will not level well. Misting occurs when fine droplets are ejected from the coating nip in high-speed operations.

Misting is often related to *cobwebbing*. The coating liquid forms threads as the gravure cylinder separates from the web. These threads look like cobwebs. Similar thread formations are often seen when painting a wall with latex paint using a roller. It is related to the viscoelastic properties of the coating liquid. Longer polymer molecules (higher-molecular-weight polymers), as indicated by higher-viscosity grades of the polymer, tend to show this when in solution or as a melt. Thus misting is related to the properties of the coating liquid. To a lesser extent it is also related to the geometry and speed of the process, with higher speeds showing greater misting. When the coating liquid cannot be altered and the system has to be run at high speed, one may have to live with the misting and control it by adequate shielding and controlled airflows and ventilation. Misting should not affect the quality of the coating, although it would reduce the coverage somewhat.

The stability of pickout depends mainly on the gravure pattern. As discussed above, the volume factor—the cell volume per unit area—controls the wet coverage, with about 59% of the cell volume being transferred. The pitch—the number of cells per inch or per centimeter perpendicular to the pattern angle—and the cell design control the pickout. Pulkrabek and Munter (1983) found that multiple-line pickout occurs when the natural ribbing frequency differs from the cell pitch multiplied by the sine of the helix angle. They worked with a number of knurl rolls and found that when the knurl roll line frequency, equal to the knurl roll pitch times the sine of the helix angle, equals the natural ribbing frequency of the system, multiple-line pickout does not occur. The severity of multiple-line pickout increases with the degree of mismatch. Figure 4-18 shows the results of their work. Larkin (1984) recommends the quadrangular pattern for direct gravure, the pyramid pattern (like the quadrangular pattern but with a pointed bottom) for offset gravure, and the trihelical for heavy coatings or for easy flushing and cleaning (see Figure 4-19).

4.5 STREAKS AND BANDS IN ROLL COATING

Streaks are long narrow lines of higher or lower coverage, usually running downweb, and bands are similar lines but much, much wider. In those devices using doctor blades, knives, or bars, such as blade or knife coating, gravure coating, and so on, particles of dirt caught against the blade, knife, or bar will cause a streak. Such particles should not be present. They should be removed by filtration before the coating solution reaches the coating station. In addition, a small filter should be installed just before the coating station to remove particles that may come from the feed lines. Also, it is advisable to have the coating operation under at least partial clean room

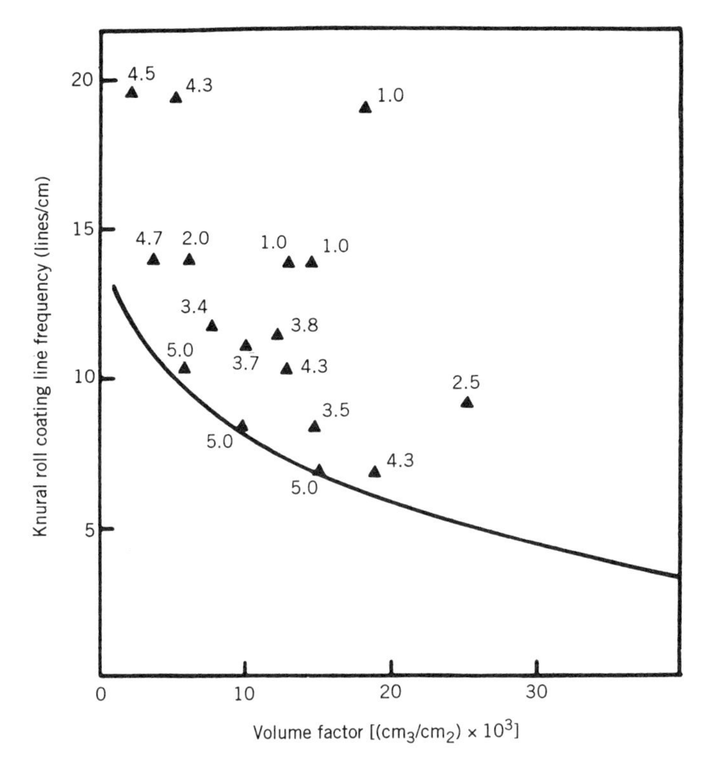

Figure 4-18. Knurl roll line frequency (equal to pitch times the sine of the helix angle) versus volume factor for stable pickout. The solid line is the natural ribbing frequency. Quality numbers range from 1, very poor, to 5, very good. Each number is the average of 12 tests covering the range of coating conditions. [Reprinted from W. W. Pulkrabek and J. D. Munter, "Knurl roll design for stable rotogravure coating," *Chem. Eng. Sci.*, **38**, pp. 1309–1314. Copyright (1983), with kind permission from Elsevier Science Ltd., The Boulevard, Langford Lane, Kidlington, UK.]

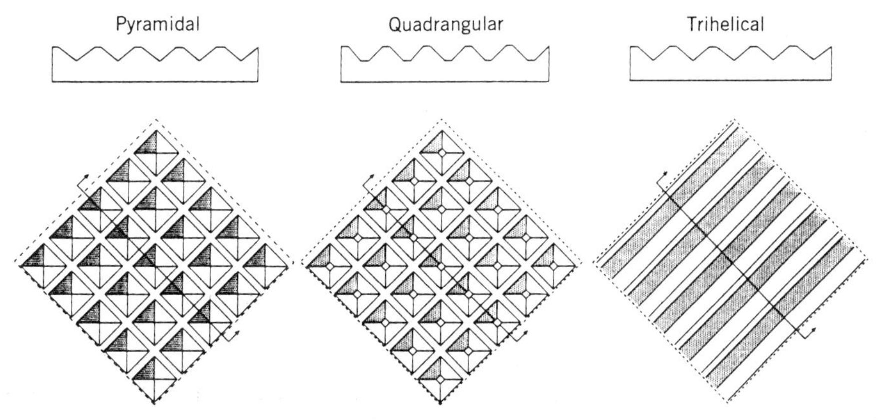

Figure 4-19. Gravure cell patterns. [From Coyle (1992). Reprinted with permission by VCH Publishers © (1992).]

conditions, perhaps on the order of class 1000 or at least class 10,000, to reduce the incidence of particles of dirt settling out of the air onto the coating liquid.

Bubbles can sometimes get stuck in a coating device, remaining there to cause streaks. The coating liquid should be thoroughly debubbled before it reaches the coating stand and precautions taken to avoid bubble formation in the lines to the coating stand. Dissolved air can come out of solution to form bubbles where the pressure is low due to high velocities in constrictions, as in valves.

Nicks or any unevenness in the blade, knife, or bar will cause a streak or a band. In high-speed blade coating, coating liquid can accumulate on the downstream side of the blade to form column structures or "stalagmites," and these can cause streaks at those locations.

Unevenness in a roll surface can cause a band. The rolls may be perfectly straight but can deflect under their own weight or under the forces present in the system. When coating or doctoring against unsupported web, any unevenness in web tension can cause bands. Unevenness in web tension is usually caused by poor-quality web. Poor-quality web can also be puckered, which will cause uneven coverage and perhaps bands. A puckered web can also cause flow by gravity in the wet coating, resulting in higher coverages in the low spots and lower coverages in the high spots on the web.

4.6 CHATTER IN ROLL COATING

Chatter is appearance of uniform cross-web bars or bands of different coverage. It may ruin the utility of the product, but sometimes it does not affect the usefulness but just ruins the appearance. Chatter in roll coating, as chatter in all coating operations, is usually caused by mechanical problems. Mechanical vibrations from the building, the air handling system, the drives, and so on, are one source of chatter. Fluctuations in web speed or roll speed caused by a speed controller or a tension controller is another source. These sources will be discussed further in Chapter 10. In addition, rolls that are out of round and operate at a narrow gap, such as in forward- and reverse-roll coating, and in certain knife coating operations, are a possible further source of chatter.

If all mechanical sources of chatter have been ruled out, the chatter may be caused by the hydrodynamics of the system. This seems to be more pronounced near the limits of coatability, that is, near the maximum coating speed or the minimum coating thickness. We are unaware of any discussion of chatter in roll coating caused by system hydrodynamics.

4.7 SAGGING

The maximum thickness of a coating is often set by the maximum thickness that the web can carry away. This obviously depends on the angle the web makes with the horizontal. A web traveling vertically upward cannot carry away as thick a layer as a horizontal web coated on its upper surface. Also, the maximum thickness increases with web speed and with the viscosity of the liquid, and decreases with increasing surface tension of the liquid. A simple theory for dip coating says that the maximum thickness that a web moving vertically upward can carry away is proportional to the $\frac{2}{3}$ power of the viscosity, to the $\frac{2}{3}$ power of the coating speed, and inversely proportional to the $\frac{1}{6}$ power of the surface tension (Groenveld, 1970).

When one tries to coat a layer thicker than this maximum thickness, sagging will occur, similar to the sagging we see when painting a vertical surface, such as a door, with too heavy a coat of oil-based enamel. From the discussion above, the cure for sagging is to coat thinner so that all the liquid coated will be carried away, or to carry away a thicker layer by coating faster or using a more viscous coating liquid. Of course, just increasing the coating speed or viscosity in these non-premetered coating systems will change the coated thickness, and then adjustments to the geometry or roll speeds would also have to be made.

From the discussion on the thickness of liquid carried away, lowering the surface tension would also have an beneficial effect in reducing sagging, but because the relationship involves only the $\frac{1}{6}$ power of surface tension, the effect would be small. Decreasing the surface tension from 40 dyn/cm to 30 dyn/cm (0.040 to 0.030 mN/m) would only increase the thickness carried away by 5%. As we already seen, for good leveling we want high surface tension. We do not recommend lowering the surface tension.

REFERENCES

Coyle, D. J., "Roll coating," pp. 63–116 in *Modern Coating and Drying Technology*, E. D. Cohen and E. B. Gutoff, eds., VCH Publishers, New York, 1992.

Coyle, D. J., C. W. Macosko, and L. E. Scriven, "The fluid dynamics of reverse roll coating," *AIChE J.* **36**, 161–174 (1990).

Eklund, D., "Blade and air knife coating," in *Web Processing and Converting Technology and Equipment*, D. Satas, ed., Van Nostrand Reinhold, New York, 1984.

Grant, O. W., and D. Satas, "Other knife and roll coaters," in *Web Processing and Converting Technology and Equipment*, D. Satas, ed., Van Nostrand Reinhold, New York, 1984.

Groenveld, P., "Low capillary number withdrawal," *Chem. Eng. Sci.* **25**, 1259–1266 (1970).

Hanumanthu, R., L. E. Scriven, and M. R. Strenger, "Operational characteristics of a film-fed, rigid-knife coater," abstract of a paper presented at the *Coating Process Science and Technology Symposium, AIChE Spring National Meeting*, New Orleans, LA, Mar. 1992.

Hasegawa, T., and K. Sorimachi, "Wavelength and depth of ribbing in roll coating and its elimination," *AIChE J.* **39**, 935–945 (1993).

Hull, M., "Visualization of wire-wound-rod coating defects," *Tappi J.* **74**, 215–218 (Apr. 1991).

Larkin, M. J., "Gravure coaters," pp. 15–33 in *Web Processing and Converting Technology and Equipment*, D. Satas, ed., Van Nostrand Reinhold, New York, 1984.

Pulkrabek, W. W., and J. D. Munter, "Knurl roll design for stable rotogravure coating," *Chem. Eng. Sci.* **38**, 1309–1314 (1983).

Steinberg, N. I., "Air-knife coating," pp. 169–192 in *Modern Coating and Drying Technology*, E. D. Cohen and E. B. Gutoff, eds., VCH Publishers, New York, 1992.

Weiss, H. L., *Coating and Laminating Machines*, Converting Technology Co., Milwaukee, WI, 1977.

CHAPTER V

Problems in Slot, Extrusion, Slide, and Curtain Coating

Slot, extrusion, slide, and curtain coating are all premetered coating systems, in that all the material metered to the coating die gets coated onto the web. Thus once we know the web speed, the coating width, and the desired wet coverage it is simple to calculate the necessary flow rate. This assumes, of course, that all the material gets coated on the web to give a uniform, smooth coating. This does not always happen, and in this chapter we address the problems that can occur. But when a smooth coating is obtained, it should not be difficult to maintain coverage control to within 2% of aim in slot, slide, and curtain coating, both in the machine direction and across the web. In extrusion coating where the slot opening is manually adjusted and in curtain coating with a weir or a downward-pointing slot, the cross-web uniformity is poorer.

In all coating operations discussed in this book, a wet layer is laid down on the web. The wet coverage is controlled by the coating operation. Unless the coating is solidified by cooling (when coating hot melts) or by curing (by radiation with ultraviolet light or with electron beams), the dry coverage is much less than the wet coverage, due to evaporation of the solvent. If the dissolved or dispersed solids are uniformly distributed in the coating liquid, and if there is no flow of the coating after it is on the web, the dry coverage will be as uniform as the wet coverage.

5.1 DESCRIPTION OF COATING METHODS

In *slot coating* the coating fluid is fed into the manifold of the coating die, where it spreads across the width and flows out of a slot onto the moving web, usually held against a backing roll as illustrated in Figure 5-1. In slot coating the coating liquid wets both lips of the coating die, although if the die is tilted or the lower lip is swept back, only one lip may be wet with the coating liquid. As illustrated in Figure 5-1, we distinguish slot coating from *extrusion coating* in that in extrusion coating the coating liquid exits the slot as a ribbon and does not wet the lips of the die. For very

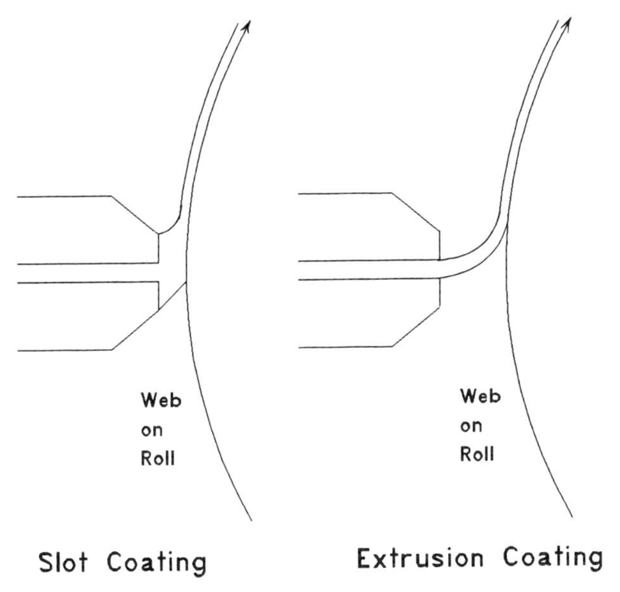

Figure 5-1. Slot and extrusion coating.

thin coatings the slot coating die may be used to coat against unsupported web, as in Figure 5-2. The coating die is usually aimed perpendicular to the web, but may be tilted up or down. Usually, the coating lips are in one plane, that is, one directly above the other. However, at times it is advantageous to have one lip protrude beyond the other, as indicated in Figure 5-3. Tilt and offset are beneficial primarily for very

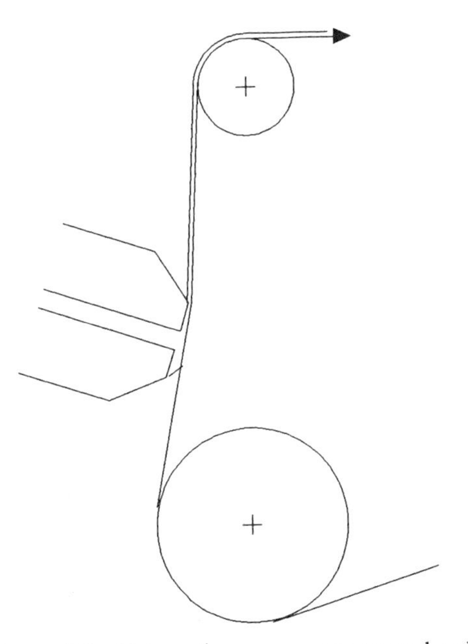

Figure 5-2. Slot coating on an unsupported web.

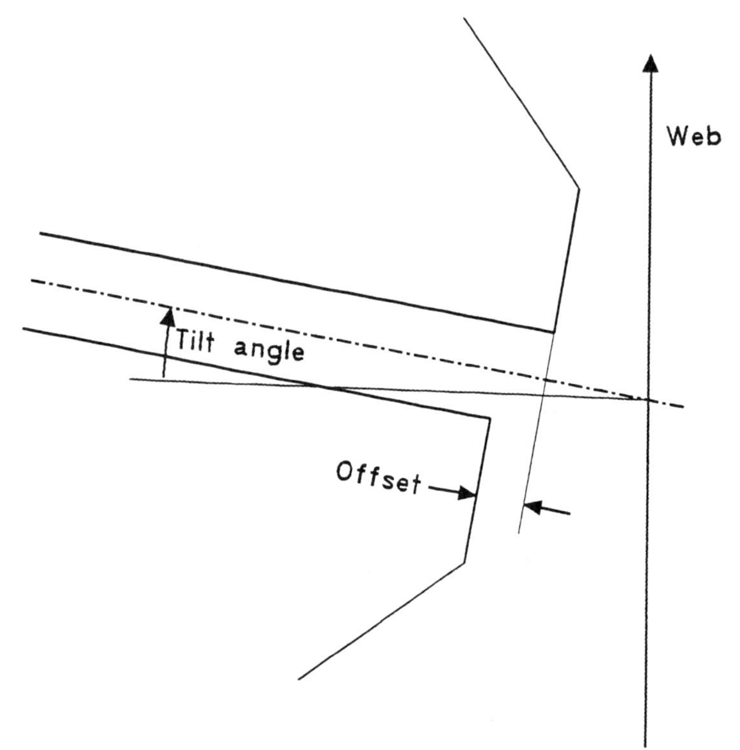

Figure 5-3. Tilt and offset in slot coating.

thin coatings, when even with the use of bead vacuum, indicated in Figure 5-4, thin-enough coating cannot be made with the standard configuration. A bead vacuum of up to 1000 Pa (4 in. of water) and even higher is often used to increase the window of coatability, that is, to allow thinner coatings to be made or to permit use of higher coating speeds.

The upper edge of the upper coating lip, where the upper meniscus is pinned, and the lower edge of the lower coating lip, where the lower meniscus is pinned, should be sharp. If it is not sharp, the meniscus may not pin at one location but wander back and forth over the curved edge, as indicated in Figure 5-5. This will lead to chatter at wider gaps. Any edge where a meniscus is pinned should be sharp. But the edges should not be too sharp, for sharp edges are easily damaged, forming nicks that cause streaks. And sharp edges can cut the hands of the operators during handling. It is suggested that the radius of curvature where a meniscus is pinned should be 50 μm (2 mils) (Fahrni and Zimmerman, 1978), but even twice that may still be satisfactory. The radius of curvature can be measured by molding the edge with dental molding compound, cutting out a thin section with a razor blade, and making a photomicrograph of the corner of the mold at low magnification, 50 to 200×. If the magnification is not known, it can be measured by including a millimeter scale in the photomicrograph. The radius of curvature of the photo can be read by eye with a millimeter scale, and then divided by the magnification. By this technique one can measure the radius of curvature to within 5 or 10 μm ($\frac{1}{4}$ or $\frac{1}{2}$ mil).

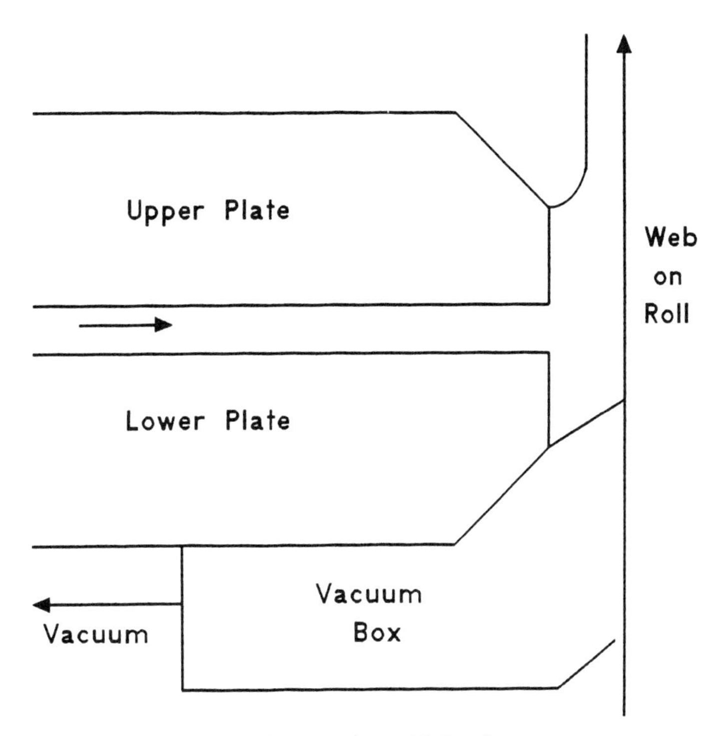

Figure 5-4. Slot coating with bead vacuum.

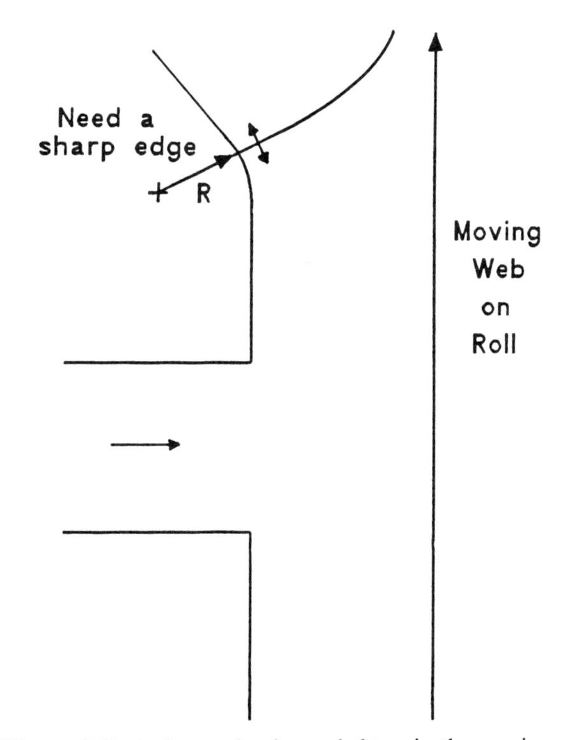

Figure 5-5. A sharp edge is needed to pin the meniscus.

Extrusion coating is similar to slot coating and is sometimes not distinguished from it. In extrusion coating the coating fluid is of much higher viscosity, often a polymer melt, and the stream exiting the slot does not spread against the lips of the die but remains as a ribbon between the die and the web, as mentioned earlier and as shown in Figure 5-1. Because of the high viscosities the pressures within the extrusion die are very high. The extruded material is often molten plastic fed by a screw feeder, and temperature control is important. To obtain a uniform coating, adjusting bolts are positioned every few inches across the slot width to adjust the slot opening or to compress internal choker bars. These adjusting bolts are sometimes under computer control, based on a scanning sensor that measures the cross-web coverage. Because of the high pressures, the adjusting bolts, and the temperature control, these extrusion dies tend to be massive, as illustrated in Figure 5-6.

Multilayer extrusion or *coextrusion* is a very common technique for obtaining the barrier or surface properties of an expensive polymer while still having a thick enough combined layer to coat, or for obtaining the desirable properties of several polymers in one multilayer coating. Several exit slots can be used in a multiple slot die, as in Figure 5-7*a*. The layers can be joined internally in a multimanifold die to exit as one layer through a single slot, as in Figure 5-7*b*. This is the most complex system. Figure 5-7*a* and *b* illustrate two-layer slot coating. The layers can also be joined together externally before entering a standard single-layer extrusion die in a combining adapter or feed block. The only change in the die is that the entry port is rectangular and not round. The is the simplest and most used system.

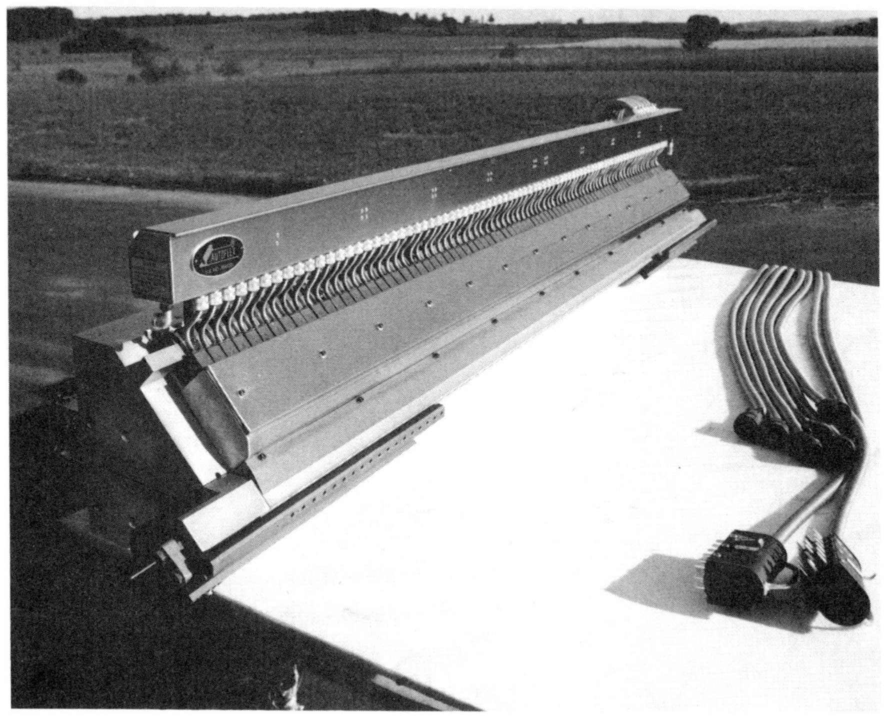

Figure 5-6. AutoFlex IV extrusion die. [Courtesy of Extrusion Dies, Incorporated, Chippewa Falls, WI 54729.]

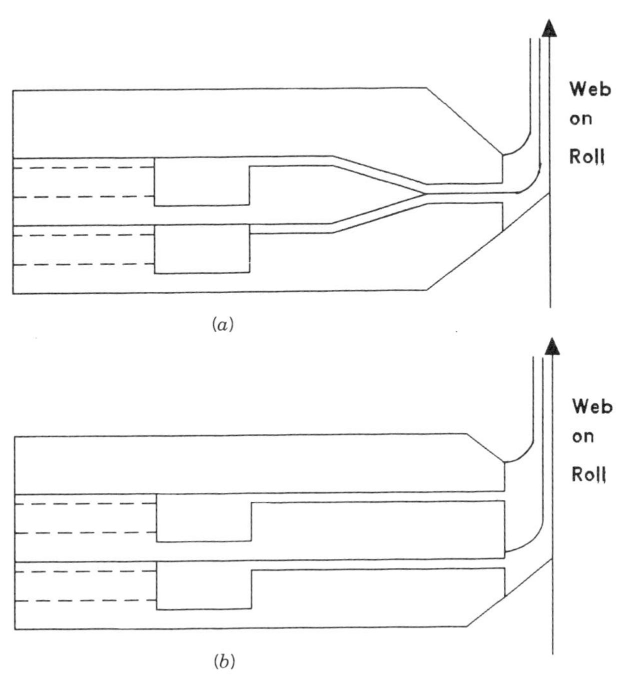

Figure 5-7. Two-layer slot coaters: (*a*) with a multiple manifold and one exit; (*b*) with multiple slots. The combining adapter system is not shown.

With the combining adapter, when the layers have similar rheologies each layer will be of uniform thickness across the full width of the extrusion die. When the layers differ in viscosity, the lower-viscosity layer will tend to be thicker toward the edges, while the higher-viscosity layer will be thicker in the center. Mainstone (1984) says that viscosity ratios between layers of as much as three or more will still give commercially acceptable thickness uniformity of the individual layers. The adjusting bolts are used to ensure that the total package is of uniform thickness.

In *slide coating* (Figure 5-8) multiple streams exit onto an inclined plane, flow down the plane without mixing, across a gap, and onto an upward-moving web. The streams remain separate and distinct from the moment they emerge onto the slide, cross the gap, get coated on the web, and in the final dried coating. This technique is commonly used in coating photographic films and papers. The edge of the bottom plate should be sharp to pin the bottom meniscus, as in slot coating. And as in slot coating, bead vacuum is often used to extend the window of coatability.

Curtain coating is an old technique that has long been used in the converting industry for coating sheets of paperboard, with the sheet passing through a falling curtain of fluid. The curtain was often formed by liquid flowing over a weir, or by an extrusion die aimed downward. In recent years curtain coating has been used as a precision coating technique for coating multilayer photographic products, with the fluid flowing off the rounded edge of a slide coater and falling vertically downward onto a horizontally moving web, as in Figure 5-9. To keep the coating from necking in edge guides need to be used (Greiller, 1972). These can be metal or plastic rods going from the coating die to the base. The bottom of the rods can be bent and can ride

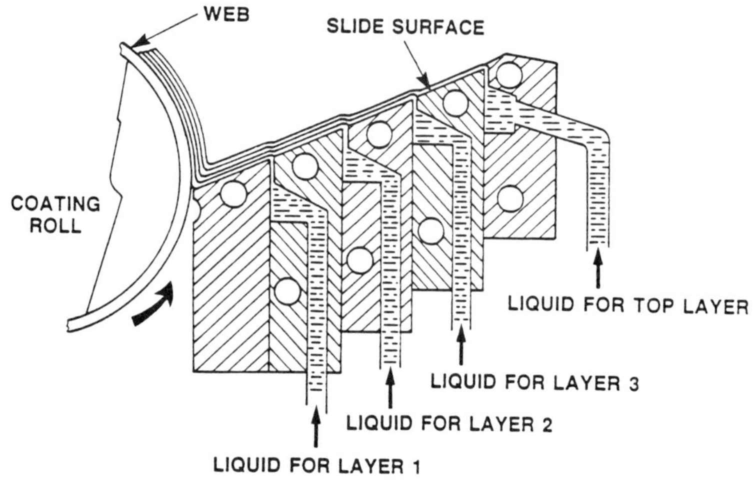

Figure 5-8. Slide coating. [After Mercier et al. (1956).]

on the web. The bent ends can be pointed slightly outward to spread out the edges to avoid heavy edges. In multilayer coating the individual layers retain their identity throughout this process.

Curtain coating has two significant advantages over slide coating:

1. In curtain coating the gap is wide, allowing splices to pass without withdrawing the coating die, as sometimes must be done in slide coating.
2. In slide coating a bubble may stay in the coating bead, floating on the surface and causing a streak until removed, while in curtain coating a bubble would flow off the slide, down the falling curtain, and onto the web to give a point defect. Point defects are less objectionable than streaks.

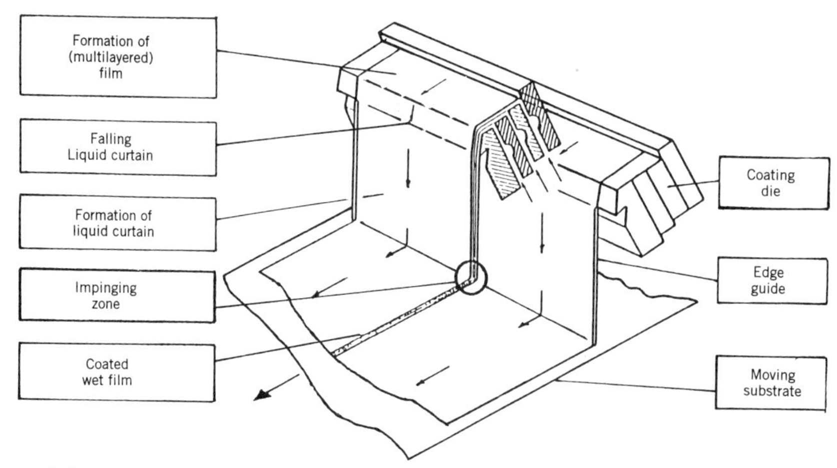

Figure 5-9. Curtain coating. [From Kistler and Scriven (1982), with permission of the authors.]

The rear lower edge of the curtain coating die, where the rear meniscus from the bottom layer is pinned, should have a sharp edge, just as in slot and slide coating. As before, by a sharp edge we mean one with a radius of curvature of about 50 μm (2 mils), and certainly not over 100 μm (4 mils).

In discussing the problems related to premetered coating, we first concern ourselves with the limits of coatability—that is, the limiting conditions under which coatings can still be made—and then discuss the problems that can arise when coatings are made within these limits of coatability.

5.2 LIMITS OF COATABILITY

When one coats too slowly or tries to apply too heavy a coverage on a web that is moving upward, the wet layer may run back. The web cannot carry away all the wet liquid layer and *sagging* occurs. The appearance is similar to sagging on a door painted with too thick a coat of an oil-based enamel. The sag lines can be far apart, as on the painted door, or very close together. If sagging occurs, the cure is simple: coat faster or make the web path more horizontal. This runback or sagging causes the high-flow limit or low-speed limit of coatability. This is rarely a problem.

On the other hand, if we are making a good coating and then try to coat faster at the same coverage, we sometimes find that we cannot do so because the coating will no longer spread uniformly across the web. And if we are making a good coating and then try to coat thinner, we sometimes find that we cannot do so, for the same reason. Both of these limits are often one and the same, and are called the low-flow limit of coatability or the high-speed limit of coatability.

5.2.1 The Window of Coatability

The low- and high-flow limits of coatability lead to a window of coatability within which one can coat the desired product. Determining this window of coatability is a useful and practical technique for optimizing the coating formulation and the coating process. We recommend that the window of coatability be determined for all new products, or at least the portions of it relevant to the product. Thus, if operations are near the low-flow limit, it may not be worthwhile determining the high-flow limits, and vice versa. The extent of the coatability window can be used to aid in determining the best of several alternative formulations. It can also be used in problem solving for production problems.

The coatability window can be found on the production coater or, preferably, on a pilot coater if it can coat at the production speeds and has been shown to duplicate production conditions. Although it is preferable to examine the dried coating for defects, the coatability can be determined just by looking at the coating as it leaves the coating stand, so drying is not necessary. A narrow pilot coater with no drying capability but able to cover the full speed range of the production machine is ideal for this purpose. Most of the coating can be doctored (squeegeed) off the web before it is wound up wet.

In the case of a slide coater there are several different types of data that can be obtained to generate the window of coatability. Ideally, the maximum and minimum wet coverages would be determined over a wide range of coating speeds, coating gaps, and bead vacuums. Practically, the following measurements are desirable:

1. The maximum and/or minimum wet coverages over a range of coating speeds at several coating gaps, at several levels of bead vacuum, including no bead vacuum
2. The range of bead vacuum that can be used to stabilize the coating bead at the desired wet coverage over a range of coating speeds and at several coating gaps
3. The range of bead vacuum that can be used to stabilize the coating bead over a range of coating gaps at the desired wet coverage and coating speed

When a series of data points are obtained, repetition is not necessary, but if only a few points are obtained, repeat measurements should be made. It is also useful to vary the direction in approaching a limit. It is desirable to operate near the center of the coatability window, as defects are more likely to occur near the boundaries.

Making measurements on the production coater is expensive and imposes limits on the conditions that can be run. Coaters often cannot be run as fast as needed to determine the high-speed limits, and dryers often cannot dry high-speed coatings. It is necessary to ensure that wet coatings do not contaminate face rolls in the dryer and in the windup, and that wet coatings do not reach the windup. One can always run the dryer as hot as possible, but this may still not dry high-speed coatings. In this case one should coat just a short section and then stop the coating machine, letting the coated section dry in place.

When the low-flow or high-speed limits of coatability are exceeded, one of several conditions can occur in slide coating (Gutoff, 1992):

1. The coating bead breaks down completely and the coating liquid cannot cross the gap to the web.
2. Ribs are formed and, at still higher speeds, the spaces between the ribs receive no liquid to become dry lanes and the ribs become rivulets.
3. Air is entrained in the coating to appear as bubbles in the coating.
4. Dry patches where air is entrained are interspersed with the coated areas. This has been noted with more viscous liquids.
5. One or both of the edges neck in. The neck-in can be considerable, a matter of centimeters or inches, and the remaining coating is then thicker.

We now examine the low-flow limits of coatability for slide and for slot coating and both the low- and high-flow limits for curtain coating.

5.2.2 Slide Coating

The window of coatability in slide coating can be shown in a plot of bead vacuum versus coating speed for a given coverage of a particular solution, as has been done by Chen (1992) (Figure 5-10). There is an upper coating speed, above which air entrainment occurs, and this is not shown in these plots. But the regions in which various defects occur are shown. Many of these defects are discussed later in this chapter. *Barring* is another term for chatter. *Gap bubbles* refers to entrained air, and *air pockets* refers to bubbles covering a larger area. Rivulets occur when the coating consists of narrow stripes of coating running down-web, separated by dry lanes. Bleeding indicates that liquid is sucked down into the vacuum pan. Chen (1992) found that at too low a vacuum, except at low speed, a good coating could not be made. Air entrainment or rivulets would occur. On the other hand, at too high a bead vacuum, ribbing or chatter would occur. In Figure 5-10*b* no defect-free coating could be made. The term *uniform coating* probably should not be used in this drawing, since an oscillating cross-web wave was present, which seemed to originate on the slide.

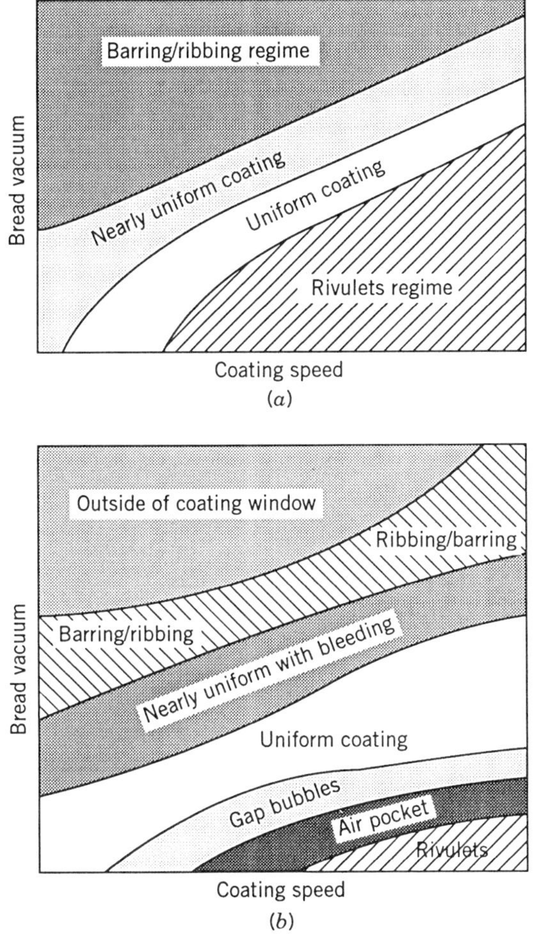

Figure 5-10. Operating diagrams in slide coating: (*a*) with glycerin water; (*b*) with an aqueous cellulosic polymer. The high speed limit is not shown. [From Chen (1992), with permission of the author.]

Vortices and Eddies

Vortices and eddies can form within the coating bead in slide coating. The bead profile in a good coating would appear as in Figure 5-11*a*. Vortex formation should be avoided, as the liquid in a vortex can remain for long periods to degrade with time and cause coating defects as the material slowly leaves the vortex. When the bead is too thick, perhaps because the coating speed is too low for this coverage, a vortex can form in the bead, as indicated in Figure 5-11*b*. When the bead vacuum is too high the bead can be sucked below the edge of the die lip, as in Figure 5-11*c*. A vortex can also form in this region. Figure 5-11*d* is a computer-generated profile showing both vortices.

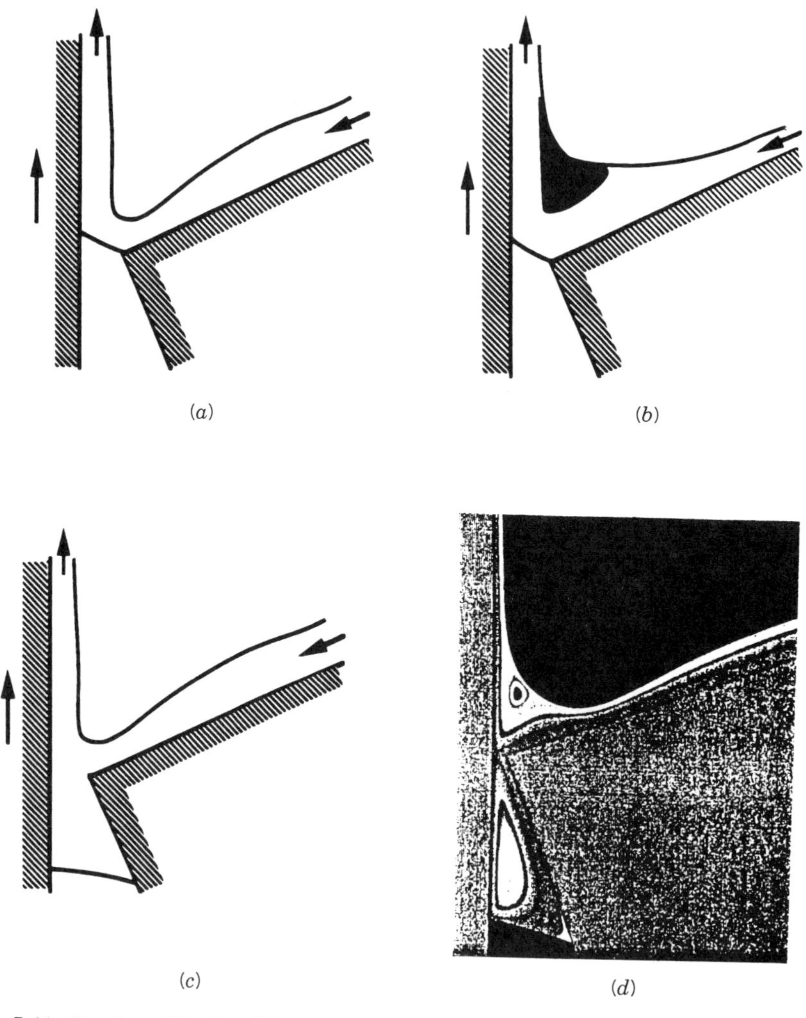

Figure 5-11. Bead profiles in slide coating: (*a*) the desirable condition; (*b*) with an upper vortex; (*c*) with die-face wetting; (*d*) with an upper vortex, die-face wetting, and a lower vortex. [From Chen (1992), with permission of the author.]

Chen (1992) studied conditions under which vortices form. Figure 5-12 shows that an upper vortex forms when the drawdown ratio, the ratio of the coating speed to the average speed on the slide, is below a certain value, depending on the capillary number, $\bar{\mu}U/\sigma$, where U is the web speed, $\bar{\mu}$ the average viscosity, and σ the surface tension. For a given system the capillary number is just a measure of the coating speed. The drawdown ratio is not easy to calculate and therefore not too easy to use. However, it does increase with coating speed and with viscosity, and with a decrease in the angle of the slide above the horizontal.

The drawdown ratio is $U/(q/H)$, or UH/q. Now the flow rate, q, is equal to the coating thickness times the coating speed, or $q = tU$. And H, the thickness on the slide, is

$$H = \left(\frac{3Ut\mu}{\rho g \sin\theta} \right)^{1/3}$$

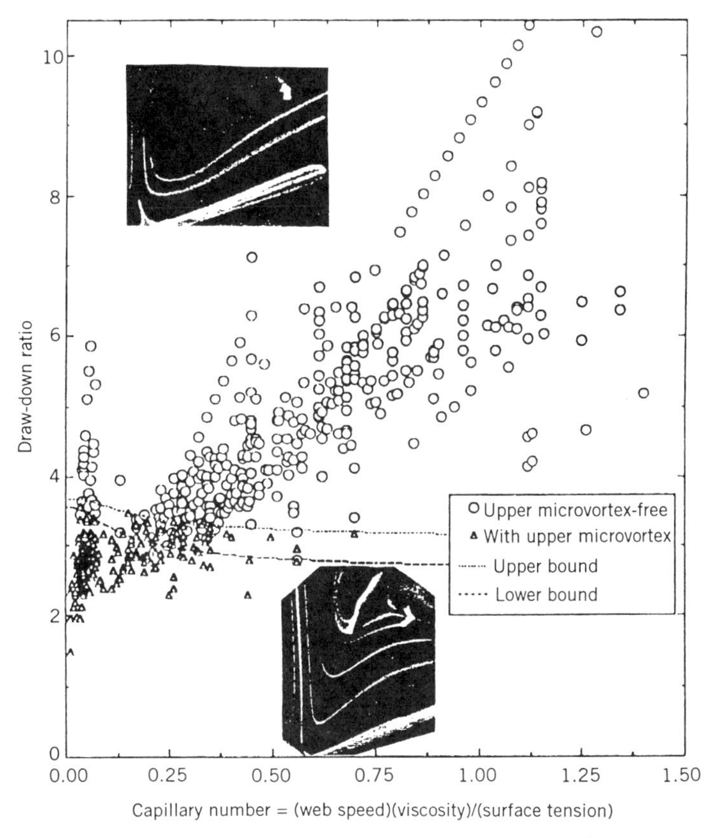

Figure 5-12. Conditions for the formation of an upper vortex, based on two- and three-layer flows of glycerin solutions. Gap: 200 to 400 μm (8 to 16 mil); bead vacuum: 0 to 900 Pa (0 to 3.6 in. water); speed: 0.05 to 2.62 m/s; wet thickness: 60 to 816 μm; viscosities: 12 to 70 mPa · s (cP). [From Chen (1992), with permission of the author.]

Therefore, the drawdown ratio can be written as

$$\frac{U}{q/H} = \frac{U}{Ut}\left(\frac{3Ut\mu}{\rho g \sin\theta}\right)^{1/3} = \frac{(3U\mu)^{1/3}}{t^{2/3}(\rho g \sin\theta)^{1/3}}$$

Figure 5-12 shows that to avoid the upper vortex one has to coat above some minimum coating speed. One should also avoid having the coating liquid wet the die face below the edge of the lip. Chen (1992) showed in Figure 5-13 that with no bead vacuum, the capillary number, $\bar{\mu}U/\sigma$, must be above some minimum value, about 0.15 to 0.2. This also sets a minimum speed for the system. Bead vacuum tends to pull the bead down and wet the die face below the lip edge. This sets a maximum bead vacuum that can be used, and this is indicated in Figure 5-14, also from Chen (1992). As seen from Figure 5-14, this maximum bead vacuum increases with coating speed.

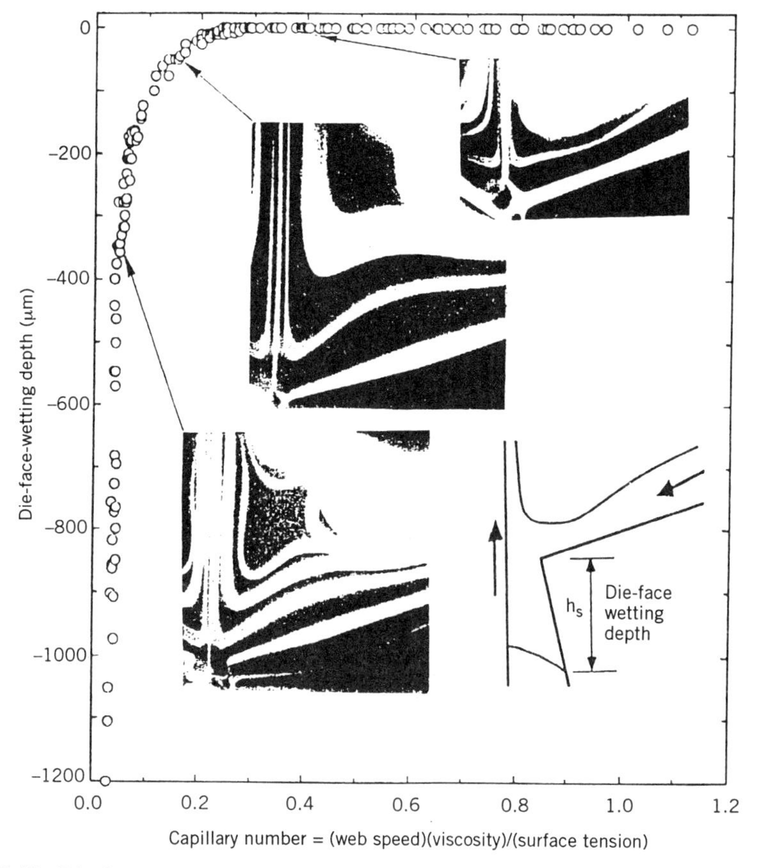

Figure 5-13. Die-face wetting as a function of the capillary number with no bead vacuum. The other conditions are the same as for Figure 5-12. [From Chen (1992), with permission of the author.]

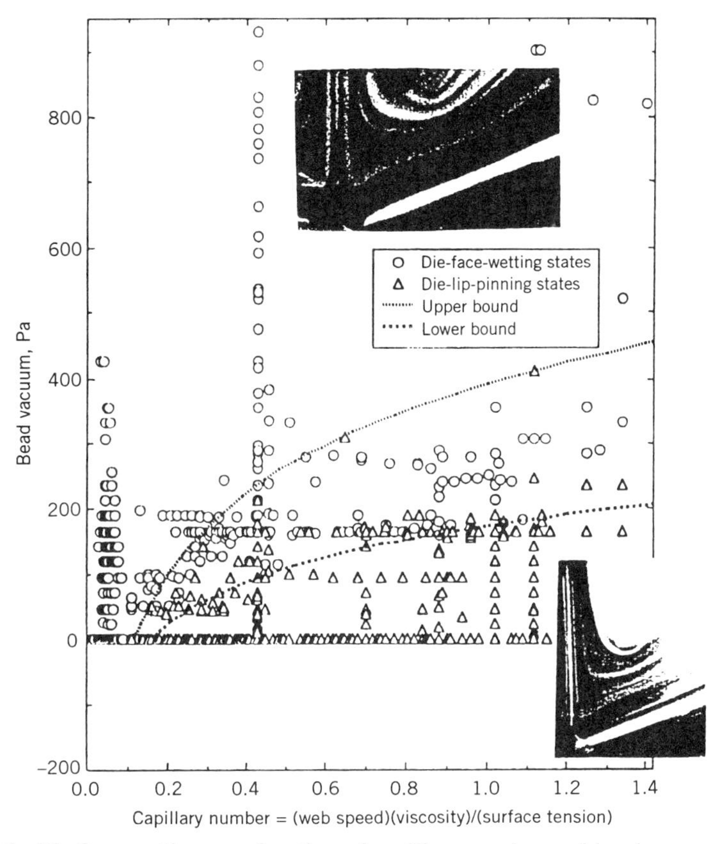

Figure 5-14. Die face wetting as a function of capillary number and bead vacuum. The conditions are the same as for Figure 5-12. [From Chen (1992), with permission of the author.]

Figures 5-12, 5-13, and 5-14 are all based on experiments with glycerin–water solutions, which are Newtonian. Polymer solutions, which are inherently shear thinning and are inherently viscoelastic, would behave differently, although the qualitative trends would be similar. It is sometimes useful to look at the window of coatability in another manner. As it is often desired to coat a given coating solution to given coating coverage at a fixed coating speed (usually, the maximum coating speed that the dryer can handle), a diagram of bead vacuum versus coating gap—the two controllable variables—showing the regions of operability can be very useful to the coating operator. Such a diagram for a particular set of conditions is shown in Figure 5-15.

In slide coating the minimum wet thickness increases with coating speed, as shown in Figure 5-16. Above some maximum speed a coating cannot be made, no matter how thick the layer. Bead vacuum lowers the minimum thickness at any speed, and slightly increases the maximum coating speed (perhaps by 25%). The data for a bead vacuum of 500 Pa (2 in. water) are not shown on this graph, as they fall almost on top of the data for a bead vacuum of 1000 Pa (4 in. water). Gap has a slight effect. In Figure 5-16 narrow gaps allow a slightly thinner coating, but gap does not affect the maximum coating speed. With some other liquids, gap has an opposite effect;

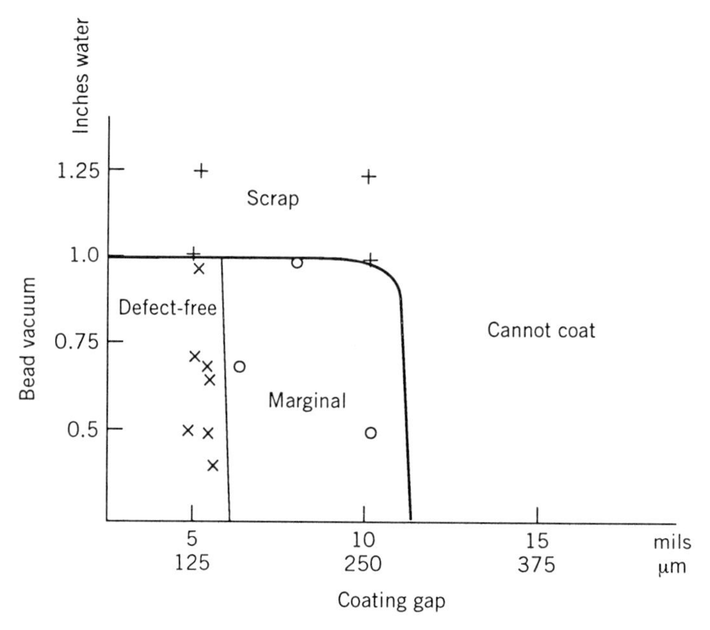

Figure 5-15. Operating window for a particular slide coating in terms of bead vacuum and coating gap.

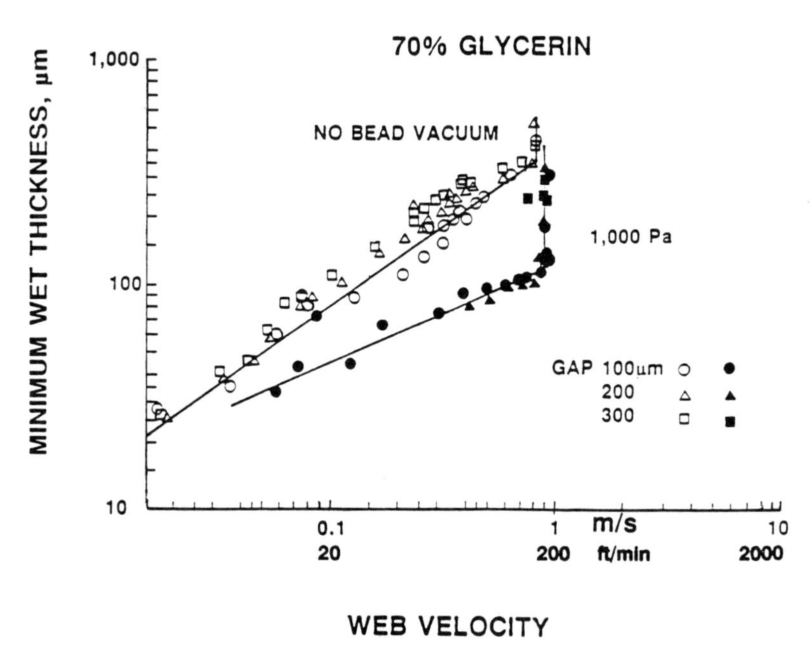

Figure 5-16. Minimum wet thickness in slide coating versus web velocity, for 70% glycerin. [From E. B. Gutoff and C. E. Kendrick, *AIChE J.*, **33**, 141–145 (1987). Reproduced with permission of the American Institute of Chemical Engineers.]

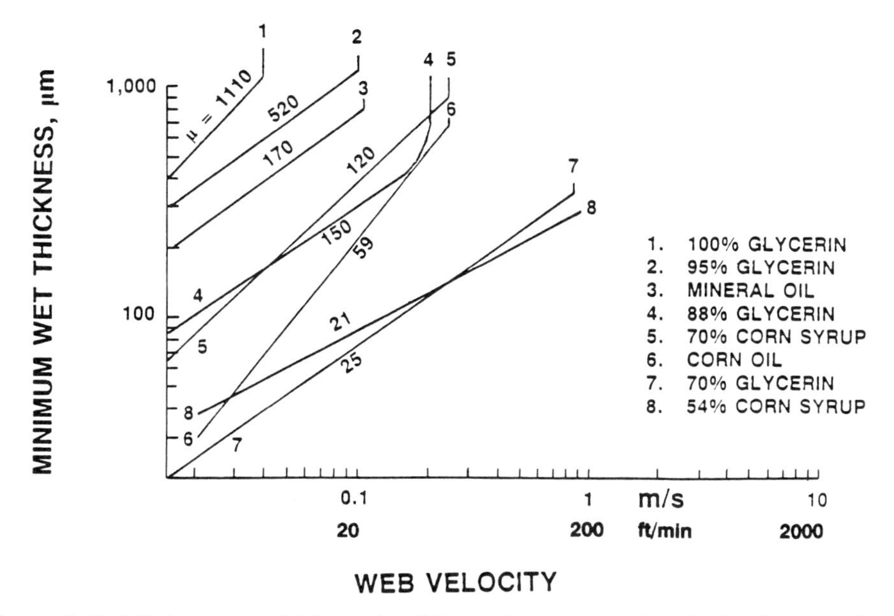

Figure 5-17. Minimum wet thickness in slide coating versus web velocity for a number of liquids. No bead vacuum was used. [From E. B. Gutoff and C. E. Kendrick, *AIChE J.*, **33**, 141–145 (1987). Reproduced with permission of the American Institute of Chemical Engineers. © 1987 AIChE. All rights reserved.]

that is, at a wider gap one is able to coat thinner, but in almost all cases the gap effect is small. Note that with the bead vacuum of 1000 Pa (4 in. water) there are very few data at the wide gap of 300 μm. At wide gaps only a low bead vacuum can be used; otherwise, the vacuum will suck the bead down. The best lines for similar data for a number of simple liquids are shown in Figure 5-17, where it is seen that as the viscosity increases, the minimum coating thickness increases and that the maximum coating speed decreases. With bead vacuum (Figure 5-18) the minimum coating thickness decreases and the maximum coating speed increases slightly. This is why bead vacuum is usually used.

Polymer solutions are easier to coat, in that one can coat thinner layers and at higher speeds than simple liquids, as shown in Figure 5-19. We see that a polyvinyl alcohol solution can be coated at about an order of magnitude (10×) faster than a glycerin solution of the same viscosity.

It is not difficult to give a qualitative explanation for all these effects (Gutoff, 1992). Let us look at the forces involved in the slide coating bead, as sketched in Figure 5-20. There are stabilizing forces, which tend to hold the bead against the web, and destabilizing forces, which tend to tear the bead apart.

The stabilizing forces are:

1. *Bead Vacuum.* The force per unit width is the bead vacuum times the length of the bead; thus at wider gaps, less vacuum should be needed. The bead vacuum tends to push the bead against the web, and so is stabilizing.
2. *Electrostatic Forces.* If the backing roll is charged and the coating die is

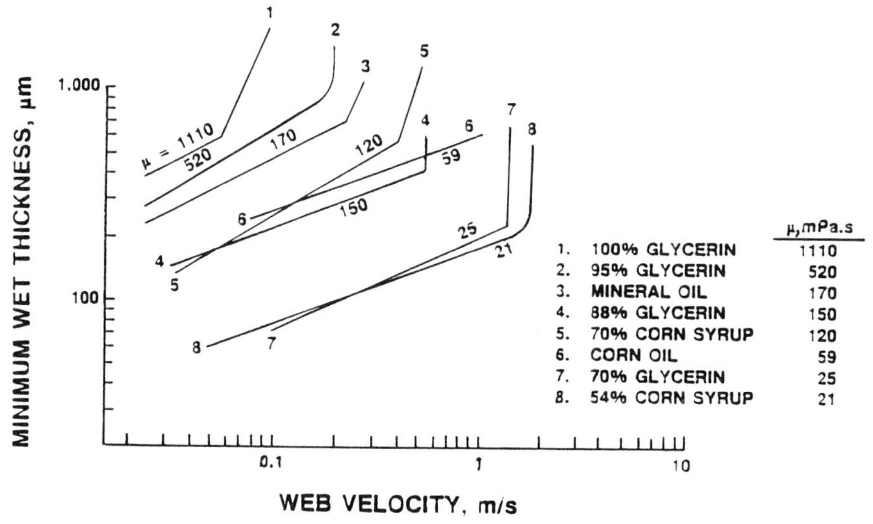

Figure 5-18. Minimum wet thickness in slide coating versus web velocity for a number of liquids, with a bead vacuum of 1000 Pa (4 in. of water). [From E. B. Gutoff and C. E. Kendrick, *AIChE J.*, **33**, 141–145 (1987). Reproduced with permission of the American Institute of Chemical Engineers. © 1987 AIChE. All rights reserved.]

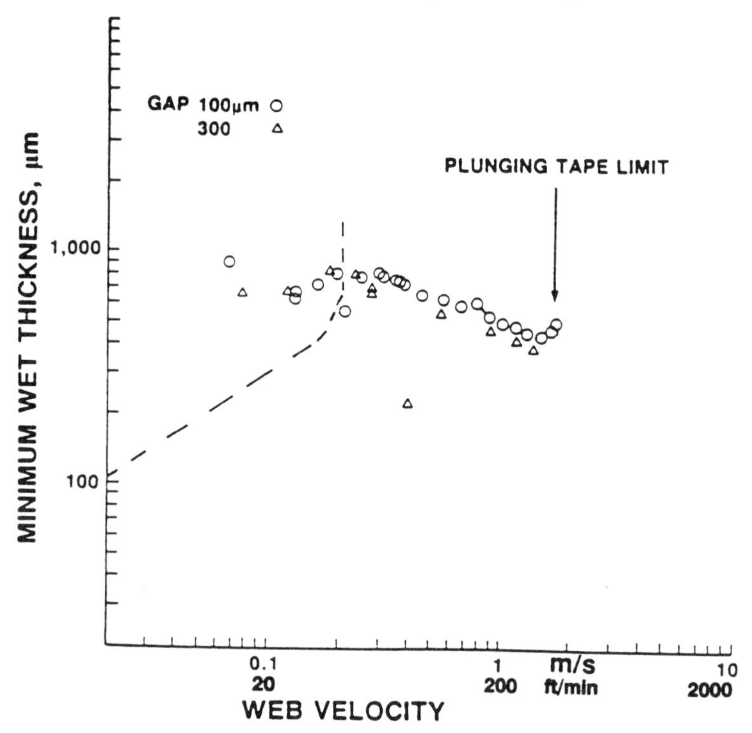

Figure 5-19. Minimum wet thickness versus web velocity for slide coating a polyvinyl alcohol solution. The viscosity is 146 mPa · s (146 cP), and no bead vacuum was used. The dashed curve is for a glycerin solution of the same viscosity from Figure 5-16. [From E. B. Gutoff and C. E. Kendrick, *AIChE J.*, **33**, 141–145 (1987). Reproduced with permission of the American Institute of Chemical Engineers. © 1987 AIChE. All rights reserved.]

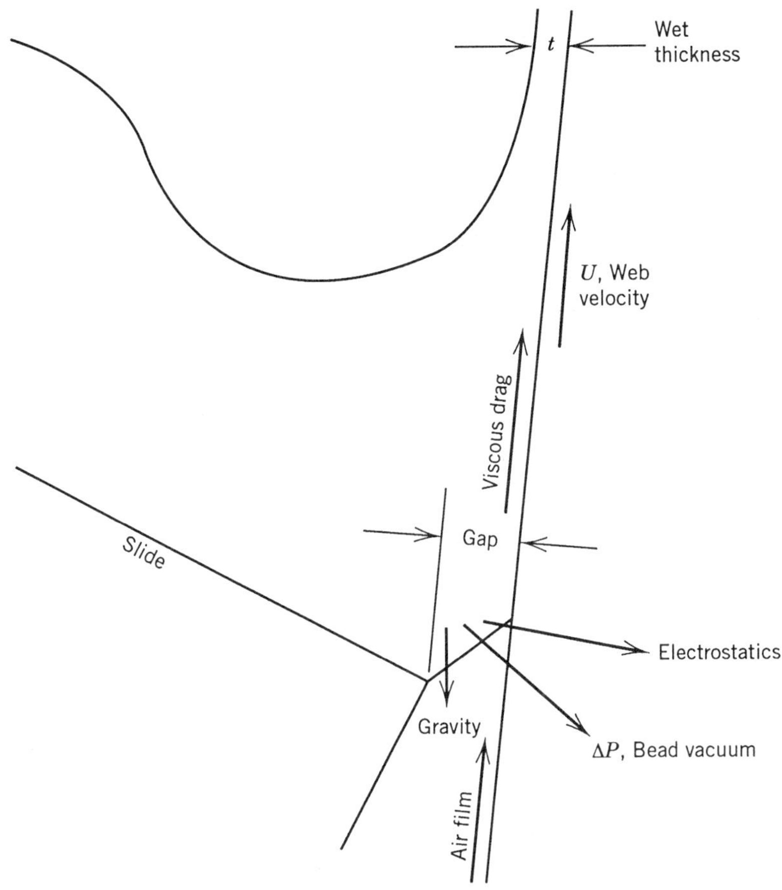

Figure 5-20. Slide coating bead showing some of the forces acting on it. [From Gutoff (1992). Reprinted with permission by VCH Publishers © 1992.]

grounded, or if the web has a uniform static charge, the coating fluid will be attracted to the web. Electrostatic assist is not commonly used. At best it is as good as bead vacuum, and in combination with bead vacuum is often no better than bead vacuum alone.

3. *Gravity.* Gravity acts downward, thus tends to keep the liquid from being pulled out from the bead. But in these thin layers the gravitational force is very weak and has no significant effect on the system.
4. *Fluid Inertia.* The momentum of the fluid flowing down the slide tends to force the bead against the web, but this force, like gravity, is very weak and usually can be neglected.

The destabilizing forces are:

1. *Shear Stress in the Liquid, Arising from the Motion of the Web.* The web tends to pull the fluid along with it. This shear stress is proportional to the fluid

viscosity and to the shear rate. The shear rate is the rate of change of velocity as one moves away from the surface. It is approximately proportional to the web speed. Thus the shear stress is approximately proportional to the viscosity times the web speed. It pulls the liquid along and tends to act as a tensile stress in the bead, trying to pull the liquid apart.

2. *Air Film Momentum.* The web carries along with it a very thin film of air, which has to be removed for the liquid to wet the web. This air pushes against the bead and tends to push the bead away from the web. Thus it is a destabilizing force.
3. *Centrifugal Force of the Rotating Backing Roll.* This tends to throw the coating fluid off the web, but like gravity and fluid inertia, it is a very weak force.

Thus there exist strong destabilizing forces of the shear stress and the air film momentum acting to tear the coating bead apart. If no bead vacuum or electrostatic assist is used, the remaining stabilizing forces of gravity and fluid inertia are too weak to have any appreciable effect. Then what resists the destabilizing forces? It is only the inherent properties of the coating fluid itself that can resist the destabilizing forces. It is the cohesive strength of the liquid, or, if you will, the elongational viscosity of the liquid, that allows a liquid in motion to sustain a tensile stress.

Now everyone knows that liquids cannot be in tension—at equilibrium. But less well known is that liquids in motion can be in tension. If we dip a finger in a jar of honey and then lift the finger up, some honey will tag along. The strand of honey between the jar and the finger is in tension. If we hold the finger still, much of the honey will drop back into the jar.

The tensile stress that a liquid can sustain is equal to the elongational viscosity times the extension rate—a measure of how fast the liquid is being pulled. The elongational viscosity of a simple fluid is three times its normal viscosity. Thus honey, which is a thick or viscous fluid, will sustain a greater tensile stress than water.

Polymer solutions, by their very nature, tend to have high elongational viscosities at high elongation rates. When a polymer solution is pulled slowly, the polymer molecules just slide past each other and the elongational viscosity is just about what it would be for a simple liquid of the same viscosity. But when a polymer solution is pulled quickly, the molecules do not have time to "relax," and they remain entangled and hard to pull apart. The elongational viscosity goes way up and the solution can sustain high tensile stresses. This would occur with all polymer solutions at the high pull rates (or elongation rates) that exist in the coating bead. This is why the polyvinyl alcohol solution can be coated about 10 times faster than the glycerin solution of the same viscosity, as we saw in Figure 5-19. With some polymer solutions, such as polyacrylamide, this ability to withstand tensile stresses can be demonstrated at the relatively low pull rates that exist in a tubeless syphon, shown in Figure 5-21. The vacuum tube is dipped into the polymer solution and is then pulled up above the surface, and the liquid continues to be sucked up.

Polymer solutions also coat better than simple liquids because their viscosity tends to decrease at high shear rates. Imagine a polymer solution in a beaker, and imagine that we are stirring the beaker with a mixer. When we stir at a very slow rate, such that the polymer molecules remain undisturbed and retain their normal shape of a random coil—a sphere—then at a slightly higher stirring rate the resistance of these

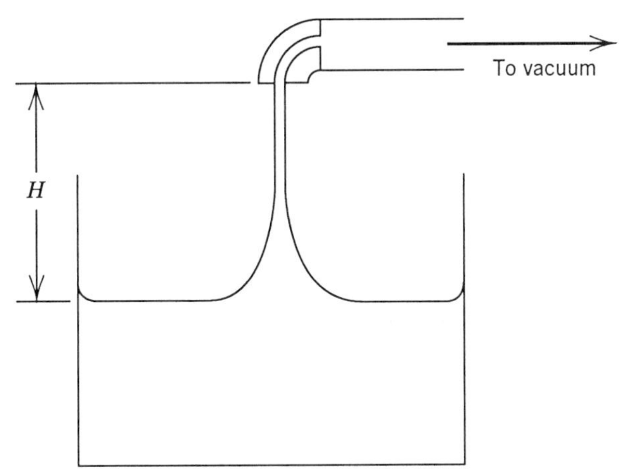

Figure 5-21. The tubeless syphon with a polymer solution such as polyacrylamide in water.

coils to flow will remain unchanged, and the viscosity will not change. This is called the zero shear viscosity. But as the stirring rate increases, the spherical coils will become elongated—football shaped—and their smaller cross section will present less resistance to flow. The viscosity will then drop, as sketched in Figure 5-22. After further increases in stirring rate, the polymer molecules will be as stretched as much they can be, and further increases in stirring will not reduce the resistance to flow. Thus the viscosity will again be constant. This is called the infinite shear viscosity. In many cases, on a double-log plot, as in Figure 5-22, there will be a straight-line region where the viscosity decreases with shear. This is called the power law region, and the fluid would be called a power law fluid. The mathematics of power law fluids are relatively simple.

When we coat a fluid at its minimum thickness and try to coat faster, we cannot get a good coating. As we increase the coating speed, the shear stress, approximately proportional to the viscosity times the coating speed, increases. This shearing force

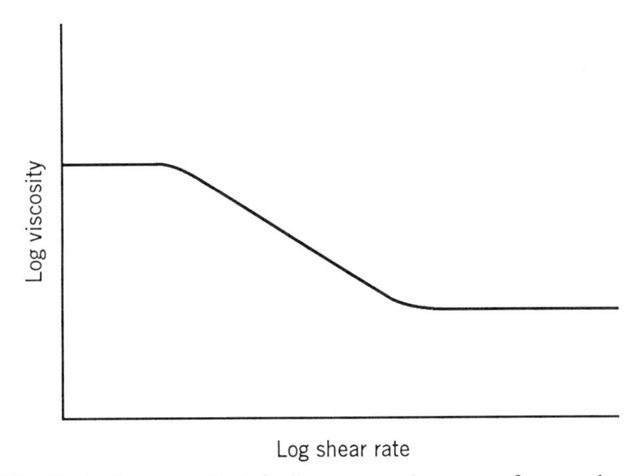

Figure 5-22. Typical curve of viscosity versus shear rate for a polymer solution.

on the web acts as a tensile force in the bead. When the shear stress increases, the tensile stress in the bead will increase, and it will pull apart the coating bead, which was already at its maximum tensile stress. What happens when we coat thicker? The tensile force remains the same. But as the flow down the slide is now greater in order to coat thicker, the bead will also be thicker. With the thicker bead the tensile stress, equal to the tensile force divided by the area of the bead, will now be less and may be within the cohesive strength of the liquid. Thus as we coat faster, we have to coat thicker.

Similarly, as we coat with higher viscosities, the shear stress and therefore the tensile forces and stresses will increase. If initially we are at the limits of coatability and then we switch to a higher viscosity liquid, we will no longer be able to coat because the higher tensile stress will exceed the cohesive strength of the liquid at that particular elongation rate. By coating thicker the tensile force will be spread over a larger area, the tensile stress will be reduced, and we may now be within the limits of coatability for this liquid.

However, we should note that the tensile stress in the liquid is greatest at the contact line where the bead meets the web. The tensile stress will be lowest at the upper surface of the bead. As the bead becomes thicker and thicker, there will come a time when the surface of the bead is relatively unstressed. Thus, increasing the thickness of the bead will no longer reduce the tensile stress in the bead. At this condition we have reached the maximum coating speed, where increasing the wet thickness, and thus the bead thickness, no longer allows an increase in speed. Bead vacuum and electrostatic assist, as they are stabilizing forces, lower the minimum wet coverage and also slightly increase the maximum coating speed.

The ability to coat lower-viscosity liquids at higher speeds and in thinner layers explains why, in multilayer slide coatings, the bottom layer is often of low viscosity. The bottom layer is most highly stressed. It had been suggested that one can use an inert, additional layer as a carrier or bottom layer, to make coating easier (Dittman and Rozzi, 1977; Choinski, 1978; Ishizaki and Fuchigami, 1985; Ishizuka, 1988).

5.2.3 Slot Coating

The window of coatability can be illustrated by a graph of bead vacuum versus the coating speed, as has been done by Sartor (1990) for a slot coating (Figure 5-23). We see in this figure that within a certain range of coating speeds and bead vacuums good, defect-free coatings can be made. Outside this region, on the low-vacuum side, ribs will form. At too high a bead vacuum the coating liquid will be sucked out of the slot. In the figure this is called swelling of the upstream meniscus. Swelling means that the meniscus is no longer pinned at the lower edge of the bottom lip but is on the swept-back section below the lip. With swelling a vortex will form there, leading to a long residence time in the vortex. A long residence time can lead to degradation of the liquid with some materials. A vortex may also cause a streak or a band. With a slight additional increase in vacuum, the bead will be sucked down out of the gap and a coating will no longer be possible.

When the coating speed is too high, air is entrained and no coating can be made. In Figure 5-23 this high-speed limit was neither measured nor calculated, but just sketched in by this author to indicate that this limit does exist. One should also emphasize two points illustrated by Figure 5-23. The first is that at very low speeds

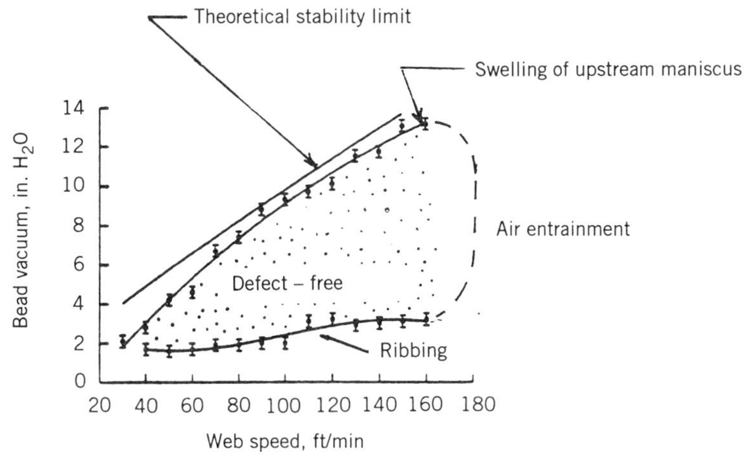

Figure 5-23. Window of coatabilty plot for slot coating a 25 mPa·s (25 cP) solution 85 μm thick with a gap of 250 μm (10 mil). [After Sartor (1990), with permission of the author.]

a defect-free coating may not be able to be made. Either ribbing will occur, or, at a higher bead vacuum, the upstream meniscus will swell with formation of a vortex and the consequent long residence time and the possibility of streaks and bands. The second is that, in this case, some bead vacuum is needed to obtain a defect-free coating. In many cases bead vacuum is not necessary, but as the coating thickness decreases or the coating speed increases, it is more likely that bead vacuum will be necessary.

Lee et al. (1992) found that at relatively low speeds and with low-viscosity liquids, the limits of coatability are very similar to the limits in slide coating, as seen in Figure 5-24. This agreement between slot and slide coating occurs when a dimensionless grouping of physical properties called the capillary number (equal to $\mu V/\sigma$, the product of the viscosity and the coating speed divided by the surface tension) is below a critical value. However, above the critical value of the capillary number the slot coater can coat at higher speeds and in thinner layers than the slide coater. Thus, as with slide coaters, at low values of the capillary number the gap has almost no effect, and the minimum wet thickness increases with viscosity and with coating speed. These effects can also be seen in Figures 5-25 and 5-26. At the higher levels of viscosity and speed, that is, above the critical capillary number, viscosity has no effect, as seen clearly in Figure 5-26 at the tighter gap of 200 μm (8 mils). At these high capillary numbers the minimum wet thickness divided by the gap is constant at a value of about 0.6 to 0.7, as seen in Figure 5-27. Thus, from Figure 5-27, at high capillary numbers, as the thickness divided by gap is constant, the minimum coating thickness is directly proportional to the gap. For thin coating one would have to use tight gaps. Figure 5-27 also shows that the critical capillary number, above which the minimum wet thickness divided by the gap is constant, increases with gap, and Figure 5-28 shows that the relationship is linear. Table 5-1 summarizes some of these high-capillary-number data.

In slot coating gravity normally has no effect, and the slot die, aimed at the center of the backing roll, can be pointed up, or down, or horizontally, and it makes no

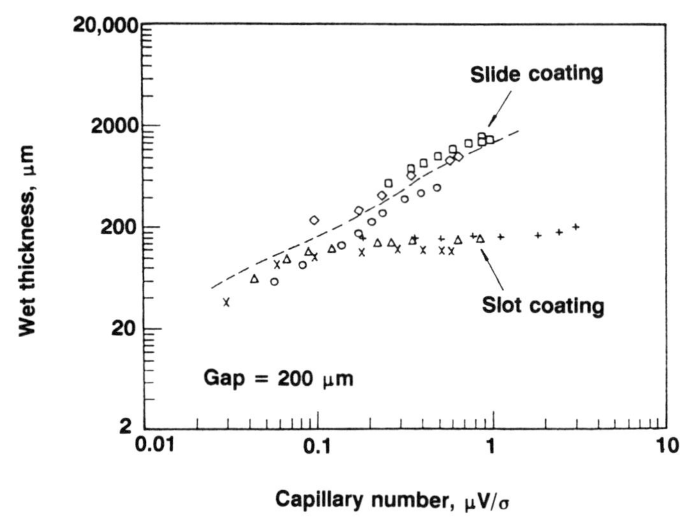

Figure 5-24. Comparison of slot coating with slide coating, with a 200-μm gap. The dashed line is from Gutoff and Kendrick (1987). $\underline{\Delta}$ and ○, silicone oil, 50 mPa · s (50 cP); +, 95% glycerin-water, 420 mPa · s (420 cP); ×, silicone oil, 10 mPa · s (10 cP); ◇, 88% glycerin-water, 133 mPa · s (133 cP). [Reprinted from K. Y. Lee et al., "Minimum wet thickness in extrusion slot coating," *Chem. Eng. Sci.*, **47**, pp. 1703–1713. Copyright (1992), with kind permission from Elsevier Science Ltd., The Boulevard, Langford Lane, Kidlington, UK, and from the authors.]

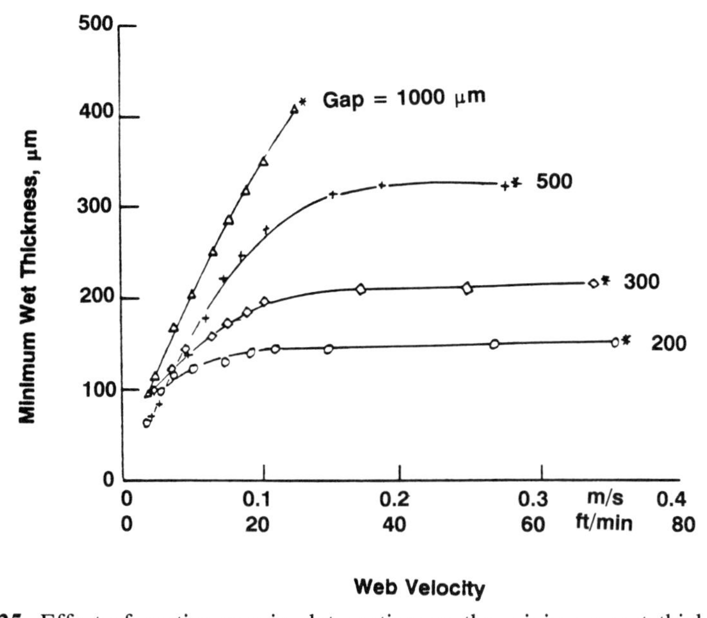

Figure 5-25. Effect of coating gap in slot coating on the minimum wet thickness, with a 50 mPa · s (50 cP) liquid. The asterisks indicate the maximum coating speeds. [Reprinted from K. Y. Lee et al., "Minimum wet thickness in extrusion slot coating," *Chem. Eng. Sci.*, **47**, pp. 1703–1713. Copyright (1992), with kind permission from Elsevier Science Ltd., The Boulevard, Langford Lane, Kidlington, UK, and from the authors.]

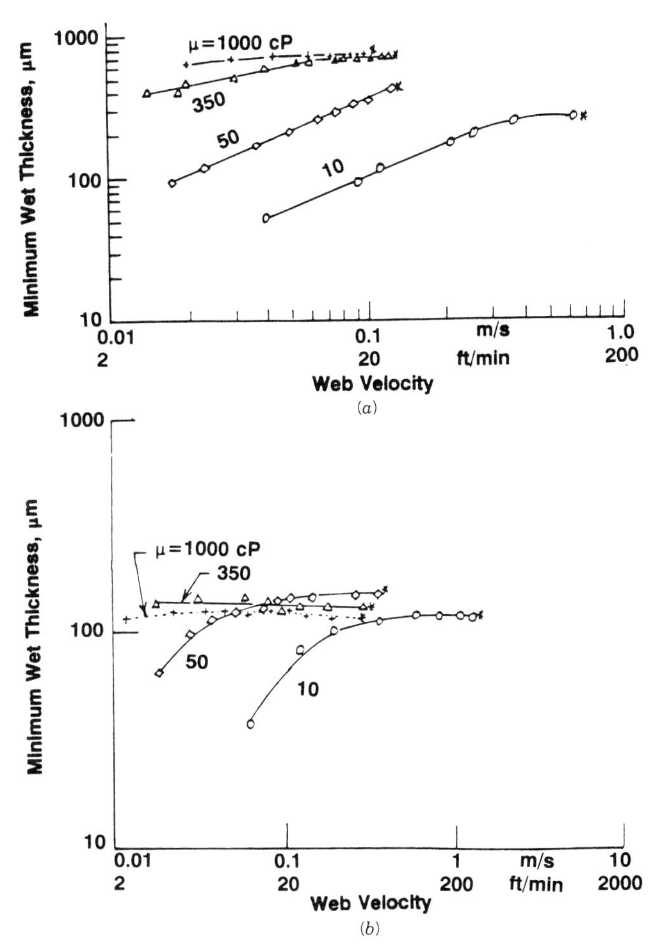

Figure 5-26. Effect of viscosity in slot coating on the minimum wet thickness, at gaps of (*a*) 1000 and (*b*) 200 μm. The asterisks indicate the maximum coating speeds. [Reprinted from K. Y. Lee et al., "Minimum wet thickness in extrusion slot coating," *Chem. Eng. Sci.*, **47**, pp. 1703–1713. Copyright (1992), with kind permission from Elsevier Science Ltd., The Boulevard, Langford Lane, Kidlington, UK, and from the authors.]

difference. The horizontal position is most common, but when aimed upward it is easier to get rid of air in the system, and there is less likelihood of bubbles building up to disrupt the flow. The slot opening also has no effect. All the slot opening does is control the liquid momentum against the web, and as we discussed in slide coating, this is normally insignificant. The slot opening does help determine the pressure within the coating die, and there are upper and lower limits on this, as we discuss in Chapter 6.

In Figures 5-25 and 5-26 the maximum coating speed is indicated by an asterisk. This is the highest speed at which a good coating can be obtained. At higher speeds either air is entrained and a complete coating cannot be had, or ribbing occurs and

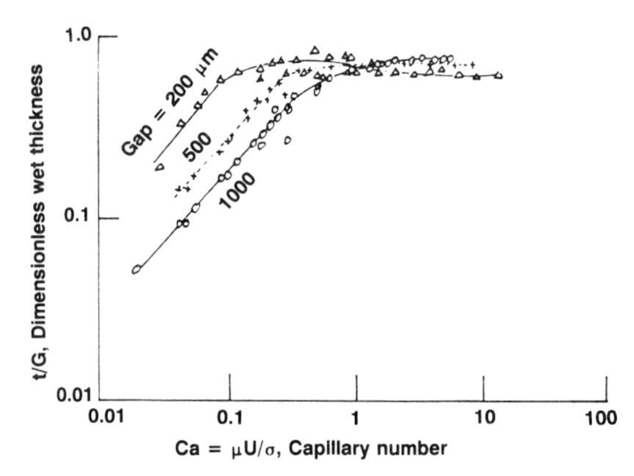

Figure 5-27. Minimum wet thickness divided by the gap in slot coating as a function of the capillary number for various gaps. [Reprinted from K. Y. Lee et al., "Minimum wet thickness in extrusion slot coating," *Chem. Eng. Sci.*, **47**, pp. 1703–1713. Copyright (1992), with kind permission from Elsevier Science Ltd., The Boulevard, Langford Lane, Kidlington, UK, and from the authors.]

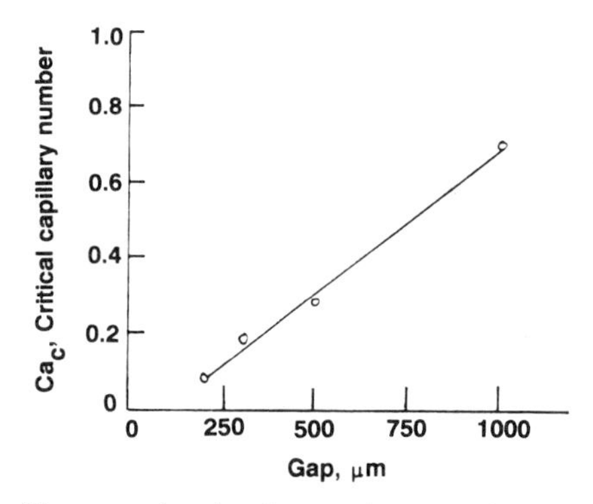

Figure 5-28. Critical capillary number in slot coating as a function of gap. [Reprinted from K. Y. Lee et al., "Minimum wet thickness in extrusion slot coating," *Chem. Eng. Sci.*, **47**, pp. 1703–1713. Copyright (1992), with kind permission from Elsevier Science Ltd., The Boulevard, Langford Lane, Kidlington, UK, and from the authors.]

Table 5-1 Dimensionless Minimum Wet Thickness, t/G, for Ca > Ca*

Viscosity, μ (mPa · s)	Gap, $G(\mu m)$			
	200	300	500	1000
10	0.61	0.55	0.44	Defects
50	0.70	0.70	0.65	Defects
350	0.67	0.71	0.66	0.70
1000	0.65	0.70	0.65	0.72

Source: K.-Y. Lee et al., *Chem. Eng. Sci.* **47**, 1703–1713, (1992), with kind permission from Elsevier Science Ltd., The Boulevard, Langford Lane, Kidlington, UK, and from the authors.

Ca* is the critical capillary number.

then at still higher speeds, stripes separated by dry regions, or rivulets, would be coated.

As with slide coating, bead vacuum lowers the minimum wet thickness (Figure 5-29). Also as with slide coating, at wide gaps the bead is unstable at higher bead vacuums. Bead vacuum seems to be less effective with high viscosities, over 500 mPa · s (500 cP) (Lee et al., 1992). At the lower viscosities bead vacuum can easily reduce the minimum wet thickness by 20 to 30%.

A comparison of slide and slot coating can be made by referring to Figure 5-30, as has been done by Lee et al. (1992). The slide coater has the lower meniscus pinned to the coating die with the upper meniscus completely free, while the slot coater has both the upper and lower menisci pinned. When the coated layer is thin compared to the gap, as it can be at low viscosities or low speeds (low capillary numbers), the upper lip of the slot coater has little influence on the meniscus that is pinned to it, except very close to the lip. The upper menisci of both the slot and the slide coaters are then similar, and the limits of coatability are similar. However, at higher capillary numbers the minimum wet thickness increases up to about two-thirds of the gap. The upper lip then acts as a doctor blade to prevent the coated layer from getting thicker (Lee et al., 1992).

5.2.4 Extrusion Coating

The upper speed limit in extrusion coating occurs when the extruded material is no longer smooth or uniform. There are many types of instabilities that can occur in extrusion coating, and they are generally classified under the terms *melt fracture* or

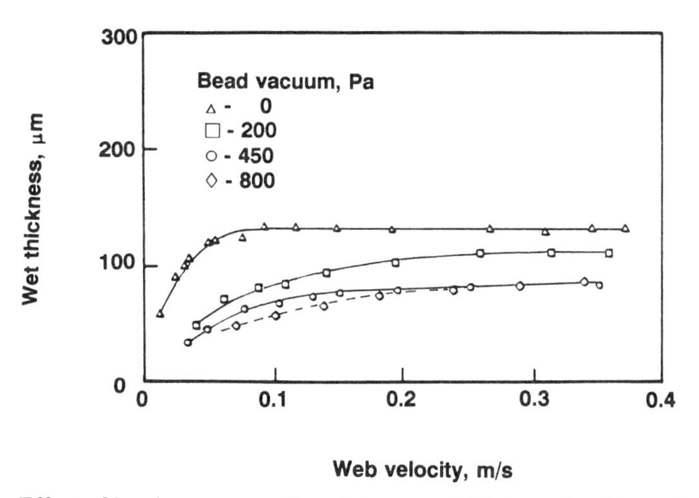

Figure 5-29. Effect of bead vacuum on the minimum wet thickness in slot coating, with a 50 mPa · s (50 cP) liquid, at a 200-μm gap. The bead vacuum is: Δ, 0; □, 200 Pa; ○, 450 Pa; ◇, 800 Pa (3.2 in. water). [Reprinted from K. Y. Lee et al., "Minimum wet thickness in extrusion slot coating," *Chem. Eng. Sci.*, **47**, pp. 1703–1713. Copyright (1992), with kind permission from Elsevier Science Ltd., The Boulevard, Langford Lane, Kidlington, UK, and from the authors.]

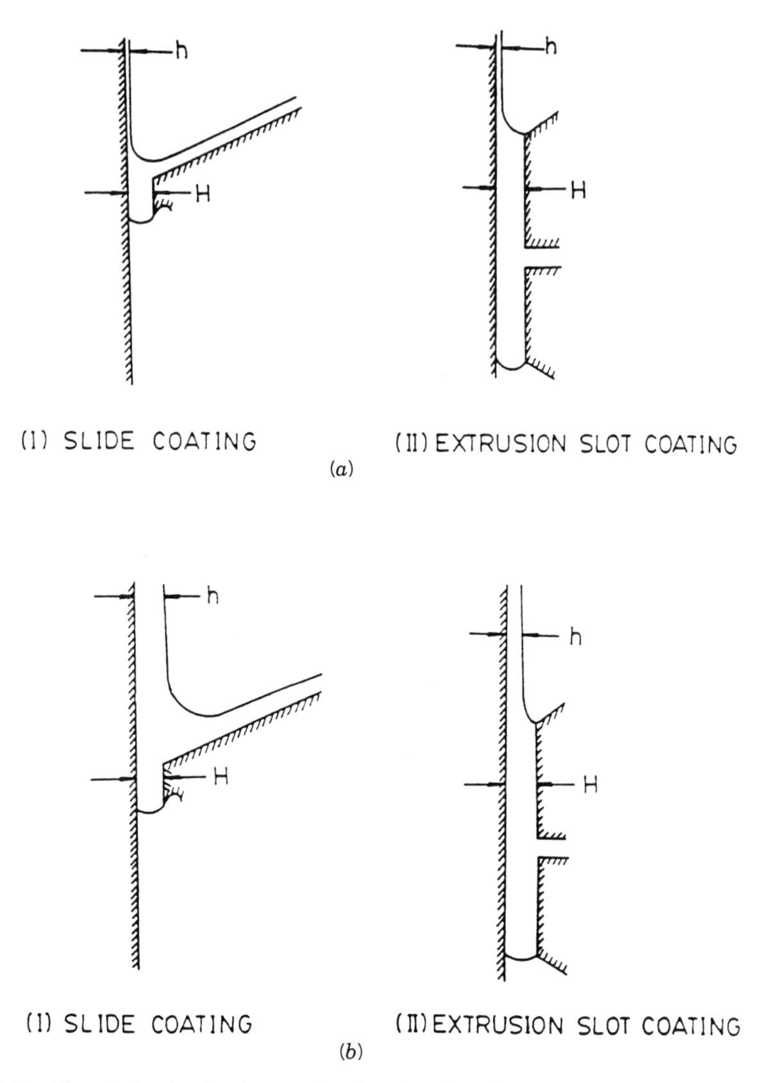

Figure 5-30. Flow behavior in the coating bead region for a slot coater and a slide coater for (*a*) thin coatings and (*b*) thick coatings. [Reprinted from K. Y. Lee et al., "Minimum wet thickness in extrusion slot coating," *Chem. Eng. Sci.*, **47**, pp. 1703–1713. Copyright (1992), with kind permission from Elsevier Science Ltd., The Boulevard, Langford Lane, Kidlington, UK, and from the authors.]

draw resonance. Melt fracture occurs when the material is extruded at too high a rate, and can result in haze, surface roughness (also called sharkskin), or even gross surface distortions. In extrusion coating the extruded film is often stretched by coating at a higher speed than the material is extruded. The ratio of the coating speed to the extrusion speed is called the *draw ratio.* When a certain critical draw ratio is exceeded, the film thickness varies sinusoidally. This instability is called *draw resonance.* Melt fracture is highly dependent on the viscoelasticity of the material, and draw resonance is somewhat related to viscoelasticity.

There are a number of possible phenomena that may result in the instabilities known as *melt fracture*. Agassant et al. (1991) lists the following:

1. *Vortex Formation in Sudden Contractions.* As indicated in Figure 5-31*a*, when a simple liquid enters a contraction, all the streamlines flow toward the exit region. However, viscoelastic liquids often do not flow completely toward the outlet but form vortices in the corners as indicated in Figure 5-31*b*. When liquid in a vortex on either side exits into the contraction, there is a pressure pulse, and this can cause the instability. Polymer melts (usually with branched chains) such as polystyrene, polymethyl methacrylate, polyamide, and most low-density polyethylenes tend to form these vortices, while other polymer melts (usually linear) such as high-density polyethylene and linear low-density polyethylene tend to have all the streamlines converge into the contraction (Figure 5-31*a*) with no vortex formation.
2. *High Entry Pressure Drops.* When a liquid enters a straight pipe or a slit, the velocity profile rearranges to form the typical parabolic shape of well-developed flow. This involves a higher pressure drop. There is some evidence that when this excess pressure drop becomes too high, flow instabilities occur.
3. *High Wall Shear Stress.* With some polymer melts, instabilities occur when the wall shear stress exceeds some critical value. In one model fluid this appears as a pulsating flow. This may be similar to the occurrence of turbulence in pipe flow when the Reynolds number, $DV\rho/\mu$, exceeds a critical value of 2100.
4. *Stick-Slip Conditions at the Wall.* For a melt of high-density polyethylene flowing in a capillary, melt fracture seems to occur at high values of the length-to-diameter ratio. It is thought that stick-slip phenomena are occurring. Stick-slip conditions may occur when the shear stress is sufficiently high that the molecular entanglements do not have time to rearrange and molecular motion becomes

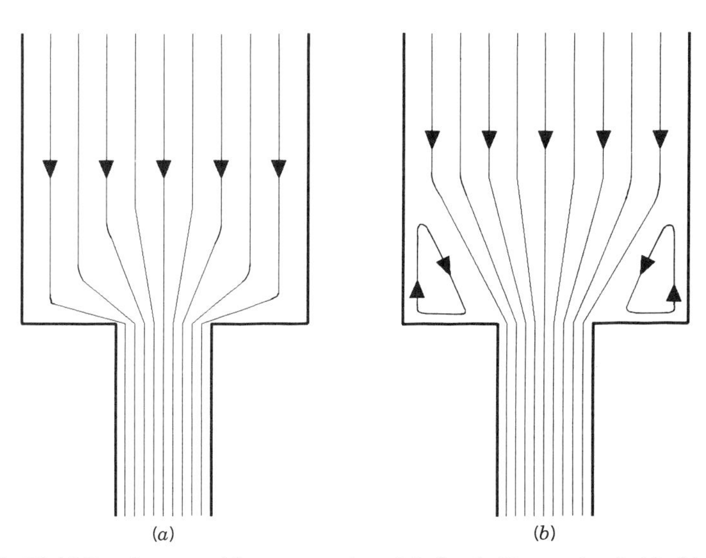

Figure 5-31. Fluid flow into a sudden contraction: (*a*) simple Newtonian fluid; (*b*) viscoelastic liquid.

restricted. When the yield stress is exceeded, melt fracture occurs, the material "gives," and the process repeats. It is also possible that the materials loses adhesion to the wall above a critical shear stress and start to slip. The wall shear stress then drops below the critical value, adhesion to the wall reoccurs, and the process repeats.

5. *Elastic Energy Stored in the Deformed Polymer Chains.* The stress increases rapidly with increasing rates of stretching. At some point the material "gives."
6. *Melt Fracture in Simple Shear Flow.* This has been demonstrated in a cone-and-plate viscometer, where the cone rotates and the shear rate is the same throughout the unit. When the shear rate exceeds a critical value, the shear stress or torque decreases with time to a new, much lower value. When the rotation is stopped, the liquid melt will heal itself to its original state. The lower stress is possibly due to disentanglement of the polymer chains along a "fracture" surface, which may be within the polymer or at the metal surface.
7. *Flow Rearrangement at the Die Exit.* Sharkskin has been attributed to this.

Many authors believe that a critical shear stress has to be exceeded for melt fracture to occur, and this critical value decreases with increasing molecular weight of the polymer. For a flow in the slit of a die, the shear stress increases with flow rate, with viscosity, and even more rapidly with decreasing gap. Thus if melt fracture does occur, one may be able to correct the problem by opening up the exit slot, by switching to a lower-molecular-weight polymer, or by operating at a lower production rate.

Other authors suggest that melt fracture occurs when a critical ratio of elastic stress to viscous stress (this ratio is called the Wiessenberg number) is exceeded. The viscous stresses are proportional to shear rate, and the elastic stresses increase much more rapidly with shear rate. Thus this ratio increases rapidly with shear rate.

The other instability in extrusion coating, draw resonance, has been analyzed mathematically. It will occur even for simple, Newtonian liquids, when the draw ratio exceeds about 20. Thus, when the final film thickness is less than one-twentieth of the film thickness at the exit of the extrusion die, we would expect draw resonance to occur. Middleman (1977) quotes one case in the casting of a polypropylene film, where, when draw resonance occurred, the film thickness varied sinusoidally between 20 and 36 μm with a wavelength of about 1 m. The thickness variations can be even more extreme, and the width of the extruded film can also vary when draw resonance occurs. The distance between the peak thicknesses increases with the stretching distance and with the draw ratio. By comparison with draw resonance in fiber spinning, we might expect that viscoelasticity would have an effect on the critical draw ratio.

5.2.5 Curtain Coating

In curtain coating there are not only the limits of coatability on the coating, but there are also limits on the curtain. A certain flow rate is needed in order to form and maintain the falling curtain. This minimum flow rate is fairly high, approximately 0.5 $cm^3/s \cdot cm$ width (30 $cm^3/min \cdot cm$ width), and it is recommended that at least double the minimum should be used. With this fixed minimum flow rate for the curtain, the minimum coverage will decrease as the coating speed goes up. This means that for thin coatings high coating speeds have to be used. Curtain coating is inherently a high-speed coating operation. Thus, with a curtain flow of 1.0 $cm^3/s \cdot cm$ width, twice the minimum, and for a wet coverage of 25 μm, the coating speed would be

$$\text{Speed} = \frac{1\ \text{cm}^3}{\text{s} \cdot \text{cm}} \frac{1}{25\ \mu\text{m}} \frac{10^4\ \mu\text{m}}{\text{cm}} \frac{\text{m}}{100\ \text{cm}} = 4\frac{\text{m}}{\text{s}}$$

or 800 ft/min. Surfactants are needed in every layer in curtain coating. The surface tensions of all the layers should be about the same, except for the top layer, which should be slightly lower than the rest.

In curtain coating the momentum of the falling curtain aids in holding the coating liquid against the web and opposing air entrainment. Thus the velocity of the falling curtain, determined by the height of the curtain and also by the velocity of the liquid film leaving the slide, helps determine the maximum coating speed. Also, as with slide and as with slot coating, the flow rate is very important, aside from the need for a minimum flow to form the curtain. If the web were stationary, the curtain on striking the web would split into two streams, with half going in the direction of the web movement and half going in the opposite direction. As the web starts moving, more of the flow goes in the direction of the web. Above some speed all the liquid moves in the web direction. But even when all the liquid moves in the web direction, at speeds just above the minimum some liquid still goes backward a short distance before turning forward, to form a heel, as shown by Kistler (1984) in Figure 5-32. In

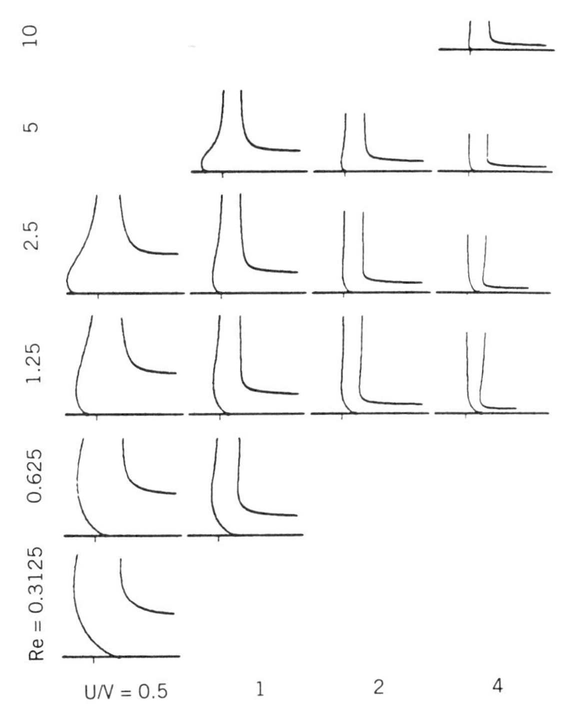

Figure 5-32. Heel formation in curtain coating. Re, the Reynolds number, equal to $q\rho/\mu$, is proportional to the flow rate. U/V is the coating speed divided by the speed of the falling curtain. [From Kistler (1984), with permission of the author.]

this figure the flow rate is plotted on the y-axis as a Reynolds number, $q\rho/\mu$, where q is the flow rate in $cm^3/s \cdot cm$ of width, ρ is the density in g/cm^3, and μ is the viscosity in poise. The x-axis is the ratio of the web speed to the speed of the falling curtain.

We see in Figure 5-32 that with higher flow rates or lower viscosities (higher Reynolds numbers), we tend to get more of a heel, and also with higher curtain speeds—because of a higher curtain—or with lower web speeds (lower ratio of web speed to curtain speed), we tend to get more of a heel.

The heel is objectionable because when it is large enough, there is a recirculating eddy there and the liquid in the eddy can remain a long time. The eddy can cause a streak or a band, and in many cases the coating liquid degrades with time, and this too can cause a defect.

The coating window, plotted as flow rate versus web speed for several different conditions of viscosity and curtain height, is given by Kistler (1984) in Figure 5-33. The lower limit of flow is given by the need to maintain a stable curtain, which requires a flow of a little over 1 $cm^3/s \cdot cm$ width. The highest coating speeds are given by the need to avoid air entrainment. With increasing flow rates the coating speeds can be increased while still avoiding air entrainment. At high flow rate both air entrainment and heel formation occur. The low-speed limit is due to heel formation.

Figure 5-33 shows the low-flow limit (the dashed horizontal line), the low-speed limit due to heel formation and, at slightly lower coating speeds, heel formation with recirculating eddies, and the high-speed limit due to air entrainment. The high-speed limit is shown both for incipient air entrainment and, at slightly higher speeds, for air entrainment over the entire width of the coating.

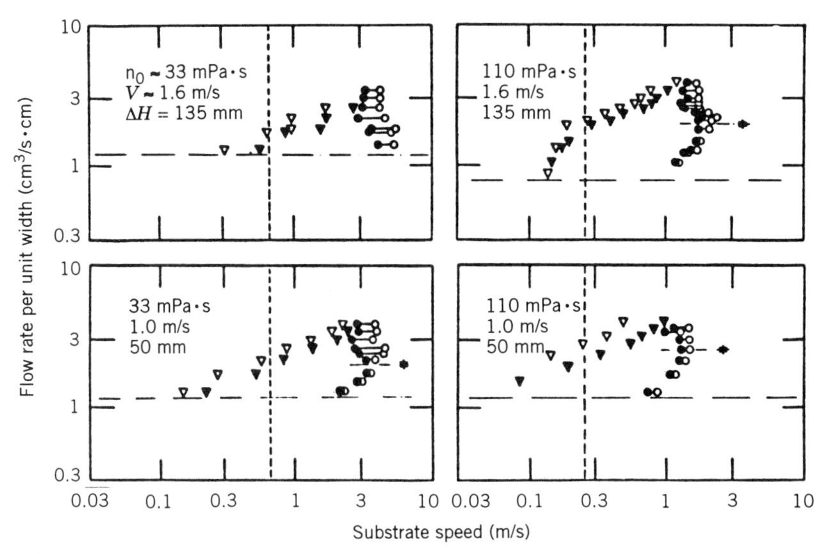

Figure 5-33. Coating window in curtain coating. $\triangledown$, heel formation with recirculation; $\blacktriangledown$, onset of heel formation; $\bullet$, incipient air entrainment; $\circ$, air entrainment over the entire width. The dashed vertical line represents air entrainment with a plunging tape. [From Kistler (1984), with permission of the author.]

The vertical dashed line in Figure 5-33 is the air entrainment velocity for a plunging tape, taken from published data. In slide coating this is the maximum coating speed with no bead vacuum. We saw earlier that bead vacuum increased the maximum coating speed by about 25%. Here we see that curtain coating can coat about an order of magnitude (10 times) faster than the air entrainment velocity for a plunging tape, or almost an order of magnitude faster than slide coating, even with bead vacuum.

Comparing the low and high curtain velocities in the lower and upper portions of Figure 5-33, we see that the high-speed air entrainment limit is unchanged. Thus, at least in this region, the upper-speed limit in curtain coating is unaffected by the curtain height. Comparing the effect of viscosity in the left and right portions of Figure 5-33, we see that, as with slide coating, lower-viscosity liquids can be coated faster.

The data in Figure 5-33 can be replotted as coating thickness versus coating speed, as has been done for one condition in Figure 5-34. This can be compared with the similar plots, shown earlier, for slide and for slot coating. The minimum flow to form a curtain curtain gives a low-speed, thin-film limit. This is the straight line with the 45° slope in the figure. Heel formation gives the thick-film limit. And air entrainment gives the high-speed limits.

Figures 5-33 and 5-34 demonstrate that the coating window in curtain coating is limited on all sides. The coating window may cover a wide speed range (in Figure 5-34 from 0.1 to 3 m/s (20 to 600 ft/min), but for most coating thicknesses the range is much narrower. The thickness range is more limited than in slot or slide coating. However, with polymer solutions we would expect the coating window to be much wider than for the glycerin–water mixture coated in Figure 5-34.

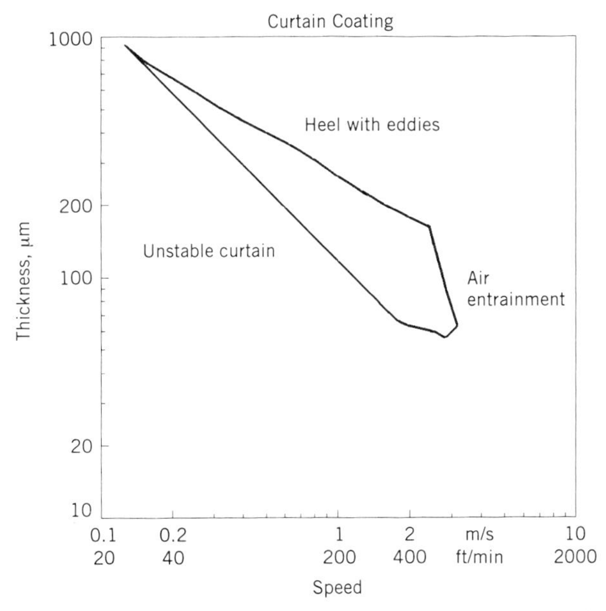

Figure 5-34. Coating window in curtain coating for a 33 mPa · s (33 cP) liquid and a 50-mm (2 in.)-high curtain. The data are taken from Figure 5-33.

5.3 INSTABILITIES IN PREMETERED COATING

5.3.1 Chatter

Chatter is the appearance of cross-web line or bars at regular intervals down-web in the coating, arising from periodic coverage variations. It is also referred to as barring. In a significant number of cases these coverage variations—chatter—results from mechanical causes, such as vibrations, speed fluctuations, an eccentric backing roll, or a roll with too great a total indicated runout in the bearings, fluctuations in the flow of the coating fluid, bead vacuum fluctuations, and so on. Some of these are discussed in Chapter 10.

Fluctuations in flow rate can occur from pump pulsations that remain undamped. Plastic lines from the pump to the coating die are usually adequate to damp the slight pulsations arising from gear pumps. Pulsations in flow have also been known to occur with jacketed plastic lines, when the jacket water experienced pressure pulsations.

One should always check whether the frequency of any of the mechanical disturbances is the same as the chatter frequency. The chatter frequency in cycles per second (hertz) is found by dividing the coating speed (in cm/s, for example) by the spatial frequency of the chatter (in cm). Using a vibration analyzer the frequency of any vibrations can be determined. One should also compare the spatial frequency of the chatter with the circumference of the backing roll, as a bad bearing or an out-of-round roll can cause chatter.

The frequency of bead vacuum fluctuations can be determined with a rapid-response pressure sensor. Bead vacuum fluctuations often arise because the bead vacuum system can act as a giant wind instrument. As in a wind instrument, the greater the airflow, the greater the amplitude of the pressure waves (equivalent to the sound level in a musical instrument) and the more likely that visible chatter will result. Therefore, one should reduce the air leakage as much as possible. Bead vacuum fluctuations cause chatter mainly at wide coating gaps.

It is impossible to eliminate all air leakage into the vacuum system, because the web has to enter the vacuum chamber, and rubbing contact against the web must be avoided to prevent scratching the web. However, the clearance between the vacuum seal and the web should be as small as possible, and all other seals—mainly against the face or sides of the backing roll—should be with rubbing contact. The coating liquid itself forms the seal between the web and the coating die.

If all mechanical causes of chatter have been eliminated and the problem remains, it is likely that the chatter is caused by the hydrodynamics of the system. Sartor (1990) has reported chatter occurring near the limits of coatability, and in some systems hydrodynamic-caused chatter is not uncommon.

5.3.2 Ribbing

Ribbing appears as evenly spaced, uniform, down-web lines in the coating, as if one ran a comb down the wet coating. Ribbing is a common defect in all coating operations. It arises from the hydrodynamics of the coating system. Even when painting a wall with a paintbrush the lines that appear are not from the bristles on the brush but from rib formation. Bixler (1982) studied ribbing in slot coating mathematically, and from his analysis one sees that to avoid or reduce ribbing one should (1) have a low-viscosity liquid, (2) coat at low speeds, (3) have high wet coverage, and (4) operate with a tight gap.

In most coating systems a diverging included angle, as shown in Figure 5-35, tends to promote ribbing. Either the lips should be parallel to the web or the included angle should be slightly converging. In slot coating a thin layer, ribs can also form when the upper or downstream lip of the coating die is too rounded, thus presenting a diverging included angle to the exiting flow.

Sartor (1990) showed that ribbing appears near the limits of coatability in slot coating, and Christodoulou (1990) showed mathematically that in slide coating ribbing is more severe at high bead vacuums, high coating speeds, and low flows. Earlier, Tallmadge et al. (1979) showed that ribbing appears near the limits of coatability in slide coating. And Valentini et al. (1991) showed that the choice and level of surfactant affects the bead vacuum in the coatability window and thus the bead vacuum at which ribbing occurs.

Although some viscoelasticity is desirable in coating, too high a viscoelasticity promotes ribbing. In extreme cases the ribs can be transformed into filaments or strings. If adjustments of bead vacuum, gap, or included angle cannot eliminate ribbing, the cure may be to use a less viscoelastic (lower-molecular-weight) polymer. However, viscoelasticity can also be caused by some surfactants forming temporary structures at higher concentrations (above the critical micelle concentration). Thus the type and concentration of surfactant may have an effect on ribbing. Viscoelasticity may also be caused by some suspensions at high concentrations tending to develop structures.

Ribs can originate in the top or in the bottom of the coating bead. The rib spacing tends to increase in thicker coatings and with wider gaps. Ribs can also form when

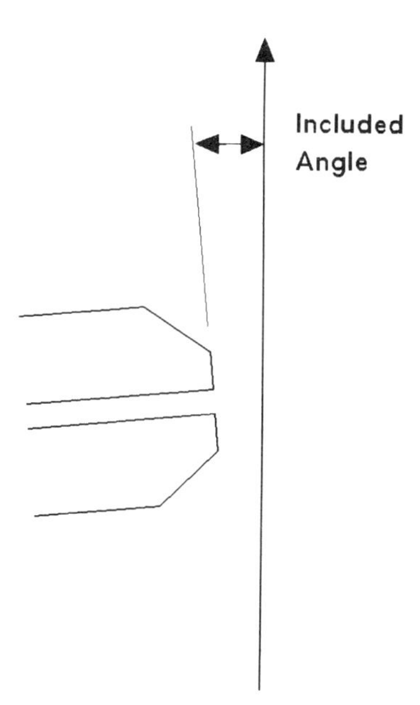

Figure 5-35. Diverging included angle between the slot coating die and the web tends to promote ribbing.

a viscoelastic fluid is forced through too narrow a slot opening in the coating die. If this occurs, the slot opening in slot, slide, or curtain coating should be increased.

5.3.3 Neck-In

In all forms of premetered coating, whenever there is a gap the liquid tends to neck-in. In curtain coating this neck-in is prevented by the edge guides. In slot, slide, and extrusion coating, it can and does occur. However, with low-viscosity liquids that are commonly used in slot and slide coating, edge guides have been used effectively to prevent neck-in.

Neck-in is most severe when one approaches the limits of coatability, where it is difficult to coat unless the coverage is increased. When an edge necks-in, the coverage near that edge is increased. Neck-in seems to be more severe with more viscous and also more viscoelastic liquids, and is more severe at wider gaps. High surface tensions would also contribute to neck-in. Therefore, using surfactants to lower the surface tension should be tried.

In extrusion coating the gap tends to be much greater than in slot coating, and neck-in is correspondingly more severe. In addition, the polymer melts tend to be of high viscosity and also viscoelastic. This also tends to increase neck-in. Neck-in is also more severe the more one stretches the film—the higher the draw ratio. Thus, to reduce neck-in one should try to reduce the draw ratio by using a narrower slot opening in the extrusion die, thus giving a smaller initial thickness but with a higher exit velocity, and one should try to use a smaller gap or draw distance.

An edge can move in and out in a smooth, cyclic manner. This may be caused by an out-of-round backing roll or by bearings with too much runout. The gap would then cycle between a tighter and wider opening with each revolution of the roll. At the wider gaps, neck-in would tend to be more pronounced, and this would cause a smooth cycling of the edge with a wavelength equal to the circumference of the backing roll.

5.3.4 Scalloped Edges

Scalloped edges are a higher-frequency defect where the edge of the coating moves in and out with a spacing of a few centimeters or inches or less, with pointed peaks and valleys. Scalloped edges seem to be more common at wide gaps and near the limits of coatability at low coverages and high speeds. The origin of this cyclic condition is not known, but it is much more common with viscoelastic liquids. The viscoelasticity is usually caused by dissolved polymers, but it can also result from some suspensions tending to form three-dimensional networks, and from some surfactants above the critical micelle concentration, tending to form temporary structures.

5.3.5 Edge Beads and Heavy Edges

In almost all coating operations the edges of the coating tend to be thicker then the broad central area. The heavy edges are objectionable for several reasons. First, a roll of coating with heavy edges cannot be wound properly. The web in the central portion of the roll will not be in contact with the wrap underneath it. One suggestion is to oscillate the roll from side to side so that the heavy buildup will not be in the same location throughout the roll.

Second, unless the coating is solidified by curing (by ultraviolet light or an electron beam, for example), the heavy edges will dry slower. If the coating encounters a face roll before the edges are dry, the wet coating will transfer to the roll, get it dirty, and this could cause defects subsequently in the coating run. The wet edges in a wound roll will cause it to stick together such that it might tear during unwinding.

In Chapter 7 it is pointed out that surface forces may be the cause of fat edges that occur during drying. But in many cases the thick edges occur in the coating operation before the wet coating gets to the dryer. In curtain coating it has been suggested that the edge guides be bent to ride on the web, and that the bent sections point outward to thin down the heavy edges. Air jets have been suggested to blow the heavy edges outward. And it has also been suggested to dilute the edge with solvent (usually water) to reduce the viscosity, and then suck it off with a vacuum tube.

In premetered coating, edge beads can be caused in the coating operation by surface tension effects, by die swell, and by stresses due to the stretching of the film (Dobroth and Erwin, 1986). These authors show that surface tension is most likely to cause edge beads when the viscosity is low, the coating speed is low, the film thickness is low, and the gap is large. In slot and slide coating the coating speed is usually high, the gap is small, and the viscosity is fairly low. In extrusion coating the coating speed is high, the gap is large, and the viscosity is high. In most cases surface tension contributes only slightly to the edge bead. When surface tension is a major contributor, the bead would normally extend no more than about 5 mm in from the edge.

Die swell can also contribute to the edge bead. Viscoelastic liquids swell when they are extruded to give an initial thickness larger than the die opening. Because of the extra wall at the edges, there can be additional stresses at the edges, and this can lead to additional swelling. It is unlikely that the thickness of an edge bead due to die swell would be more than twice the thickness in the broad central area, and it is unlikely that such an edge bead would extend more than 5 mm in from the edges (Dobroth and Erwin, 1986).

The most common cause of edge beads in extrusion coating—and probably in slot and slide coating as well—is from the stresses that build up in stretching the film, with the web or coating speed being much higher than the speed at which the material exits the coating die. This ratio of coating speed to the die exit speed is the draw ratio. Dobroth and Erwin (1986) show that the ratio of the bead thickness to the thickness at the center—the bead ratio—is equal to the square root of the draw ratio irrespective of the properties of the coating liquid. It does not matter whether or not the material is viscoelastic or what its viscosity is. It depends only on the draw ratio. The draw ratios for typical coatings are about 10 and range from about 2 to 14. Therefore, one can expect that the bead will be thicker than the central portion by a factor ranging from about 1.4 to a little under 4.

The extent of the edge bead due to the stresses caused by stretching will vary with the gap or draw distance. In extrusion operations with gaps as large as 15 cm (6 in.), the edge bead can extend about 5 cm (2 in.) in from the edge. In slot or slide coating, with gaps of 100 to 400 μm (4 to 16 mils) the edge bead would extend well under 1 cm.

There is not much one can do to prevent the formation of this edge bead. Narrowing the slot opening of the extrusion or slot die would increase the die exit speed and thus decrease the draw ratio, but it would increase the pressure drop in the die, and the narrower opening might well require more precise machining (lapping rather than grinding, for example) to maintain the cross-web uniformity of the coating. The narrower slot would increase the stresses in the coating liquid, which, if viscoelastic, would increase the die swell, and may also result in rib formation. Higher die swell may be undesirable, as it is possible that the draw ratio should be defined as the ratio of the maximum thickness at the exit (in the die swell region rather than the slot thickness) to the coated thickness, and thus this might be counterproductive and increase the bead ratio.

However, as the extent of the edge increases with gap or draw distance, decreasing the gap should help in reducing the extent of the edge. This may be possible to do in extrusion coating, but in slot or slide coating one usually wants a wide gap of over 200 to 300 μm (8 to 12 mils) to allow splices to pass without withdrawing the coating die.

In slide coating one could coat at a steeper angle to increase the velocity on the slide. This would reduce the draw ratio and thus the edge ratio. But this is not recommended, for with slide coating the edge extent is only about 5 mm, and a steeper slide is more likely to cause interfacial waves (see below).

5.3.6 Waves in Slide and in Curtain Coating

When a liquid flows down an inclined plane, waves can form on the surface. Many of us have seen such waves when rainwater runs down a hill on a paved street. The waves often have a wavelength of about 30 cm (1 ft). These are surface waves. Such waves can form on the slide in slide or curtain coating. In multilayer slide or curtain coating, both surface and interfacial waves can form. Figure 5-36 gives an example of interfacial waves on a slide, with a wavelength of about 1 mm. The wavelength of surface waves, although much shorter than for rainwater on a hill, is still fairly long, on the order of 10 cm. Interfacial waves have shorter wavelengths, on the order of a millimeter to a centimeter. The longer interfacial waves can appear as a series of jagged V's.

In pure liquids surface waves form above some critical flow rate that is proportional to the viscosity and to the cotangent of the angle of inclination of the slide to the horizontal. Thus with higher viscosities the critical flow rate is higher and waves are less likely to form. Also, because of the cotangent function, at lower angles the critical flow rate is higher and waves are less likely to form. Surfactants tend to stabilize the flow, and when surfactants are present, much higher flow rates are required for surface waves to form.

The use of surfactants to dampen waves was known to the ancient Greeks. When a ship was caught on the rocks and in danger of breaking up due to the wave action, the sailors would slowly pour vegetable oils over the windward side to calm the waters. This is the origin of the expression "pouring oil on troubled waters."

Because surfactants are always used in aqueous systems, and multilayer coatings are most often aqueous systems, surface waves are rarely seen in slide or curtain coating. However, in multilayer flow there are interfaces between adjacent layers,

Figure 5-36. Interfacial waves in a three-layer slide coating.

and interfacial waves can form. But if all the layers have the same properties, the system would behave as if it were one large layer with no interfaces (as long as there are no interfacial tensions). In this case there could be no interfacial waves. Density and viscosity are the main physical properties of concern. Densities of aqueous layers do not vary greatly from that of water, and in addition we have little control over them. Kobayashi (1992) concluded that the viscosities of adjacent layers should either be the same, or the upper layer could be more viscous by a factor of up to 1.5 or at most 2, to avoid interfacial waves. Also, one should avoid using an upper layer with a viscosity under 70% of the viscosity of the lower layer. When the upper layer is very much more viscous, well above a factor of 10, the interface again becomes stable. This explains the stability of the flow suggested by Ishizuka (1988), who uses water as a carrier layer with a second layer having a viscosity of 50 mPa · s (50 cP), giving a ratio of the upper-layer viscosity to lower-layer viscosity of about 60.

Just as with surface waves, there is greater stability at lower angles of inclination to the horizontal. At lower angles the layers are thicker, and this also gives increased stability to interfacial waves. Coating at higher speeds gives thicker layers because of the higher flow rates, and thus interfacial waves are less likely to occur at higher coating speeds. Interfacial tensions, where they do exist, tend to stabilize the system against interfacial waves (Kobayashi, 1990).

Although we are unaware of any reports of waves forming in a coating bead, they presumably could form there as well as on a slide. Presumably, the cure would be the same as for waves forming on the slide. Slot coaters and extrusion coaters can also be used for multilayer coatings. We are unaware of reports of waves forming in these systems.

5.3.7 Streaks and Bands in Premetered Coating

Streaks are lines running in the direction of the coating, and appear in many forms, but are visible because they usually have a significantly lighter or significantly heavier coverage than the average value. One example of a streak is shown in Figure 5-37. They can have a variety of widths. Bands are wider than streaks and often differ only slightly in coverage from the rest of the coating. Bands can be caused by broad fluctuations in the slot opening in the coating die, by incorrect tension in the

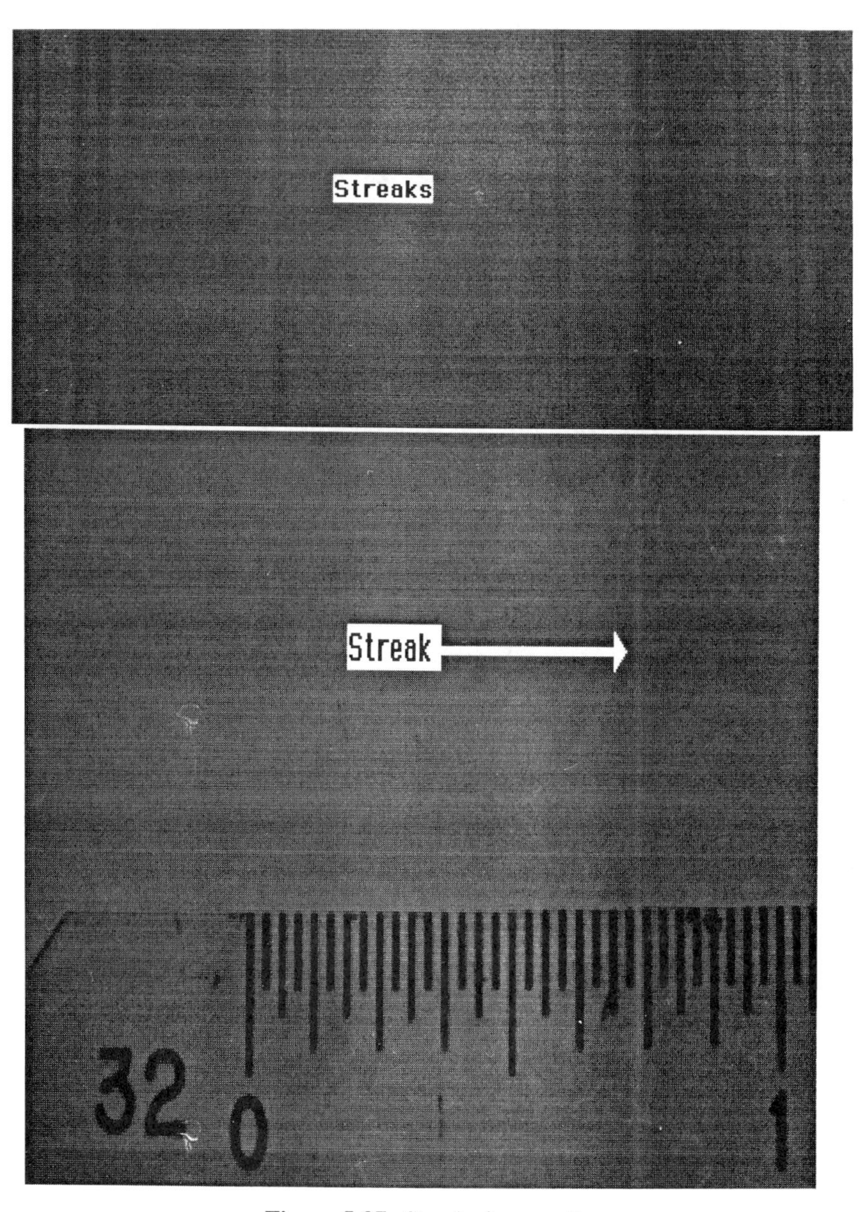

Figure 5-37. Streaks in a coating.

web when coating against unsupported web, by uneven tension in the web or by poor-quality web with "valleys" and "ridges" causing flow of the wet coated layer and, perhaps, even by the system hydrodynamics. Eddies and vortices are also thought to cause streaks and bands.

Particles of dirt carried along by the web and caught in the gap between the coating die and the web can cause streaks. This is less likely to occur with a wide coating gap. If it does occur, a piece of plastic shim stock can be used to sweep the particles to the side. The web used should be clean and free of dirt, and a good web cleaner should be used just before the coating stand.

Particles of dirt and even bubbles that are carried in the coating liquid can get caught inside the coating die—usually in the slot—there to disrupt the flow and cause streaks. If this occurs, the slot should be "shimmed," a piece of plastic shim stock inserted into the slot and moved across the width to sweep any particles to one side. With a slot coater the die would have to be pulled back to do this. The coating liquid should be filtered before being delivered to the coating die and an additional coarse, small filter should be placed just before the coating die, to remove any particles that break free from the line or from the fittings. Coating dies often have two channels, with the front slot having a greater clearance than the rear slot. Thus any particles large enough to be trapped in the die are more likely to get caught in the narrower rear slot than in the wider front slot. A flow disturbance in the rear slot, such as that caused by a trapped particle, is more likely to heal itself by the time it reaches the slot exit than would a disturbance to the front slot.

Particles lighter than the coating liquid, especially bubbles, can float in the bead of a slide coater and not get carried off onto the web by the coating liquid, causing streaks. If this occurs, one should "shim" the gap and sweep the bubbles or particles to the side with a piece of plastic shim stock.

Nicks in the coating die can cause streaks. Even though it may be made of stainless steel, the coating die is very delicate and should be handled as if it were made of the finest crystal. The edges, especially, are easily damaged, and a damaged edge will cause a streak. Nicks can be "stoned" out with a whetstone (an abrasive stone), but this rounds the edge. A round edge does not pin the meniscus tightly and can cause chatter at wider gaps. When the edge gets too rounded, above about 100 μm (4 mils) radius of curvature, the coating die has to be refinished (by grinding or lapping).

Coating liquid flowing under the front edge of the slide coater can cause streaks, as shown by Ishizuka (1988) (see Figure 5-38). Liquid in this region is likely to form vortices or eddies. This disturbance to the flow may be the cause of these streaks. Liquid will enter this region when the bead vacuum is too high or the gap is too wide for the given coating liquid and speed. Less bead vacuum must be used for low-viscosity liquids, for wide gaps, and for low coating speeds.

Streaks can also be formed in slide and in curtain coating by liquid from an upper slot flowing into an incompletely filled lower slot, as in Figure 5-39 (Ishizuka, 1988). This rarely occurs, because to obtain a reasonable minimum pressure drop across the slot, the slot opening would be narrow enough to prevent this from occurring. However, in very thin layers of low-viscosity liquids, such as carrier layers, one could inadvertently use a wide slot designed for higher flow rates of more viscous liquids.

The coating fluid can dry up under the coating bead, and also on the top plate of a slide or curtain coater, there to disturb the flow and cause a streak. If this occurs,

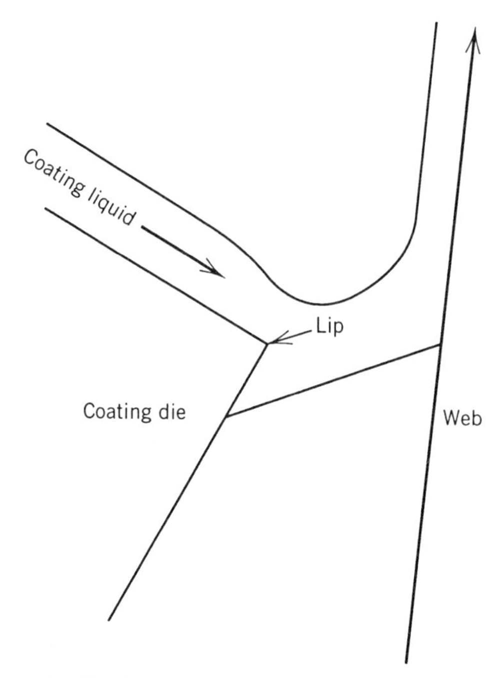

Figure 5-38. In slide coating liquid flowing below the lip can cause streaks (Ishizuka, 1988).

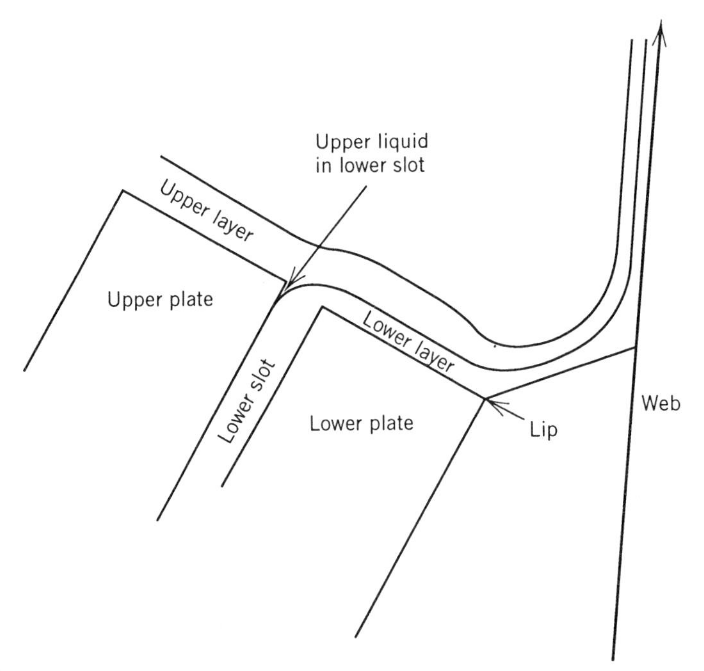

Figure 5-39. In slide and curtain coating, liquid flowing into an incompletely filled lower slot can cause streaks. [From Ishizuka (1988).]

the buildup has to be washed off. If the buildup is under the bead, the coating die first has to be withdrawn. Laticies are often coated, and they are known to build up aggregates under the coating lip and cause streaks. Huang and Nandy (1993) found that latex particles which have more carboxylic acid groups on them, and therefore swell more as the pH is raised to 7.0, have greater colloidal stability, and have a greatly reduced tendency to form streaks.

Buildup on the top plate can be prevented by elevating the top plate to about the height of the flowing layer, so that the liquid does not run up over the edge (Figure 5-40). Ade (1978) suggests chamfering the lower edge of the top slot (Figure 5-40*c*) to get the same effect.

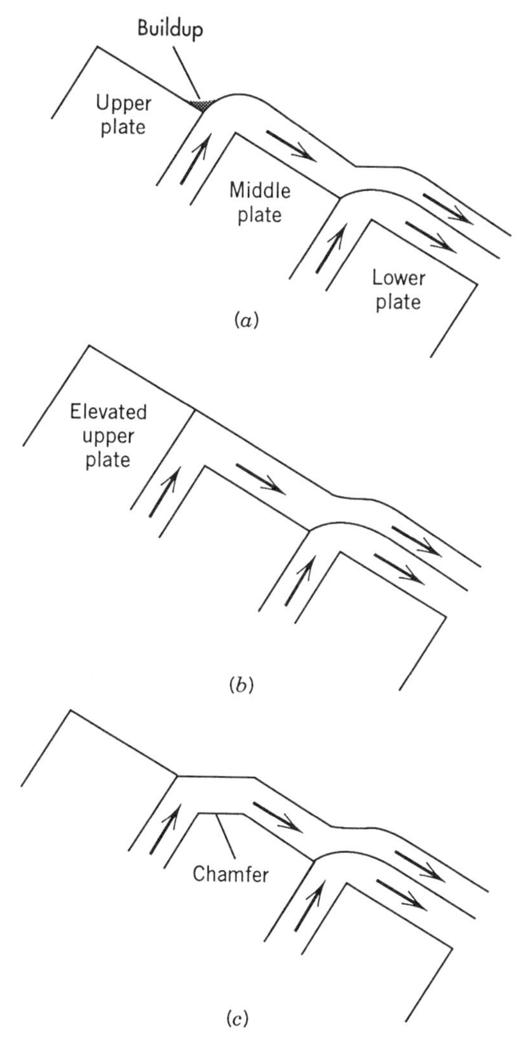

Figure 5-40. Preventing runback and buildup on the top plate of a side or curtain coater. (*a*) Buildup formation. Buildup prevented by (*b*) elevating the top plate, or (*c*) chamfering the lower edge of the top slot. [From Ade (1978).]

REFERENCES

Ade, F., "Cascade or curtain coater," Brit. Patent 1,509,646, May 4, 1978.

Agassant, J.-F., P. Avenas, J.-P. Sergent, and P. J. Carreau, *Polymer Processing*, Hanser Publishers, New York, 1991.

Bixler, N. E., "Stability of a coating flow," Ph.D. dissertation, University of Minnesota, 1982; University Microfilms, Ann Arbor, MI.

Chen, K. S. A., "Studies of multilayer slide coating and related processes," Ph.D. dissertation, University of Minnesota, 1992; University Microfilms, Ann Arbor, MI.

Choinski, E. "Method of multilayer coating," U.S. Patent 4,113,903, Sept. 12, 1978.

Christodoulou, K. N., "Computational physics of slide coating flow," Ph.D dissertation, University of Minnesota, 1990; University Microfilms, Ann Arbor, MI.

Dittman, D. A., and F. A. Rozzi, "Method of multilayer coating," U.S. Patent 4,001,024, Jan. 4, 1977.

Dobroth, T., and L. Erwin, "Causes of edge beads in cast films," *Polym. Eng. Sci.* **26**, 462–467 (1986).

Fahrni, E., and A. Zimmerman, "Coating device," U.S. Patent 4,109,611, Apr. 29, 1978.

Greiller, J. F., "Method and apparatus for curtain coating," U.S. Patent 3,632,403, Jan. 4, 1972; "Method of making photographic elements," U.S. Patent 3,632,374, Jan. 4, 1972.

Gutoff, E. B., "Pre-metered coating," pp. 117–167 in *Modern Coating and Drying Technology*, E. D. Cohen and E. B. Gutoff, eds., VCH Publishers, New York, 1992.

Gutoff, E. B., and C. E. Kendrick, *AIChE J.* **33**, 141–145 (1987).

Huang, D., and S. Nandy, "Effects of latex colloidal stability on coating performance," pp. 325–329 in *Paper Summaries, IS&T's 46th Annual Conference*, May 9–14, 1993, Society for Imaging Science and Technology, Springfield, VA, 1993.

Ishizaki, K., and S. Fuchigami, "Method of simultaneously applying multiple layers to web," U.S. Patent 4,525,392, Jan. 25, 1985.

Ishizuka, S., "Method of simultaneous multilayer application," European Patent Appl. No. 88,117,484.1, Pub. No. 0 313 043, Oct. 20, 1988.

Kistler, S. F., "The fluid mechanics of curtain coating and related viscous free surface flows with contact lines," Ph.D. Thesis, University of Minnesota, 1984; University Microfilms, Ann Arbor, MI.

Kistler, S. F., and L. E. Scriven, "Finite element analysis of dynamic wetting for curtain coating at high capillary numbers," presented at *AIChE Spring Meeting*, Orlando, FL, Mar. 1982; see also Kistler (1984).

Kobayashi, C., "Stability analysis of multilayer stratified film flow on an inclined plane," presented at *AIChE Spring Meeting*, New Orleans, LA, Apr. 1990.

Kobayashi, C., "Stability analysis of film flow on an inclined plane. I. One layer, two layer flow," *Ind. Coat. Res.* **2**, 65–88 (1992).

Lee, K.-Y., L.-D. Liu, and T.-J. Liu, "Minimum wet thickness in extrusion slot coating," *Chem. Eng. Sci.* **47**, 1703–1713 (1992).

Mainstone, K. A., "Extrusion," pp. 122–146 in *Web Processing and Converting Technology and Equipment*, D. Satas, ed., Van Nostrand Reinhold, New York, 1984.

Mercier, J. A., W. Torpey, and T. A. Russell, "Multiple coating apparatus," U.S. Patent 2,761,419, Sept. 4, 1956.

Middleman, S., *Fundamentals of Polymer Processing*, McGraw-Hill, New York, 1977.

Sartor, L., "Slot coating: fluid mechanics and die design," Ph.D. dissertation, University of Minnesota, 1990; University Microfilms, Ann Arbor, MI.

Tallmadge, J. S., C. B. Weinberger, and H. L. Faust, "Bead coating instability: a comparison of speed limit data with theory," *AIChE J.* **25**, 1065–1072 (1979).

Valentini, J. E., W. L. Thomas, P. Sevenhuysen, T. S. Jiang, H. O. Lee, Y Liu, and S.-C. Yen, "Role of dynamic surface tension in slide coating," *Ind. Eng. Chem. Res.* **30**, 453–461 (1991).

CHAPTER VI

Coating Problems Associated with Coating Die Design

The assumption made in all coating operations is that the coating device will lay down a wet coating that is completely uniform across the web and is free of all defects. Unfortunately, this is not always the case. In this chapter we explore some of the problems with coating dies: more specifically, with coating dies for slot, extrusion, slide, and curtain coating. With slot, slide, and curtain coating a coverage variation of under ±2% can be achieved without too much difficulty, and we explore what is necessary to achieve this uniformity. In extrusion coating with the slot opening adjusted manually this level of uniformity normally cannot be attained. We also explore how to achieve a more uniform residence-time distribution and when it may be desirable, how to maintain a minimum wall shear stress and why one might want to do this, how to clean the coating head, and how to change the coating width without changing the coating head. Down-web or machine direction variability was discussed in Section 5.3.1.

6.1 CROSS-WEB UNIFORMITY

The internals of slot, slide, and curtain coating dies are all similar. We will make reference to the internals of a slot coating head, as in Figure 6-1. In slide coating the plates are tilted upward and the slot exits onto an inclined plane (Figure 6-2). In curtain coating the coating fluid flows off the inclined plane to form a curtain that falls down onto a horizontally moving web, as in Figure 6-3. In all these systems the die internals should be built to give a uniform flow across the width of the coating, not with the crown in the profile shown in Figure 6-1. The die should not be adjustable because adjustable coating dies are very difficult to adjust manually, and uniformities of ±2% are almost impossible to attain with them. By a ±2% uniformity we mean that the coefficient of variation of the coverage across the web—that is, the standard deviation of the cross-web coverage divided by the mean—is no greater than 2%. The nonadjustable dies are used routinely for low-viscosity liquids, under some thousands of mPa · s (cP). The coating dies should be made so that they can be assembled only

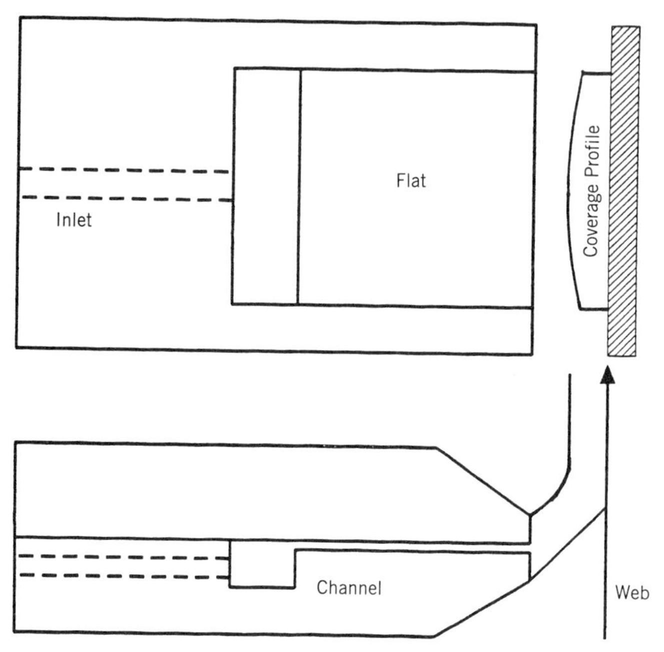

Figure 6-1. Slot coating showing the coverage profile with uncorrected internals. [From E. B. Gutoff, *J. Imag. Sci. Technol*, **37**, 615–627 (1993), with permission of the Society of Imaging Science and Technology.]

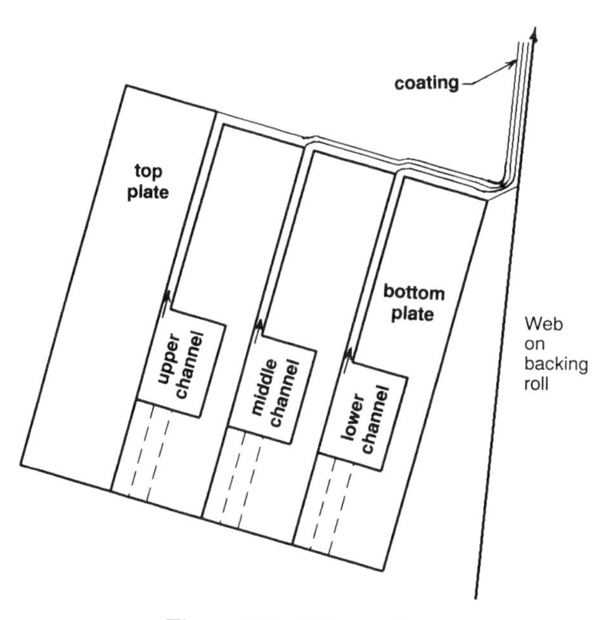

Figure 6-2. Slide coating.

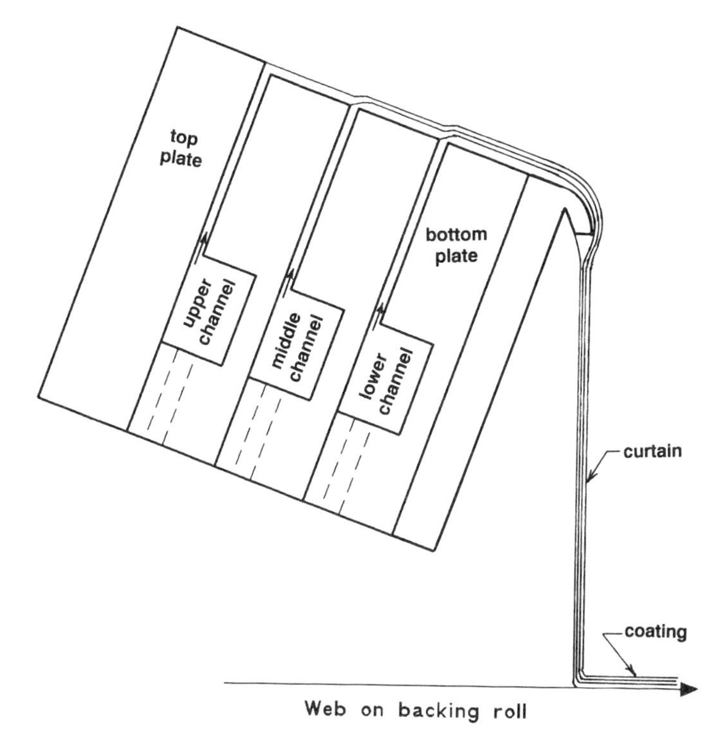

Figure 6-3. Curtain coating.

one way, thus ensuring that they are always assembled correctly. Extrusion coating dies, because they are normally used for many different high-viscosity liquids with varying rheologies, such as polymer melts, do use adjustable coating dies, and these are discussed separately.

Let us examine the slot coater in Figure 6-1. Coating liquid enters the center of the distribution channel, flows along it toward both edges, and flows out through the slot clearance to exit at the coating lips. If the slot length—the distance from the channel to the front lips—is constant, and the slot clearance is everywhere constant, the flow through the slot at any point along the width will be proportional to the pressure in the channel. But the pressure at the feed entrance, the center of the channel, is higher than at the ends of the channel to cause flow in the channel from the center to the edges. Because of this the flow out of the slot at the center will be greater than the flow out of the slot at the edges, and the wet coating will be thicker at the center, as shown in Figure 6-1. This crown in the profile is sometimes called a "frown" profile. The opposite situation, when the coverage is low in the center and higher near the edges, gives a "smile" profile. Now the die internals should be altered or corrected to give a uniform profile, and usually are, but it is important to know when an uncorrected die can be used and still give a sufficiently uniform coverage across the web—that is, to know the coverage variation in an uncorrected die.

Carley (1954) studied the coverage variation in an uncorrected die, and his results can be plotted as in Figure 6-4, where the percent error is the coverage variation from

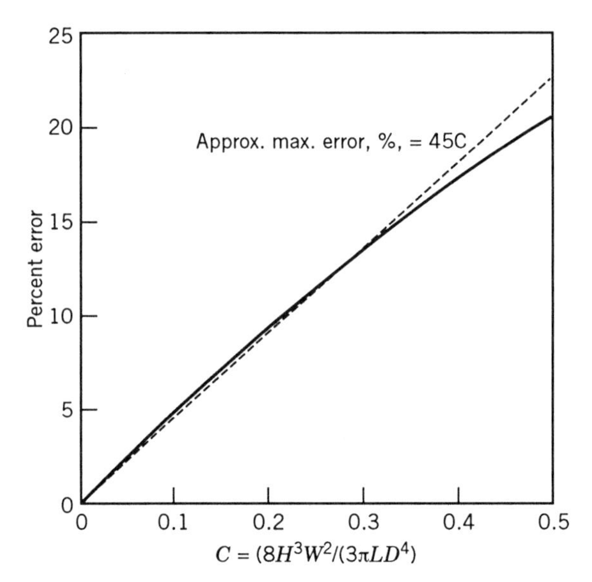

Figure 6-4. Coverage error for Newtonian liquids in an uncorrected die. See text for nomenclature. [From E. B. Gutoff, *J. Imag. Sci. Technol*, **37**, 615–627 (1993), with permission of the Society of Imaging Science and Technology.]

the center to the edge. For a simple Newtonian liquid the error is only a function of the geometric parameters,

$$C = \frac{8H^3W^2}{3\pi LD^4}$$

where D is the equivalent diameter of the channel, H the clearance in the slot, L the length of the slot from the channel to the lips, and W the width of the coating (for a side-feed die, this would be twice the width of the coating). All these measurements have to be in the same units, such as millimeters or inches. The curve in Figure 6-4 can be accurately approximated in the region of interest by the linear relationship (Gutoff, 1993).

$$\text{Fractional error} = 0.45\,C$$

Thus for simple (Newtonian) liquids it is easy to calculate whether a given uncorrected die design will give the necessary coverage uniformity. We would suggest that if a ±2% uniformity is desired, the calculated uniformity should be no worse than 0.5%, or at most 1%, to allow for other sources of error.

Note that the side-to-side coverage variation for Newtonian liquids is only a function of the geometry of the coating die and is independent of the liquid viscosity and flow rate. Now, obviously, the flow rates and the viscosity are important to the coating head design. For a given geometry the pressure drops in the coating head, and thus the pressure in the channel, are directly proportional to the viscosity and to the flow rate. If the viscosity or the flow rate is too high, the internal pressures will be excessive and the die may spread apart or the feed system may not be able to supply coating liquid at the necessary pressure. A wider slot opening or clearance would be required to lower the pressure drop in the slot and thus lower the internal pressure in the channel. As the height of the slot opening is raised to the power of 3 in the geometric grouping C above, this would greatly increase C and thus the coating nonuniformity unless the channel is also made larger.

If the viscosity or the flow is too low, the pressure in the die may be so low that gravity will become important and the liquid will just dribble out of the slot unevenly. In this case a narrower slot opening or clearance should be used. Also, if the pressure is too low, inertial effects—the kinetic energy in the moving liquid—may become significant, and this is undesirable for the coating head to be versatile and usable with different viscosities and flows. The minimum pressure needed to ensure that the inertial effects are insignificant is discussed below.

When the pressure in the die is too low in slide and in curtain coating, liquid from an upper layer can fall into the incompletely filled slot. This can result in streaks being formed, as shown by Ishizuka (1988). This was illustrated in Figure 5-39 for slide and curtain coating.

Since the channels in coating dies are usually not round but rectangular in cross section, the equivalent diameter of a rectangular channel needs to be found. If the channel width and height are a and b, where a is the larger dimension, the equivalent diameter is $D = g(b/a)a$, where $g(b/a)$ is tabulated in Table 6-1 (Gutoff, 1993).

For non-Newtonian liquids, where the viscosity is not constant but depends on the shear rate, the coverage errors are considerably worse and the simple relationship given above is no longer reliable. The shear rate can be thought of as the rate of

Table 6-1 Values of $g(b/a)$ for Calculating the Equivalent Diameter of Rectangular Channels[a]

b/a	$g(b/a)$
0.1	0.2375
0.2	0.3925
0.3	0.5222
0.4	0.6350
0.5	0.7346
0.6	0.8231
0.7	0.9021
0.8	0.9728
0.9	1.0364
1.0	1.0939

Source: Gutoff (1993), with permission of the Society for Imaging Science and Technology.

[a] a is the larger dimension, normally the channel width; b is the smaller dimension, normally the channel height.

$$D_e = g\left(\frac{b}{a}\right) a$$

relative motion of the fluid. It is defined as the rate of change of velocity in the direction perpendicular to the motion, such as the rate of change in a channel as one moves away from the channel walls. Shear rates were discussed in Chapter 5, and a common type of shear thinning behavior was shown in Figure 5-22. When the plot of the logarithm of viscosity against the logarithm of shear rate is a straight line, as in the central section of Figure 5-22, the liquid is called a power law liquid, and the viscosity can be expressed as $\mu = K\gamma^{n-1}$, where K is the consistency, equal to the viscosity when the shear rate is unity, n the power law index, γ the shear rate, and μ the viscosity. The coverage errors are now a function of the power law constant n as well as the geometric factors, but are not a function of the consistency, K.

6.2 TEMPERATURE CONTROL

The calculation of the coating nonuniformity in an uncorrected die assumes a constant viscosity, and the viscosity will be constant only if the temperature is constant. If the coating is made at room temperature, the coating liquid should be at room temperature. If the coating liquid is stored at a different temperature, it should be brought to room temperature before the start of coating. If coating is carried out at elevated temperatures, as may be necessary the coating liquid and the coating head should be brought to the coating temperature before the start of coating.

The coating liquid can be brought to the coating temperature in the feed vessel, or it can be brought to coating temperature in the lines from the feed vessel to the coating head using a heat exchanger. For moderately elevated temperatures, a simple technique is to pump the liquid through jacketed lines through which hot water is flowing. The temperature of the coating liquid should be measured as it enters the coating head, and the jacket water flow or temperature should be adjusted if needed. One should not assume that because the jacket water is at the coating temperature, the coating liquid is at the coating temperature. Also, if the coating liquid is heated to the coating temperature in the feed vessel, the circulating water in the jacket should be at the coating temperature to prevent the liquid from cooling down in the lines.

The coating die should also be at the coating temperature. When coating is done at other than room temperature, there are several techniques commonly used to do this. One is to circulate water or another heat transfer fluid through holes bored through the plates of the coating die. This works quite well but complicates the assembly of the die. Another technique for moderately elevated temperatures is to enclose the coating stand in a small, temperature-controlled room. This can be used for temperatures up to about 40° or perhaps even 50°C. However, operators do not enjoy working at such elevated temperatures, and this technique should be considered only when the coating die is normally unattended. Electrical heaters in and around the coating head can also be used. Uniform temperature control is more difficult with electrical heaters, but this method is commonly used for the high temperatures needed with polymer melts in extrusion dies.

The temperature of the coating liquid in the coating die should be the same everywhere in order for the viscosity to be the same. Because the flow rate is directly proportional to the viscosity, we can specify how much viscosity variation can be tolerated. We would suggest that the flow variation caused by viscosity variations should be no more than 0.25%, or at most, 0.5%. This can then be used to set the allowable temperature variation. But first one should determine how the viscosity varies with temperature. Figure 6-5 shows some curves of viscosity versus temperature for several coating liquids containing gelatin. One should measure the viscosity at the coating temperature, and then at 5°C (9°F) above and below the coating temperature. A smooth curve can then be drawn through the three points, and the viscosity variation found for a 1° temperature variation. It is then simple to determine the temperature variation equivalent to a viscosity variation 0.25% or 0.5%, or whatever.

If it is found that the viscosity varies 1% per degree Celsius temperature change in the vicinity of the coating temperature (this is a reasonable value), and if we want to control the viscosity to within ±0.25%, we would have to control the temperature to within ±0.25°C. This temperature control is about the most that can be attained with reasonable controls and without going to extreme measures.

As the coating liquid in the coating lines can have a different temperature at the center than at the wall, and as the liquid in the center may go to a different part of the coating head than the liquid at the wall, the liquid should be mixed just ahead of the coating die. Both static or motionless mixers and dynamic mixers can be used. These mixers will also remove any stratification that may have occurred in the lines.

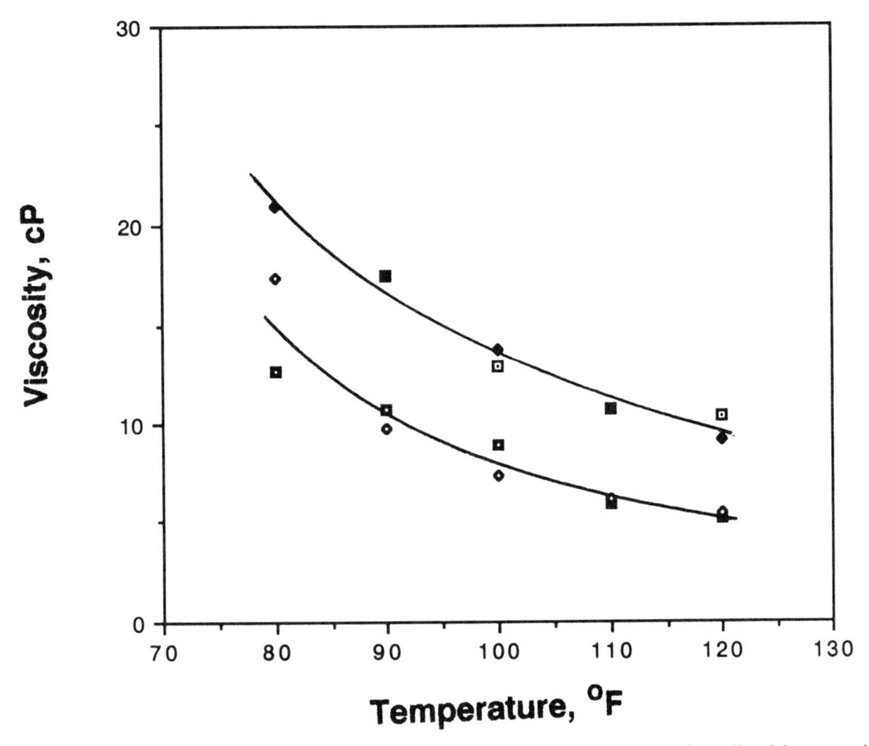

Figure 6-5. Variation of viscosity with temperature for some coating liquids containing gelatin.

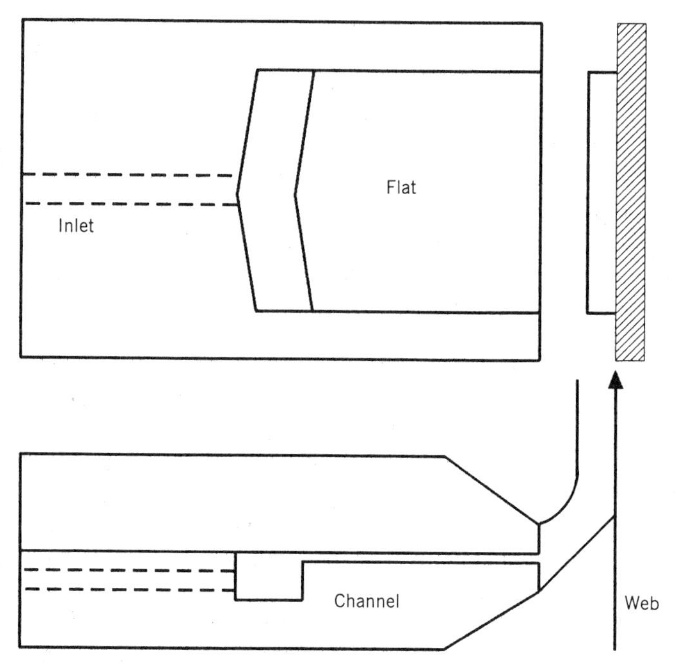

Figure 6-6. Slot coating die corrected to give a uniform flow. [From E. B. Gutoff, *J. Imag. Sci. Technol*, **37**, 615–627 (1993), with permission of the Society of Imaging Science and Technology.]

6.3 CORRECTIONS TO THE COATING DIE INTERNALS

If the error in the uncorrected die is excessive, the internal shape has to be changed so as to give a uniform flow across the width. The methods of doing this are described in texts such as that by Tadmor and Gogos (1979). In principle, the slot length from the channel to the die lips has to be changed so as to be shorter at the edges, illustrated schematically in Figure 6-6. When inertial effects can be neglected, one can use for both Newtonian and for shear-thinning liquids a simple equation for the slot length given by Gutoff (1993). The fractional correction to the slot length can also be used as an approximation to the error in an uncorrected die. A very accurate finite element technique for calculating the complete flows is described by Sartor in his Ph.D. thesis (1990) under Professor Scriven at the University of Minnesota. In a recent paper, Vrahopoulou (1991) gives a simpler finite difference technique.

6.4 DIE INTERNAL PRESSURES AND SPREADING

As mentioned earlier there are both low and high limits on the pressures in the coating head. On the low pressure side the liquid will just dribble out unevenly, and gravity will become important. Also, if inertial effects are neglected in the design then the pressure in the channel should be such that the pressure effects due to inertia are negligible. And, whenever possible, the inertial effect should be neglected in the design in order that the coating head should be versatile and usable with different liquids of different viscosities and at different flow rates. Now the pressure rise due to inertia may be about ρV^2, where V is the average velocity in the channel near the feed entrance and ρ is the density. (It is easy to show that ρV^2 has units of pressure. Thus, in SI units, and introducing Newton's second law to convert between force and mass,

$$\rho V^2 = \frac{\text{kg}}{\text{m}^3}\left(\frac{\text{m}}{\text{s}}\right)^2 \frac{\text{N}\cdot\text{s}^2}{\text{kg}\cdot\text{m}} = \frac{\text{N}}{\text{m}^2} = \text{Pa}$$

Pascal is the SI unit of pressure.) We may then specify that the pressure in the channel be high enough that the inertial pressure is only some small fraction, say 0.0025 ($\frac{1}{4}$%) of it. Thus if the inertial pressure in the channel at the feed entrance is 80 Pa, the minimum pressure in the channel should be 80 Pa/0.0025, or 32,000 Pa, or 4.6 psi.

The pressure in the channel is equal to the pressure drop in the slot, and for simple (Newtonian) liquids this is

$$\Delta P = \frac{12\eta q L}{H^3}$$

where H is the slot clearance (cm), L the length of the slot (cm), q the flow rate per unit width ($\text{cm}^3/\text{s}\cdot\text{cm}$ of width), and η the viscosity (poise). Any consistent set of units can be used; cgs units are given above, and the pressure drop will then be in dyn/cm^2. To convert dyn/cm^2 to pascal, one should divide by 10. To convert pascal to pounds per square inch, one divides pascal by 6895.

The maximum pressure is limited by the strength of the die. If the internal pressure in the channel is 1 MPa (150 psi), the die is 1 m wide with a combined length of the slot plus channel of 10 cm, and we say that the average pressure in the die is half the maximum pressure, the spreading force will be

$$\frac{10^6 \text{ Pa}}{2}\,\frac{\text{N/m}^2}{\text{Pa}}\,(1 \text{ m wide})\,(0.1 \text{ m deep})$$
$$= 50{,}000 \text{ N} \quad \text{or} \quad 11{,}000 \text{ lb of force}$$

A die has to be massive to resist this force. Even then, it is difficult to prevent the slot clearance from opening up in the center region of the die. Therefore, the dies that are not adjustable are suited only for relatively low-viscosity liquids, where the internal pressures to force the liquid out the slot are relatively low, often under 100 kPa (15 psi).

6.5 EXTRUSION DIES

Although some engineers use the terms interchangeably, we distinguish between extrusion coating and slot coating by noting that in slot coating the coating liquid wets the lips of the die (see Figure 6-1), while in extrusion coating the lips are not wet, as shown in Figure 5-1. In extrusion coating the viscosities are high, ranging up to the millions of mPa · s (cP). The pressures in the die are high and the pressure drops in the channels are high. Because of these high-pressure drops in the channels, the corrections have to be large, leading to the "coat hanger" design (Figure 6-7*b*). Because these dies are used for fluids with differing rheologies (shear-thinning behaviors) the corrections that are good for one liquid are often not good for another. For this reason, and because the high internal pressures tend to spread open the slot clearance even with massive dies, the slot openings are made adjustable at the front edge, or choker bars are used farther back. The adjustments are made with bolts at least every few inches near the front edge of the slot (Figure 6-8). It is very difficult to adjust all of these bolts manually to obtain a reasonably uniform cross-web coverage; some operations use an on-line sensor to measure the cross-web coverage and tie into a computer, to set the adjustments correctly. In some cases the adjusting bolts can be heated or cooled, and thermal expansion is used to adjust the bolt lengths.

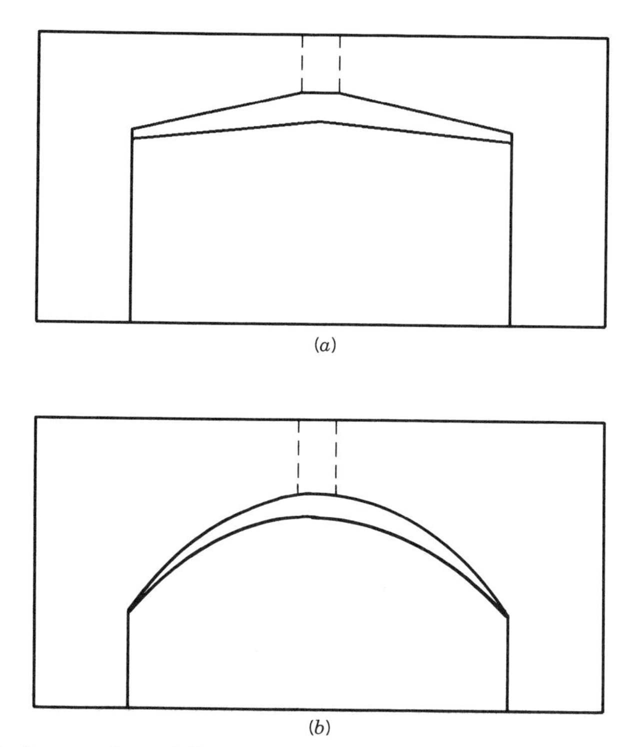

Figure 6-7. The bottom plate of dies with tapered channels: (*a*) slot coating die; (*b*) extrusion die of coat-hanger design. [(*b*) From E. B. Gutoff, *J. Imag. Sci. Technol*, **37**, 615–627 (1993), with permission of the Society of Imaging Science and Technology.]

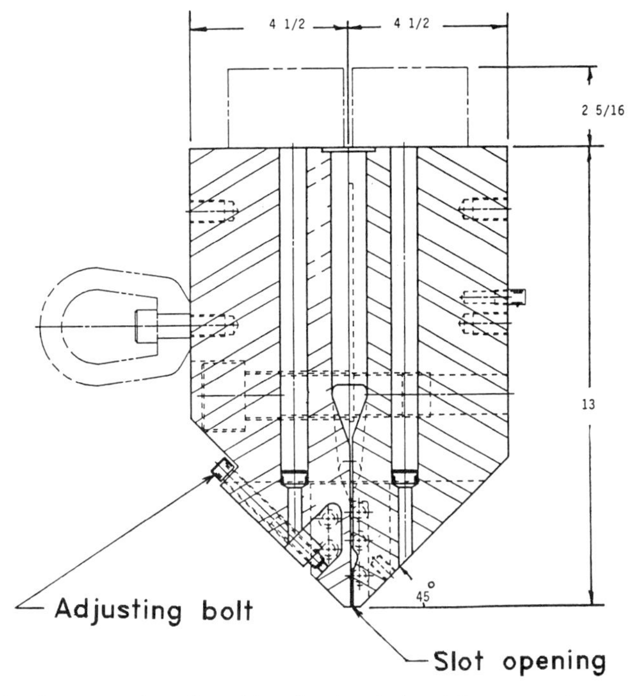

Figure 6-8. Ultraflex extrusion die with adjusting bolts. [Courtesy of Extrusion Dies, Inc., Chippewa Falls, WI 54729.]

6.6 WALL SHEAR STRESS CONTROL AND RESIDENCE-TIME CONTROL

By stress we mean the force per unit area of material. Most of us are familiar with tensile and compressive stresses. A tensile stress is illustrated in Figure 6-9*a* and is equal to the force, F, tending to pull the rectangular rod apart divided by the area, wh, perpendicular to the force. A shear stress is illustrated in Figure 6-9*b*, as the force, F, pulling the box along the surface divided by the contact area, wh, which is an area parallel to the force. If one thinks of the box in the sketch as a packet of fluid, this is the wall shear stress for the fluid.

In some operations one wants to be above a minimum wall shear stress in all parts of the coating die to prevent accumulations of material on the wall, especially if the coating liquid has a yield stress and tends to set up as a gel. Even when the liquid does not form a gel, one may want a minimum shear stress to prevent particles from settling out onto the wall. With laminar flow—in coating dies the flow is laminar and not turbulent—the wall shear stress is directly proportional to the liquid viscosity and to the flow velocity and inversely proportional to the diameter of the channel. Thus, as we move away from the feed port and down the channel toward the edges and the flow rate decreases because of the flow into the slot, the equivalent diameter of the channel will have to decrease to maintain a minimum wall shear stress. This will give rise to the tapered channel (Figure 6-7). This is sometimes used in slot coating dies, and almost always used in extrusion dies of the coat hanger design.

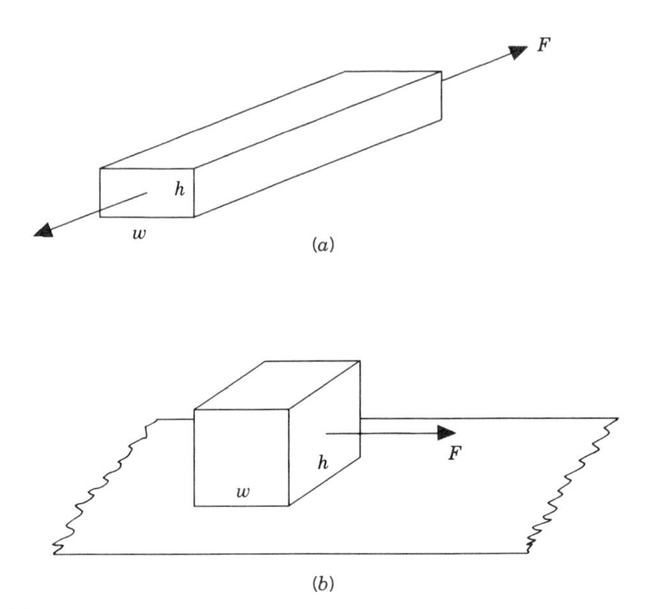

Figure 6-9. (*a*) tensile stress; (*b*) shear stress. In each case the stress is F/wh.

Residence-time control can also be important. In some but not all operations the properties of the coating liquid deteriorate with time, and thus no liquid should remain in the coating head for excessive periods of time. Especially in such cases, there should be no dead areas, where fluid does not move at all; there should be no areas where vortices form and fluid moves in closed circles and is only replaced very slowly; and the fluid velocity should not drop too low, to the point where the transit time down the channel is excessively long.

The first two of these three criteria apply to all good designs. There should be no dead areas in a good design, and there should be no vortex formation in a good design. The third or the transit-time criterion would apply only where residence-time control is important.

The avoidance of dead areas is relatively simple and is accomplished by avoiding dead-ended spaces. The avoidance of vortices is more difficult and is often not obvious. Figure 6-10 gives examples of vortex formation in some flow situations. Streamlining is the means to avoid vortices. To predict vortex formation one usually has to use involved numerical techniques such as finite element analysis. Sartor (1990) has done this for slot coating dies. Figure 6-11 shows some results of his study of vortex formation in the exit slot. The ideal situation is with no vortices. The most common type of vortex formation is along the upper meniscus, between the upper lip and the web, shown in (*a*), which occurs when the gap is too thick, more than three times the wet coated thickness. When the gap is too tight relative to the wet thickness, or the coating speed too low, the liquid can wet the shoulder of the upper die lip, as in (*b*), and a vortex will form there. At very low coating speeds vortices cannot be avoided: either the vortex as in (*a*), due to a wide gap, or the vortex as in (*b*), due to a tight gap, will form. Only at higher speeds can both types of vortices be avoided. If the coating liquid retreats from the lower edge into the slot, as in (*c*),

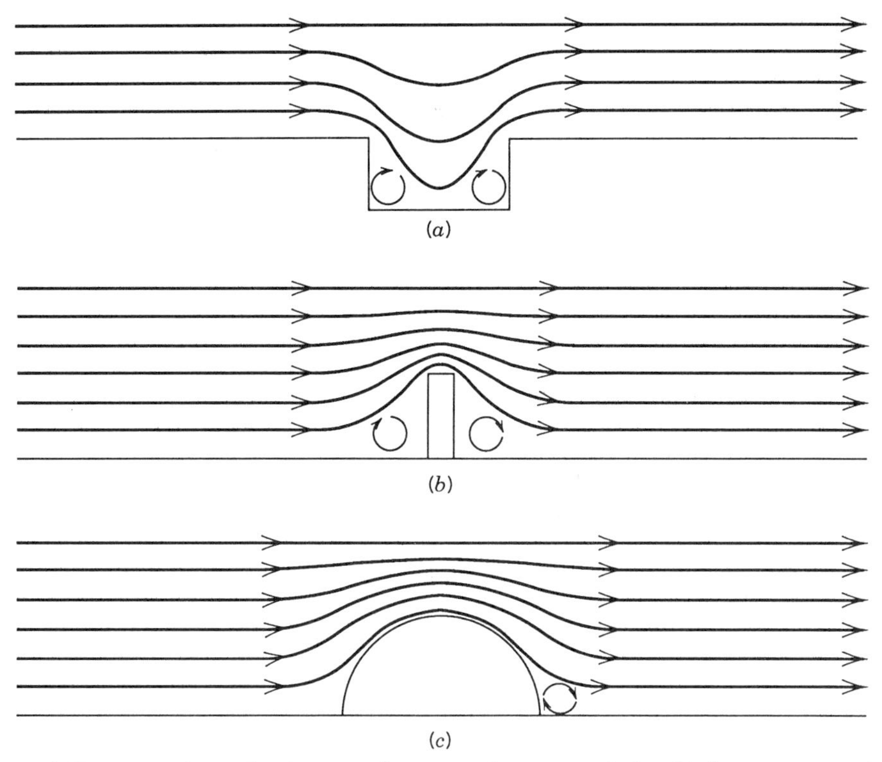

Figure 6-10. Vortex formation in some flows: (*a*) flow over a hole; (*b*) flow over a vertical plate; (*c*) flow over a hemisphere.

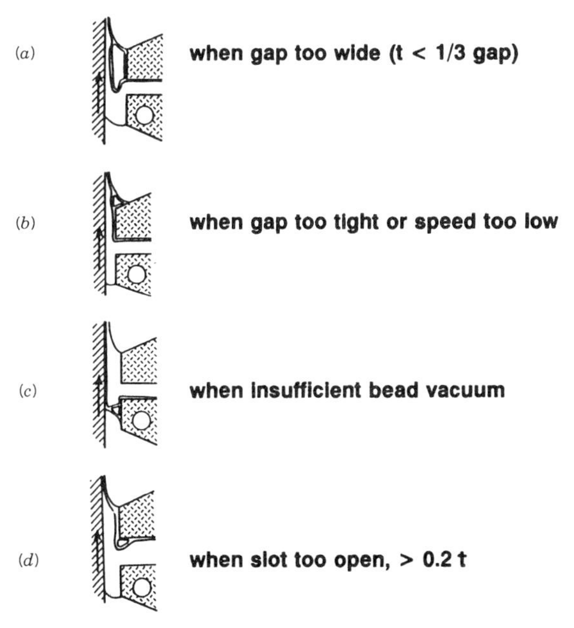

Figure 6-11. Vortex formation in slot coating dies. [From Sartor (1990), with permission of the author.]

a vortex will form near the lower meniscus. If the die surface were nonwetting, this vortex might not form. More bead vacuum will pull the meniscus back to the lower edge and eliminate this vortex. When the feed slot opening is too large, more than five times the wet film thickness, a vortex may form at the exit of the feed slot, as in (*d*). It always forms when the feed slot opening is greater than eight times the wet film thickness. The different vortices can join and combine to form one large vortex.

REFERENCES

Carley, J. F., "Flow of melts in 'crosshead'-slit dies; criteria for die design," *J. Appl. Phys.* **25**, 1118–1823 (1954).

Gutoff, E. B., "Simplified design of coating die internals," *J. Imag. Sci. Technol.* **37**, 615–627 (1993).

Ishizuka, S., "Method of simultaneous multilayer application," European Patent Appl. No. 88,117,484.1, Pub. No. 0 313 043, Oct. 20, 1988.

Sartor, L., "Slot coating: fluid mechanics and die design," Ph.D. Dissertation, University of Minnesota, 1990; University Microfilms, Ann Arbor.

Tadmor, Z., and C. G. Gogos, *Principles of Polymer Processing*, Wiley, New York, 1979.

Vrahopoulou, E. P., "A model for fluid flow in dies," *Chem. Eng. Sci.* **46**, 629–636 (1991).

CHAPTER VII

Surface-Tension-Driven Defects

In coating operations the coating liquid is spread out as a thin layer on a moving web, thus creating large surface areas. These surfaces have associated surface forces, as does every surface. These surface forces, called surface tensions, tend to reduce the areas of the surfaces. They can be the cause of a number of defects that occur in coating operations, such as convection or Bénard cells, craters, fat edges or picture framing, nonuniform edges in multilayer coating, dewetting and crawling, adhesive failure and delamination, and nonuniform coverage due to nonuniform base surface energies. In this chapter we discuss these defects and their origins, and possible remedies for them. We also explain the use of surfactants to alter surface tensions and how surfactants affect these defects.

All systems tend toward a state of lowest energy, and to minimize the surface energy the systems tend to minimize the surface area. As spheres have the smallest surface area for a given volume, small volumes of fluids tend to have spherical shapes—thus drops and bubbles tend to be spherical. Surface tension is a tension, as illustrated in Figure 7-1. The Greek lowercase letter σ (sigma) is frequently used to represent surface tension, and at times the Greek lowercase letter γ (gamma) is used. If we imagine we are pulling the surface of a bubble or drop in the direction of the surface arrows in Figure 7-1, we see that there will be created an extra force directed inward, making the pressure on the inner or concave side of the curved interface higher than on the outer or convex side. The pressure difference across a curved interface is called *capillary pressure*. Thus the pressure inside drops and bubbles is higher than that on the outside. This concept of capillary pressure is used to explain the shape of some cylindrical liquid surfaces in coating operations (Gutoff, 1992).

Surface tension and *surface energy* are sometimes used interchangeably. The term *surface tension*, in units of force per unit length, and the term *surface energy*, in units of force times distance per unit area, are equivalent. Surface tension is expressed as dyn/cm, or, equivalently, as mN/m. Now

$$\frac{\text{dyn}}{\text{cm}}\,\frac{\text{cm}}{\text{cm}} = \frac{\text{ergs}}{\text{cm}^2}$$

is energy per unit area, and

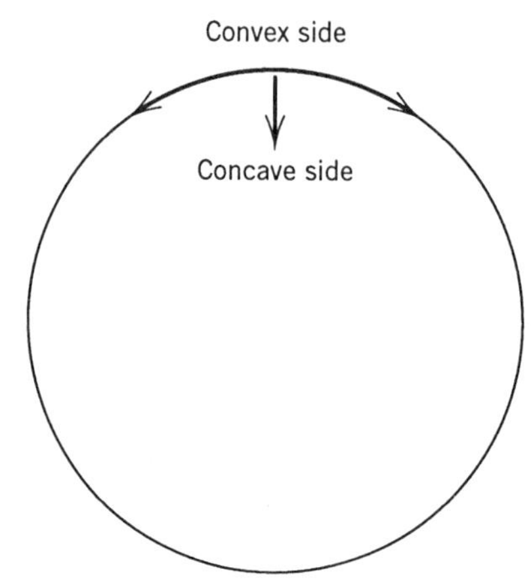

Figure 7-1. The concave side of a curved interface has a higher pressure than the convex side.

$$\frac{\text{mN}}{\text{m}}\frac{\text{m}}{\text{m}} = \frac{\text{mJ}}{\text{m}^2}$$

is also energy per unit area. Numerically, all four of these terms are the same.

When we speak of liquids, we usually refer to *surface tension*, but in reference to solids we often refer to *surface energy*, as it may be more difficult to visualize an undisturbed solid surface being in a state of tension. It may be easier to conceive of the surface of a solid having a surface energy. But this is strictly a matter of personal preference.

7.1 SURFACTANTS

We cannot discuss surface tension without discussing surfactants, which are used to lower surface tensions. *Surfactants*—the term represents a contraction of *surface-active agents*—consist of relatively long molecules. One part of the surfactant molecule is soluble in the solvent (usually water) and the other part is insoluble. Because of this dual nature, the molecules tend to concentrate at the surface of the liquid (or at an interface with a solid), where the soluble portions will remain down in the liquid and the insoluble parts will stick up in the air (or against a solid surface). Because the molecules congregate at the surface, they are called surface active. The insoluble portions of the molecules, which now cover the surface, have a lower surface tension than the solvent, so the system surface tension is lowered.

Normally, the insoluble portion of the surfactant molecule carries no electrical charge, and the soluble portion may or may not carry a charge. In *nonionic surfactants* the soluble portion has no charge; the solubility in water is conferred by nonionized oxygen groups, such as ether linkages. In *anionic surfactants*, water solubility is caused by negatively charged groups such as carboxylic or sulfonic acid groups. Electrical neutrality is maintained by positively charged ions (cations) such as hydrogen or sodium ions. Once the surfactant is in water, it ionizes, meaning that these counterions, such as sodium or hydrogen, split off and are delocalized; that is, they "swim" around in the water and are not attached to a particular molecule. In *cationic surfactants* the water-soluble portion has a positive charge, such as by salts of derivatives of ammonia (quaternary ammonium salts and salts of amines). The counterions are now negatively charged anions such as chloride or bromide ions.

Because part of the surfactant molecule is insoluble in the solvent, the bulk concentration of surfactant remains low while the surface concentration is much higher. Once the surface is fully covered with surfactant, and the surface tension is as low as it will get, additional surfactant has to go into the bulk solution. But because of the insoluble portion of the molecule, most surfactant molecules will come together to form *micelles*, with the insoluble portions of the molecules facing each other in the interior and the soluble portions "sticking out" into the solvent. The concentration at which this occurs is called the critical micelle concentration (CMC). The critical micelle concentration can be found from a plot of surface tension versus concentration. The surface tension levels off at the critical micelle concentration. We often want the surfactant concentration to be above the critical micelle concentration.

Surfactants may be added to coating liquids for other reasons than to provide good wetting of the web. Often, coating liquids contain latices or dispersed solids. Surfactants are usually present as dispersing agents, to prevent the solids from flocculating and to prevent coagulation of the latex particles. For more information on surface tension and its measurement, and on surfactants, one should consult any of the many good books available, such as that by Adamson (1982).

7.2 SURFACE TENSION EFFECTS IN COATING

Surface tension plays a very important role in wettability. For a liquid to wet a surface, its surface tension has to be low relative to the surface energy of the solid. We can demonstrate this by looking at the surface forces acting at the edge of a wet coating on a plastic web, as in Figure 7-2. At equilibrium the forces acting to the left are balanced by the forces acting to the right. With these thin films we may neglect the force of gravity. There is only one force acting to the left, the surface tension, or if you will, the surface energy, of the plastic base. Acting to the right is the horizontal

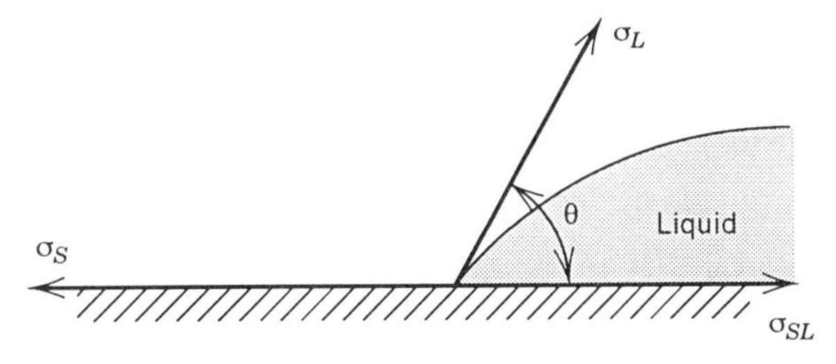

Figure 7-2. Surface forces at the edge of a wet coating. σ_L is the surface tension of the liquid, σ_S is the surface energy of the solid, and σ_{SL} is the interfacial tension. For spreading to occur, the force to the left, σ_S, must be greater than the forces to the right, $\sigma_{SL} + \sigma_L \cos \theta$.

component of the surface tension, plus the interfacial tension between the solid and the liquid. If the liquid is to spread to the left on the surface, the forces to the left must be greater than the forces to the right. Thus we want the surface energy of the plastic base to be high, the surface tension of the liquid to be low, and the interfacial tension between the solid and the liquid also to be low. With aqueous solutions we add surfactants to lower the surface tension, as the surface tension of water is very high, 72 mN/m (or dyn/cm). Surfactants also tend to lower the interfacial tension. Also, for coatings of aqueous solutions, we often increase the surface energy of the base by oxygenating the surface. The oxidized surface has a higher surface energy than the unoxidized plastic. Impurities on the surface may be oxidized, and this can contribute to improved wettability. It has been suggested that oxidizing the surface may pit it slightly, and this increased roughness can aid in wettability.

We can oxidize the surface of the web by passing it through a flame and burning the surface slightly. This is called *flame treatment.* We can treat the surface with a corona discharge, where a high-frequency, high-voltage electrical discharge breaks down the resistance of the air and ionizes the air, and the reactive ionized oxygen oxygenates the surface. The increase in surface energy by the corona discharge, however, is often transient, as in some cases the oxidized molecules diffuse into the interior. Therefore, corona treatment is often carried out on-line, directly before the coating operation. We can also increase the surface energy by coating the web with a thin, high-energy layer called a *subbing layer.* For photographic systems subbing layers are gelatin, coated from a water–alcohol dispersion.

The surface energies of various materials are given in Table 7-1, where we see that fluorinated surfaces have the lowest surface tension. This is why it is so difficult to coat-on the fluorinated polymer Teflon, and also why fluorinated surfactants are so effective in lowering surface tension. Hydrocarbon surfaces also have a low surface tension, and therefore plastic bases must be treated before coating aqueous layers on them. We see in Table 7-1 that oxygenated and chlorinated surfaces have higher surface tensions, and metals have the highest surface tension. Table 7-1 shows that the surface tensions vary widely for the various hydrocarbons and oxygenated and chlorinated solvents. The surface tension depends on which portions of the molecules are exposed at the surface. If an oxygenated molecule lies in the surface with the oxygen facing down into the liquid, the surface tension will be that of the hydrocarbon portion of the molecule. This is likely to be the case with ethanol and acetaldehyde. In *i*-pentane the methyl —CH_3 groups are more likely to be in the surface than in *n*-hexane or *n*-octane, where there are more methylene —CH_2— groups and the surface tension is lower. The aromatic hydrocarbons benzene, toluene, xylene, and styrene, with no methyl groups, have higher surface tensions. Note that metal surfaces have the highest surface tensions or energies. Thus metals, when clean, are easy to wet. However, it is very difficult to clean a metal surface thoroughly and to maintain the surface free of contamination, especially by metal oxides.

Organic solvents generally have low surface tensions, and therefore surfactants are rarely used or needed in solvent systems. Also, many surfactants are not effective in solvent systems. However, some fluorinated surfactants do lower the surface tension of solvent systems very effectively.

In multilayer coatings, such as in slide and curtain coating, the top layer has to spread over all the other layers and so has to have the lowest surface tension of all.

Table 7-1 Surface Energies or Tensions (dyn/cm or mN/m)

Fluorinated surface, close packed	6
i-Pentane	14
n-Hexane	19
n-Octane	22
Benzene	29
Toluene	28
p-Xylene	28
Styrene	32
Ethanol	22
n-Octanol	28
Phenol	41
Ethylene glycol	48
Glycerin	63
Acetaldehyde	21
Furfural	41
Acetone	24
Acetic acid	28
Benzyl alcohol	39
Carbon tetrachloride	27
Tetrachloroethylene	32
Tetrachloroethane	36
Tetrabromoethane	50
Liquid sodium	191
Liquid potassium	400
Mercury	485
Liquid aluminum	860

We illustrate this in Figure 7-3. Multilayer coatings are usually aqueous coatings, and as aqueous layers can usually mix together in all proportions, there is no interfacial tension between them. Thus for one layer to spread over another layer, we only require that the upper layer have the lower surface tension. When the top layer has to spread over the other layers on the slide and on the base, none of the other layers is

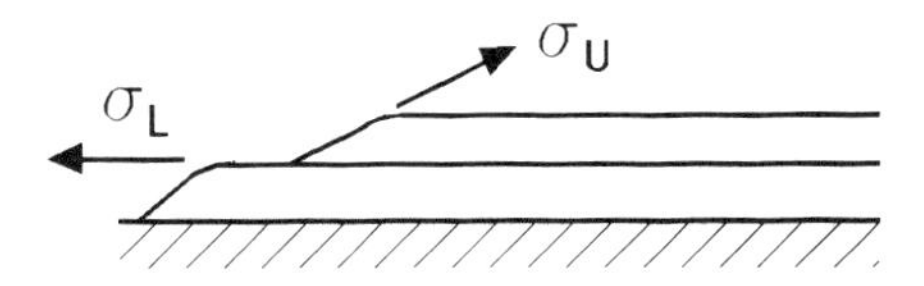

Figure 7-3. Spreading of one layer over another. For the upper layer to spread over the lower, when there is no interfacial tension, the surface tension of the lower layer, σ_L, must be greater than the surface tension of the upper layer, σ_U.

in contact with the air. In this case only interfacial tensions, not surface tensions, play a role, and there are no interfacial tensions between the various aqueous layers. With no interfacial tensions between the aqueous layers, the surface tensions of all the layers below the top can have any value whatsoever as far as spreading is concerned, as long as the top layer has the lowest surface tension. However, in curtain coating the stability of the curtain has to be considered. It appears that the curtain is most stable when all the layers have about the same surface tension, while for spreading, the surface tension of the top layer should be slightly lower than the others.

These statements need to be modified somewhat. When a fresh surface is created (and in all coating operations a fresh surface is created), this surface is not at equilibrium. It takes time for any surfactant to diffuse to the surface, there to lower the surface tension. During this time when equilibrium is being approached, the surface tension decreases with time as surfactant diffuses to the surface. The surface tension drops from that of the surfactant-free liquid, when the surface is first formed, to the equilibrium value for an aged surface. This changing surface tension is known as the dynamic surface tension. Some surfactants diffuse very rapidly and the surface tension reaches its equilibrium value in under 10 ms; others diffuse slowly and it may take hours to reach equilibrium.

We are interested in the surface tension of the coating liquid in the coating bead, on the web, and also on the slide in slide and curtain coating. The surface of the coating liquid has a certain age measured from the moment the surface was first formed at the exit of a slot. Thus we are interested in the dynamic surface tensions at certain ages. But we rarely know the surface tension as a function of time, so we usually use the equilibrium value, which often is all we know.

7.3 SURFACE-TENSION-DRIVEN DEFECTS

A number of defects are caused by surface forces. Some of these, such as craters and convection cells or Bénard cells, are more common in solvent coatings where surfactants are rarely used. In aqueous coatings surfactants are almost always used at relatively high concentrations—above the critical micelle concentrations—and the surface is saturated with surfactant. The surface tension in the surface then remains constant and rarely is the cause of defects.

7.3.1 Convection or Bénard Cells

Convection cells (Figure 7-4) often appear in the shape of hexagons (like a beehive), and were first observed when heating a thin layer of liquid from below. These cells are caused by density gradients resulting from the hotter, less dense liquid at the bottom rising through the cooler overlying liquid, as illustrated in Figure 7-4*b*. Convection cells also result from surface tension gradients, as illustrated in Figure 7-4*a*. The surface tension gradients can result from temperature gradients or from concentration gradients. When the wet layers are less than 1 mm thick, as is the case in almost all coatings, convection cells are almost always due to surface tension gradients (Koschmieder and Biggerstaff, 1986).

The liquid motions illustrated in Figure 7-4 that cause these cells can be reduced by having thinner layers and by having higher viscosities and higher surface viscosi-

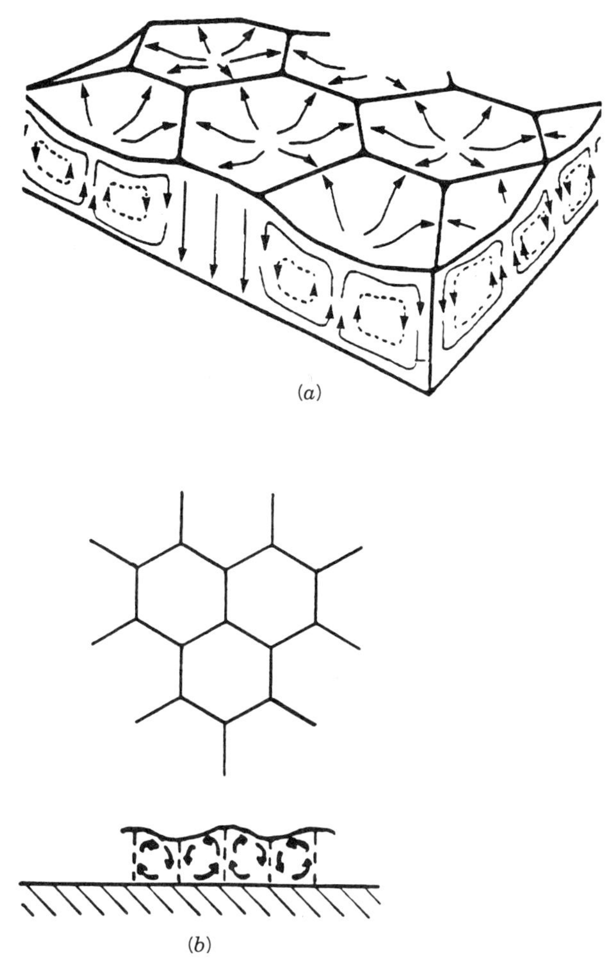

Figure 7-4. Convection or Bénard cells: (*a*) surface-tension-driven cells (From 3M Specialty Chemicals Division, Fluorad Product Bulletin, St. Paul, MN 55144); (*b*) density-driven cells.

ties, since a given driving force (a surface-tension-driven force) will have less of an effect under these conditions. The surface tension gradients are sometimes caused by using mixed solvents, which undergo concentration fluctuations during evaporation. Since the surface tensions of the several components will differ, these concentration fluctuations will lead to surface tension fluctuations. Evaporative cooling may cause temperature fluctuations in the surface. Since surface tension is a function of temperature, this will also lead to surface tension fluctuations. Drying at a slower rate is often helpful in reducing the severity of convection cells. However, we always want to increase our production rates, which means that we want to coat faster and to dry faster. Adding a lower-volatility solvent is often helpful, and just changing the solvent composition can also help. Adding a surfactant will usually help, and some of the fluorinated surfactants are very effective in solvent systems. Of course, we have to make certain that the added surfactant does not ruin the quality of the final coated product.

7.3.2 Craters

We use the term *crater* to refer to a specific defect caused by a low-surface-tension spot on the surface, such that the liquid flows out from that spot to regions of higher surface tension, as indicated in Figure 7-5. The low-surface-tension spot can be caused by a particle of dirt from the air, by an oil droplet, by a particle of gel, or by use of silicone defoamers at concentrations above their solubility. The liquid will flow out from the low-surface-tension spot at remarkably high velocities; speeds as high as 65 cm/s have been measured (Kornum and Raaschou-Nielson, 1980).

The edge of the crater becomes elevated due to the acceleration of the top liquid and the inertia or drag of the underlying liquid. The center thins and only a monolayer may remain on the web, although often a high "pip" remains at the very center.

As with convection cells, craters can be reduced by having thinner layers and higher viscosities. Use of surfactants is beneficial, and clean room conditions should be observed to avoid airborne dirt. Often, a change of surfactant, or the use of an additional surfactant, is helpful, perhaps because the new or additional surfactant better wets dirt particles or any dispersed solids. It is possible that in some cases pinholes are actually small craters. At times it can be very difficult to distinguish between craters and bubbles that have risen to the surface and burst.

7.3.3 Fat Edges or Picture Framing

Fat edges, sometimes called *picture framing*, are illustrated in Figure 7-6. Note that with uniform drying the evaporation rate should be uniform over the entire coating.

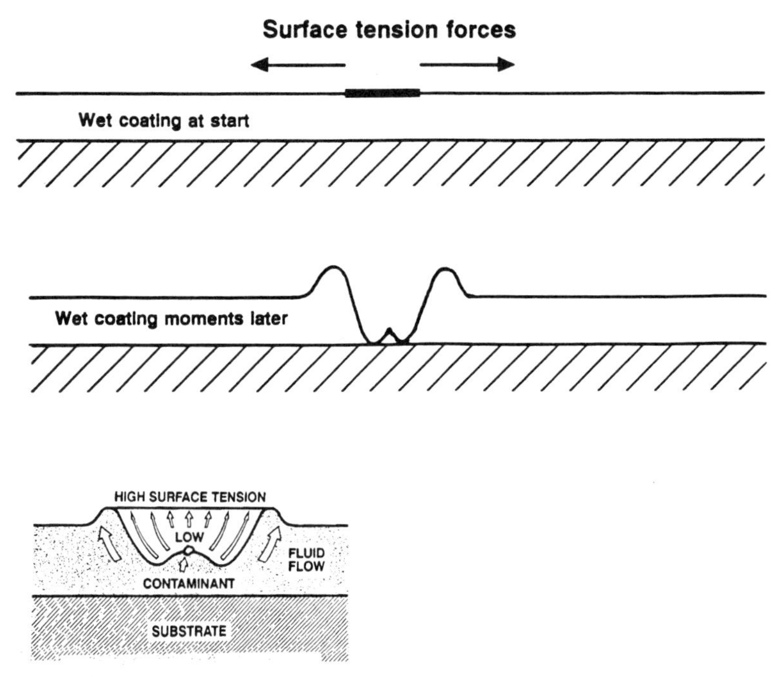

Figure 7-5. Crater formation. (From 3M Specialty Chemicals Division, Fluorad Product Bulletin, St. Paul, MN 55144).

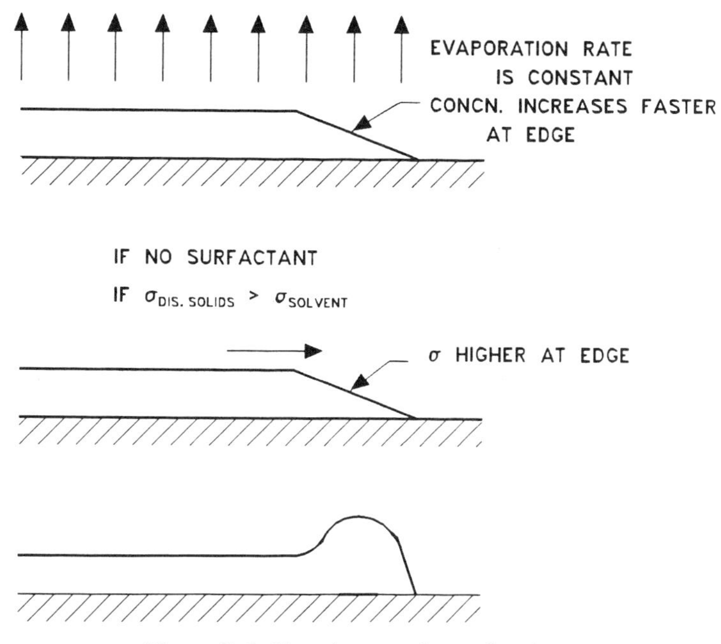

Figure 7-6. Fat edges or picture framing.

And because the edge of the coating is usually thinner, the solids concentration at the edge will increase faster during drying than in the bulk. When no surfactants are present, the surface tension at the higher solids concentration will usually be higher, because the surface tension of the dissolved solids is usually higher than that of solvent. The higher surface tension at the edge will cause flow to the edge, giving a fat edge. Surfactant will usually eliminate this problem. There are other causes of heavy edges in coatings, some of which are discussed in Section 5.3.5.

7.3.4 Nonuniform Edges in Multilayer Coatings

Figure 7-7 illustrates the boundary between two adjacent layers in a multilayer slide coating. The edge guide is used to prevent the layers from spreading beyond the

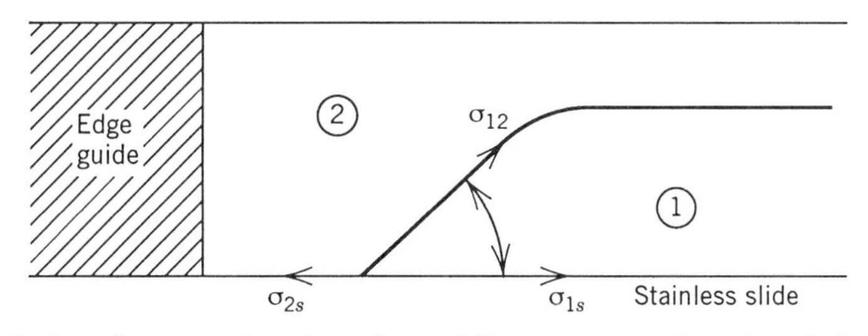

Figure 7-7. Surface forces at the edge of a multilayer system. If the interfacial tensions of the layers with the slide, σ_{1s} and σ_{2s}, are unequal, with no interfacial tension between the layers the contact line will move, causing a wide edge.

desired width. There should be no interfacial tension between the two aqueous layers. At equilibrium, the two interfacial tensions between the layers and the metal surface must be equal. However, if they are not equal, the interface will move in the direction of the higher interfacial tension. If the slide is long because there are many layers in the coating, this movement of the interface can be considerable. This means there would be a wide edge with coverage differing from that of the bulk of the coating.

Measuring interfacial tensions between a liquid and a solid is not simple. However, if the same surfactant is used in all the layers, we may assume that if all the surface tensions are equal, the interfacial tensions between the liquid layers and the metal of the coating head would also be approximately equal. Thus if there is an objectionably large edge effect in multilayer coating, one might want to adjust all the layers to the same surface tension, with the top layer having a slightly lower surface tension.

7.3.5 Dewetting and Crawling

Earlier we pointed out that for good wetting the coating fluid has to have a low surface tension and the base a high surface tension. If these are not correct, the coating can dewet or ball up: that is, retract from the coated areas. The start of this process is illustrated in Figure 7-8. When this occurs over a large area it is called *crawling*.

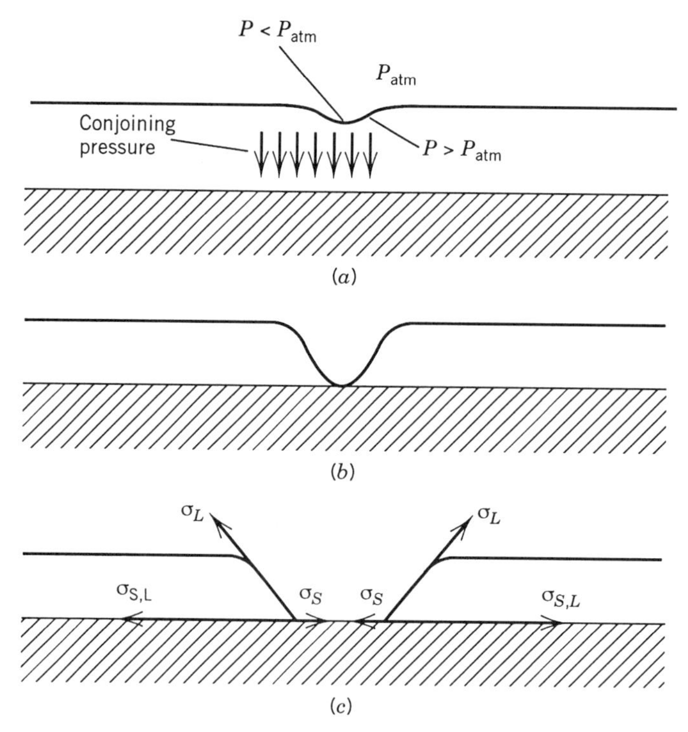

Figure 7-8. Dewetting and crawling: (*a*) surface forces tend to heal a depression while conjoining pressure tends to amplify it; (*b*) conjoining forces cause a hole to form in the coating; (*c*) surface forces then cause the system to dewet.

Dewetting can occur when a web has a low surface energy due to contamination, such as by an oil. When a layer is thick, dewetting cannot occur. Thick layers on a horizontal surface, such as water in a polyethylene beaker, will not dewet, because of the counteracting force of gravity. Surface forces (Figure 7-8*a*) aid gravity in preventing dewetting in thick layers. If a depression has formed in a liquid layer, the pressure under the concave edges will be above atmospheric pressure, as the pressure on the concave side of a curved interface is always higher than on the convex side. For the same reason the pressure under the convex center will be below atmospheric pressure. Thus liquid will flow from the high-pressure, elevated regions to the low-pressure, depressed regions due to the surface forces as well as by gravity. The surface will thus heal itself.

In thin layers, other, short-range forces come into play. They are called disjoining pressure and conjoining pressure. The *disjoining pressure* tends to keep two surfaces apart. This tends to prevent dewetting. The *conjoining pressure* tends to bring two surfaces together. This aids dewetting, for once the top surface of the liquid touches the solid, one has a dry spot (Figure 7-8*b*), and then the surface forces illustrated in Figure 7-8*c* come into play. However, the conjoining pressure normally does not play a role until the film has thinned to a thickness under 1 μm, according to Kheshgi and Scriven (1991). They point out that this is likely to occur if the film has projections from the surface, such as by bubbles in the coating, or if the surface tension at a spot on the surface is lowered by dirt from the air (as in the formation of craters) or by local temperature or concentration gradients (as in the formation of Bénard cells).

Dewetting and crawling do not take place at the point of application of the coating to the web but shortly after. Surprisingly, it is possible to coat aqueous liquids on Teflon without the use of a surfactant—but almost immediately after coating, the liquid will dewet. In the strictest sense, the surfactant in aqueous coatings is not needed for coating; it is needed to prevent dewetting and crawling immediately after coating. It is also needed to help in adhesion of the coating to the base.

Dewetting can also occur with one liquid over another. In multilayer slide and curtain coating, dewetting can occur on the slide if the surface tension of the top layer is not lower than that of all the other layers. This was discussed earlier in connection with Figure 7-3. Valentini et al. (1991) showed that this occurs when the dynamic surface tension of the upper layer is greater than about 1.2 or 1.3 times the dynamic surface tension of the lower layer.

7.3.6 Adhesive Failure and Delamination

Dewetting and crawling do not always take place when the liquid and the base surface tensions or energies are not adjusted properly. If the mismatch is small, or the layer very thick, the force to dewet may be insufficient to move the liquid significantly. But adhesion to the base will be poor, and the coating may easily be removed from the base. This has been known to occur when the proper coating liquid is coated but on the wrong side (the untreated side) of a plastic base. The dried coating looks good to the eye but comes off the base as a separate, intact sheet when a piece of coated base is crinkled or shaken vigorously.

In multilayer coatings it is also possible to get delamination between layers if the chemistry of the adjacent layers is not correct, that is, if the compatibility of the components is inadequate. This can occur when the binders in adjacent layers are at least partially incompatible. Suitable surfactants should reduce this incompatibility.

7.3.7 Nonuniform Base Surface Energies

If the surface energy of the base is nonuniform, the attractive force between the coating fluid and the web will fluctuate. The contact line of the coating liquid with the web will then oscillate back and forth, causing coverage fluctuations. These fluctuations in coverage will be random and may be enough to ruin the coating. Usually, nonuniform base surface energies are caused by contamination. Examples are contamination by oil vapors from a bearing and contamination by particles of a plastic that were abraded off in the coating area. In the latter case, pinholes were the result.

REFERENCES

Adamson, A. W., *Physical Chemistry of Surfaces*, 4th ed., Wiley, New York, 1982.

Gutoff, E. B., "Premetered coating," in *Modern Coating and Drying Technology*, E. D. Cohen and E. B. Gutoff, eds., VCH Publishers, New York, 1992.

Kheshgi, H. S., and L. E. Scriven, "Dewetting: nucleation and growth of dry regions," *Chem. Eng. Sci.* **46**, 519–526 (1991).

Kornum, L. O., and H. K. Raaschou-Nielson, "Surface defects in drying paint films," *Prog. Org. Coat.* **8**(3), 275–324 (1980).

Koschmieder, E. L., and M. I. Biggerstaff, "Onset of surface tension driven Bénard convection," *J. Fluid Mech.* **167**, 49–64 (1986).

Valentini, J., et al., "Role of dynamic surface tension in slide coating," *Ind. Eng. Chem. Res.* **30**, 453–461 (1991).

CHAPTER VIII

Problems Associated with Static Electricity

Static electricity can cause many problems in the coating industries. Besides the well-known hazard of explosions caused by static discharges in flammable vapors, static charges in the web can also cause coverage nonuniformities in coating, pickup of dirt and dust by plastic film base, problems in stacking of cut sheets, problems in film tracking, jamming of sheet ejection devices, fogging of photosensitive materials, pinholes in release coatings, and so on. Shocks to personnel can be more than annoying; the involuntary reaction to the shock can cause a major accident. The magnitudes of the static charges needed to cause these problems are shown in Table 8-1.

Especially in high-speed coating lines there are many locations that are ideal for generating static electrical charges. Charges form when two dissimilar surfaces separate, and web, which often is made of nonconducting plastic, separates from every roll over which it passes. If the surfaces are not grounded, the charges can build to high levels. Humidity in the air leads to the adsorption of moisture onto the surface of the web, raising its conductivity. Thus the static charges on the surface can leak away and dissipate. Static problems are far more prevalent in the winter when the air is dry than in the summer when the relative humidity tends to be high. Control of relative humidity is one means of controlling static.

Fire and explosion are a major concern in any solvent-handling area, and sparks caused by static electricity usually have enough energy to ignite solvent vapors. Discharge may occur whenever the charge on the web is high enough to overcome the breakdown resistance of the surrounding air. The discharge is usually to the ground. The ground might be a metal roll or a metal portion of the frame.

In addition to the static discharge as the source of ignition, a flammable solvent-air mixture must be available for a fire or explosion to occur. The solvent concentration must be above its lower explosive limit, and the oxygen content must be high enough to support combustion. Thus, to prevent the possibility of fire or explosion, one must ensure that static charges with enough energy to cause ignition cannot build up, or that sufficient ventilation is provided so that the solvent concentration in the air cannot reach the lower explosive limit, or that the oxygen content of the air is reduced to a low enough level (about 5%) that it will not support combustion. Low oxygen

Table 8-1 Static Voltages and Associated Problems

Minimum Voltage (V)	Associated Problem
30	Coating coverage nonuniformities
400	Jamming of ejection devices
500	Dust and dirt pickup
600	Fogging of photosensitive coatings
1000	Problems in registration of sheets
1500	Fire and explosion hazard
2000	Stacking problems with cut sheets
4000	Problems in packaging
6000	Problems in tracking (controlling film travel)
7000	Arcing

Source: Kisler (1994) and Keers (1984).

levels are normally found when the coating is made in a nitrogen atmosphere. A nitrogen atmosphere is sometimes used in a recirculating system where the solvent is recovered by condensation.

Like charges repel each other and unlike charges attract. A charged object can induce an opposite charge in a nearby neutral object. If this object is electrically isolated, it will remain electrically neutral, and thus one side will have one charge and the distant side will have an equal and opposite charge. This neutral object will now be attracted to the original charged object because of the opposite charge on the near side. Thus charges on the web will attract the coating liquid, and nonuniform charges will attract the coating liquid unevenly, leading to nonuniform coverage. On the other hand, a uniform charge on the web, or a uniform electrostatic field between the backing roll and the grounded coating die, will attract the coating liquid uniformly. This is the basis for electrostatic assist in coating (Nadeau, 1960; Miller and Wheeler, 1965).

When coated web, if charged, is cut into sheets at the end of the coating line, the sheets, all with the same charge, will repel each other and stacking the sheets will be difficult. The sheets will tend to slide off the top of the stack. In equipment ejection devices where the sheets are pushed rather than pulled, the charged sheets will tend to adhere to the device and cause jams. Similarly, charged web in the coating line may have difficulty tracking properly because it will be attracted to and want to adhere to the rolls and guides. This is mainly a problem on low-tension coating lines with thin or weak webs. At usual web tensions this is not normally a problem, for the web tension will tend to overcome the attractive forces due to static electricity. Problems can also arise at a laminating station, where, before two sheets with like charges enter, they can repel each other and balloon apart.

Charged web will attract dirt of an opposite charge from as far away as 15 to 25 cm (6 to 10 in.), and this dirt is very difficult to remove because of the large attractive forces. Dirt particles as small as 1 μm on the web cannot be tolerated in high-quality products such as photographic films and magnetic tapes. Web cleaners should discharge the web before trying to remove the dirt. Many web cleaners do not even try to do this, and even when they do, they are unable to remove the bound charges, which cause some of the difficulty. Even the best of web cleaners do not do a perfect job. With charges remaining on the web, it is impossible to remove dirt thoroughly.

In cases where dirt or dust pickup is undesirable—and this includes most operations—static eliminators should be installed just after the unwind stand. Charge neutralization here will reduce pickup of dust from the air. Clean room conditions will reduce the dust level in the air and will reduce the amount of dirt pickup by the web. Clean room, or at least semiclean room conditions (somewhere in the range of class 1000 to class 10,000 cleanliness), are recommended for all dirt-sensitive coating operations. This includes almost all coating operations.

In the photographic industry the blue arc of a static discharge can fog or expose light sensitive films and papers. This can occur on the coating line as well as in subsequent operations. It can even occur in the camera. Some people can build up an abnormally high static charge on their bodies. There are available meters that measure the static charge on a person. The person stands on an insulated platform and touches a probe, and the meter indicates the static charge. A person who does build up a high body charge should be equipped with the devices used to remove static charges from personnel, which are discussed later in Section 8.5.1, if they have to work in sensitive areas. Most of us do not build up such high charges.

The discharge of charged webs can have sufficient energy to burn pinholes in release coatings on the web surface. There may be a sufficient number of these pinholes to allow adhesion of the product to the underlying web, rather than permitting separation.

In many industries the quality of production decreases in the winter months. This has been tracked down to the low humidity in the winter months, allowing an increase in static charges. Moisture in the air tends to adsorb onto surfaces, causing a decrease in the surface resistivity. In winter, with its very low humidity, surface resistivities tend to be high, allowing static charges to build up to a high level. Many of us have experienced static discharges in winter when approaching a metal object after walking on a carpet. To avoid this type of static problem, it may be worthwhile to consider using a humidifier to add moisture to the air in winter. This is practical to do in small areas.

8.1 FORMATION OF CHARGES

Static charges can be developed by many different methods, such as the following:

1. *Separating Two Surfaces.* When two dissimilar surfaces are separated, one will have a greater affinity for electrons or ions than the other, and they will each become charged with respect to the other. It is difficult to distinguish between this method and charging by friction.
2. *Friction (Triboelectric Charging).* When two dissimilar materials are rubbed together a large charge can be built up, as when we use a hard rubber comb in our hair on a dry winter day. This type of charging is very common.
3. *Fracture.* When a solid is broken or an adhesive bond is broken, electrons are frequently emitted, leaving two highly charged surfaces.

4. *Induction.* If a neutral solid particle or liquid drop enters a region where there is an electric field, the end nearest the negative electrode will become positively charged and the end nearest the positive electrode will become negatively charged, while the particle overall remains neutral. The particle or drop is said to be polarized. If the electric field is nonuniform, as, for example, if one electrode is pointed, the particle will be attracted to the region of the strongest field, toward the pointed electrode. Let us assume that the pointed electrode is positive. When the particle gets close to the pointed positive electrode, the field strength may be great enough for electrons to be removed from the particle. This field emission charging will leave a net positive charge on the particle, and it will then be repelled by the pointed, positive electrode.
5. *Pressure.*
6. *Stretching.* Stretching an uncharged material can cause charge formation.
7. *Impact.* Dust particles impinging on a charged surface may become charged, depending on how long they are on the surface. Conductive particles become charged faster than nonconductors.
8. *Freezing.* During the freezing of water there has been found a large potential difference between the ice and the water. This has been suggested as a mechanism for the large charge buildup during thunderstorms.
9. *Electrostatic Field.* This will induce polarization in neutral particles, will align dipoles, and will charge particles that are in contact with the electrodes.
10. *Spraying.* When liquids are sprayed, the resulting cloud of droplets is usually charged.
11. *Corona Charging.* A corona is an electrical discharge that occurs in regions of strong electrical fields, usually produced by pointed conductors or fine wires. A reddish or violet discharge is usually seen in air. The current is carried by ionized gas molecules. These ionized molecules can transfer their charges to particles or surfaces.
12. *Ion and Electron Beams.* Particles or surfaces bombarded by a stream of ions or electrons will pick up the charges from the ions or electrons.
13. *Thermionic Emissions.* When a material is heated, the electrons move faster. If the velocity of the electron in the direction perpendicular to the surface is high enough, they will escape through the surface. This is the basis of the hot cathode in vacuum tubes. The material will be left positively charged unless it is connected to a source of electrons, such as ground.
14. *Photoelectric Charging.* Electromagnetic radiation such as light, and more particularly x-rays, when striking the surface of a material may transfer sufficient energy to the electrons in the surface to permit them to be ejected, leaving the material positively charged.

When separating two surfaces, as when unwinding a web or even when web passes over a roll, the ions and electrons that are present in the interfacial region are not evenly distributed between the separating surfaces. Positive charges will tend to accumulate on one surface, and negative charges on the other. The same effect occurs when rubbing one surface on the other. The faster the rubbing or the greater the speed of separation, the greater the charge buildup, as a slow separation allows the charges on the different surfaces to partially neutralize each other. The greater the conductivity of the surfaces, the more uniform will be the surface charge, and the less chance the surface will hold the charge. If the surface is conductive and grounded, the charge will leak away to ground. In coating operations both surfaces are usually in contact with grounded metal rolls, so there will normally be no significant surface charges in those cases where the surfaces are conductive. In most plastic webs the voltages that develop can be quite high, in the hundreds or even thousands of volts.

Static charges are of two types, free and bound. Free charges are the type we have been discussing, caused by an excess of electrons or of ions at the surface. Free charges tend to move under the influence of an electrostatic field—a voltage gradient—and will flow to ground given a conductive path. (Positive ions will not flow to ground, even though we speak of them as flowing. Instead, electrons will flow from ground to the object to neutralize the positive ions. The effect, however, is the same.) Bound charges are not mobile. They are formed by orientation of dipoles in the solid material. By a dipole we mean a molecule that while electrically neutral, has an excess of negative charges in one region and an excess of positive charges in another region. Most large organic molecules such as polymer molecules, and many small molecules such as water, fit this definition. In any solid polymer such as a sheet of film, normally the dipoles are randomly oriented as in Figure 8-1*a*. However, under the influence of a strong enough electric field the dipoles will become oriented and the sheet of film will become charged, such as in Figure 8-1*b*. These bound charges are not mobile and will decay only by disorientation of the dipoles. By itself, this

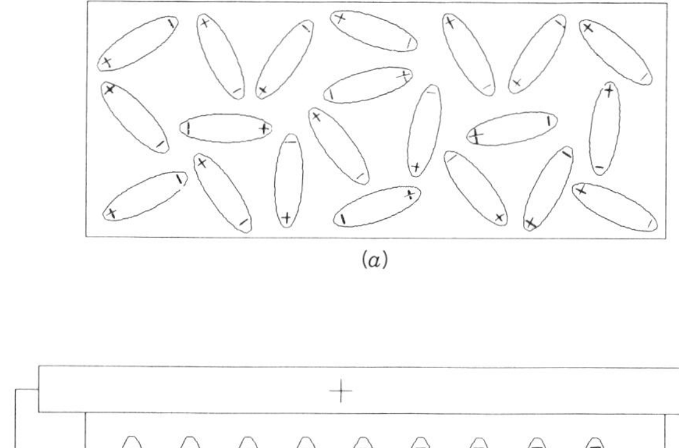

(*a*)

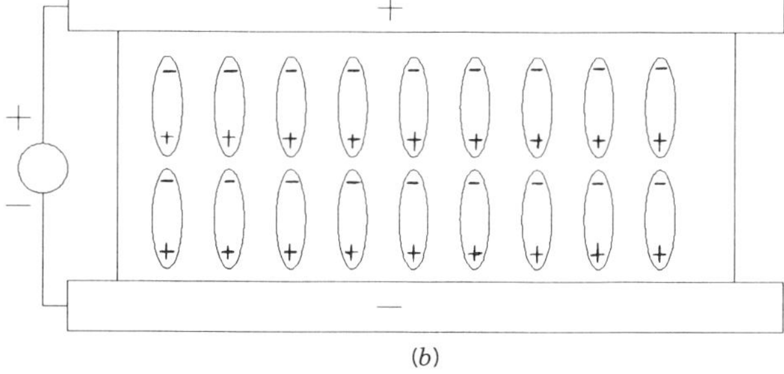

(*b*)

Figure 8-1. Orientation of dipoles in an electric field: (*a*) unoriented film; (*b*) film in a strong electric field.

may be a very slow process. Sheets of plastic may retain their bound charges for years.

8.2 ELECTROSTATIC FIELDS

A direct-current (dc) field is formed, as in Figure 8-2*a*, by having two electrodes a set distance apart. If the voltage difference between the electrodes, one of which is usually grounded, is V volts and the distance apart is d cm, the field strength is V/d V/cm. However, if one electrode is pointed, as in Figure 8-2*b*, the field strength near the pointed electrode, surprisingly enough, is greatly magnified and now depends inversely on r, the radius of the point of the electrode. The maximum field strength between the pointed electrode and the flat electrode will be directly under the pointed one. There the field strength will equal

$$\left(\frac{V}{d} + \frac{V}{r}\right)$$

Thus if the voltage difference between two flat electrodes is 1000 V and the electrodes are 1 mm (40 mils) apart, the field strength between two flat electrodes will be 1000 V/0.1 cm, or 10,000 V/cm. However, if one electrode is a needle with a tip radius of 0.01 mm (0.4 mil), the field strength would now be [(1000 V/0.1 cm) + (1000 V/0.001 cm)], or 1,010,000 V/cm. Thus, with pointed needles, very high field strengths can be attained. This is why lightning rods on the roofs of buildings are pointed. The grounded lightning rods build up the field strength between themselves and the clouds, to better attract the lightning.

Between a needle and a flat electrode the field will be very nonuniform, as indi-

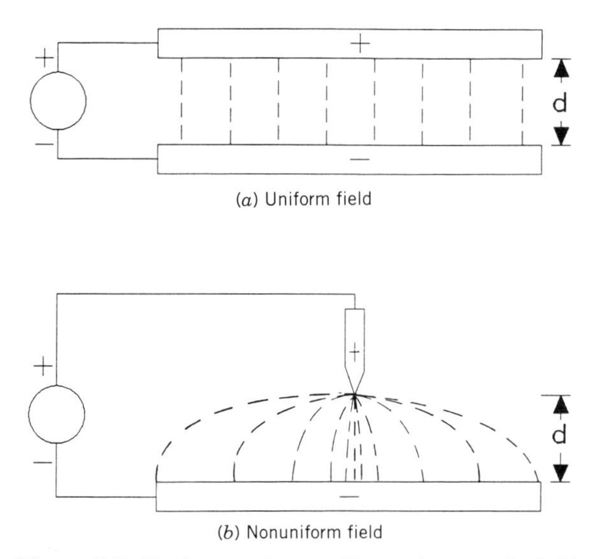

Figure 8-2. Uniform and nonuniform electrostatic fields.

cated in Figure 8-2*b*. However, if instead of one needle thousands of needles—or better yet, very fine wire brushes—are used, the field will be extremely uniform with a magnitude equal to that directly under the tip of a single needle.

8.3 SURFACE RESISTIVITY

Free charges on the surface will readily dissipate or leak away if the surface is sufficiently conductive. The resistance of a surface layer to the flow of electricity is called the *surface resistivity*. Materials that are conductive will conduct electricity on their surface as well as in the bulk. But materials that are nonconductive can still have surfaces that will conduct electricity. Surfaces with a surface resistivity under about 10^6 ohms/square (this term will be defined soon) are considered electrically conductive. Surfaces between 10^6 and 10^{10} Ω/□ are adequately conductive to dissipate static charges rapidly. Surfaces with resistivities above about 10^{13} Ω/□ will not dissipate electrical charges, and the materials can hold static charges for a significant amount of time.

Figure 8-3 shows the complete decay of charge in a period of about a week when the surface resistivity is 10^9 to 10^{11} Ω/□, and Figure 8-4 shows that while with surface resistivities of 10^{13} to 10^{16} Ω/□ there is some decay of charge, a high residual charge remains indefinitely.

How is surface resistivity measured? Let us first review the measurement of bulk

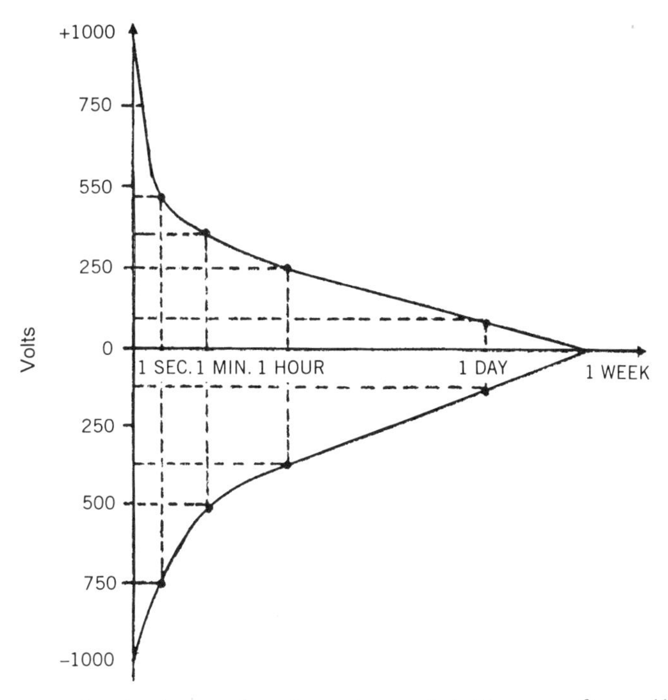

Figure 8-3. Decay of static charge when the surface resistivities are 10^9 to 10^{11} Ω/□. [From Kisler (1994), with permission of the author.]

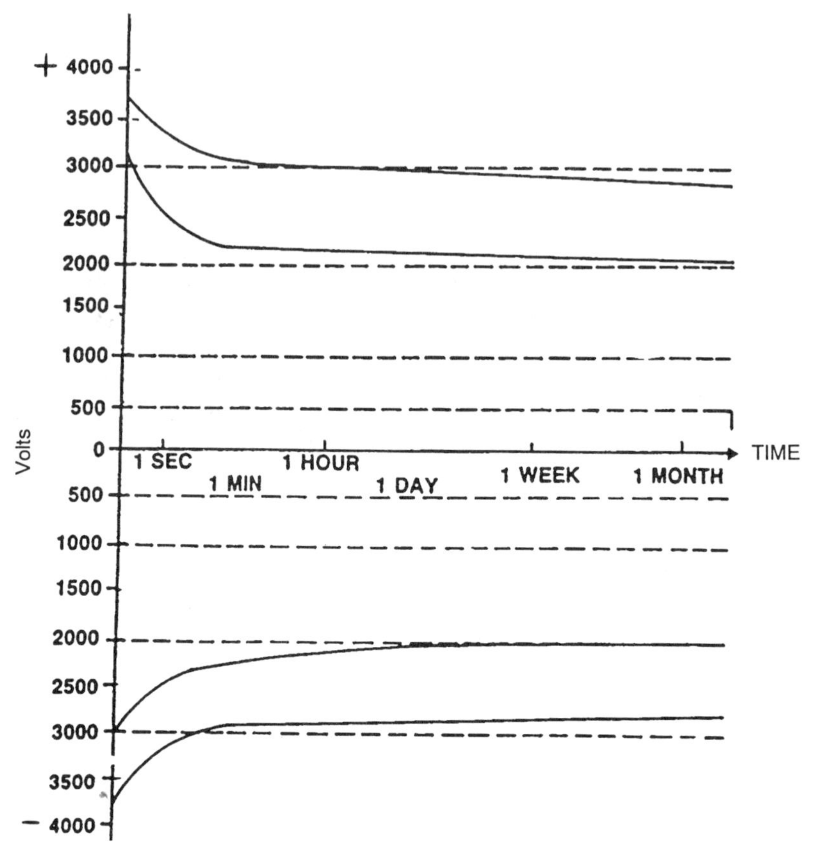

Figure 8-4. Decay of bound static charges when the surface resistivities are 10^{13} to 10^{16} $\Omega/\square$. [From Kisler (1994), with permission of the author.]

or volume resistivity. Volume resistivity is the resistance to the flow of electricity of a 1-cm cube of material. It is calculated by measuring the resistance, R, of material of cross section A between two electrodes d centimeters apart. The bulk or volume resistivity, ρ_v, is then

$$\rho_v = R\frac{A}{d}$$

with units being resistance times length, or $\Omega \cdot \text{cm}$. The volume resistivities of a number of solids are given in Table 8-2.

Surface resistivity could be measured as in Figure 8-5, with R being the electrical resistance between electrodes L centimeters long and d centimeters apart. The surface resistivity is then

$$\rho_s = R\frac{L}{d}$$

Table 8-2 Volume Resistivities (Ω · cm)

Conductors	
Silver	1.59×10^{-6}
Copper	1.72×10^{-6}
Gold	2.44×10^{-6}
Aluminum	2.82×10^{-6}
Steel	1.2×10^{-5}
Mercury	1.0×10^{-4}
Carbon	3.5×10^{-3}
Semiconductors	
Germanium	4.6×10^{1}
Silicon	2.3×10^{5}
Dielectrics	
Dry wood	3×10^{6}
Polymethyl methacrylate	1×10^{13}
Glass	2×10^{13}
Polyester	1×10^{14}
Nylon	4×10^{14}
Teflon	$> 10^{15}$
Polystyrene	$> 10^{17}$

Source: Kisler (1994), with permission of the author.

and the units would simply be ohms, the same as the units of resistance. However, when L and d are equal so that the resistance is measured over a square area, it does not matter what the size of the square is, and the units of surface resistivity have come to be called ohms/square. While the term *square* has no dimensions and could just be dropped, its use does avoid confusion with the units of resistance. Thus the

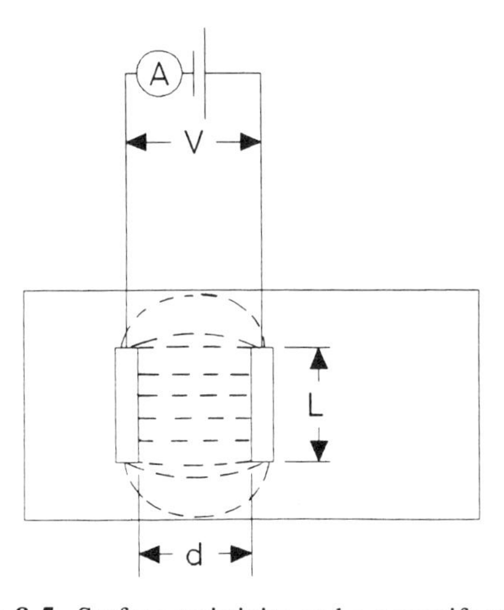

Figure 8-5. Surface resistivity and a nonuniform field.

denominator *square* has no units, and serves only to distinguish surface resistivity from resistance.

In Figure 8-5 the electric field near the edges of the electrodes is very nonuniform, as indicated in the sketch. A concentric cylindrical probe, as shown in Figure 8-6, avoids this problem completely.

As mentioned previously, the surface resistivity decreases greatly with increasing humidity. At higher humidities more water is adsorbed by the surface, greatly increasing the conductivity. As the relative humidity increases, the surface resistivity of many materials will drop greatly, with the magnitude of this decrease depending on the material.

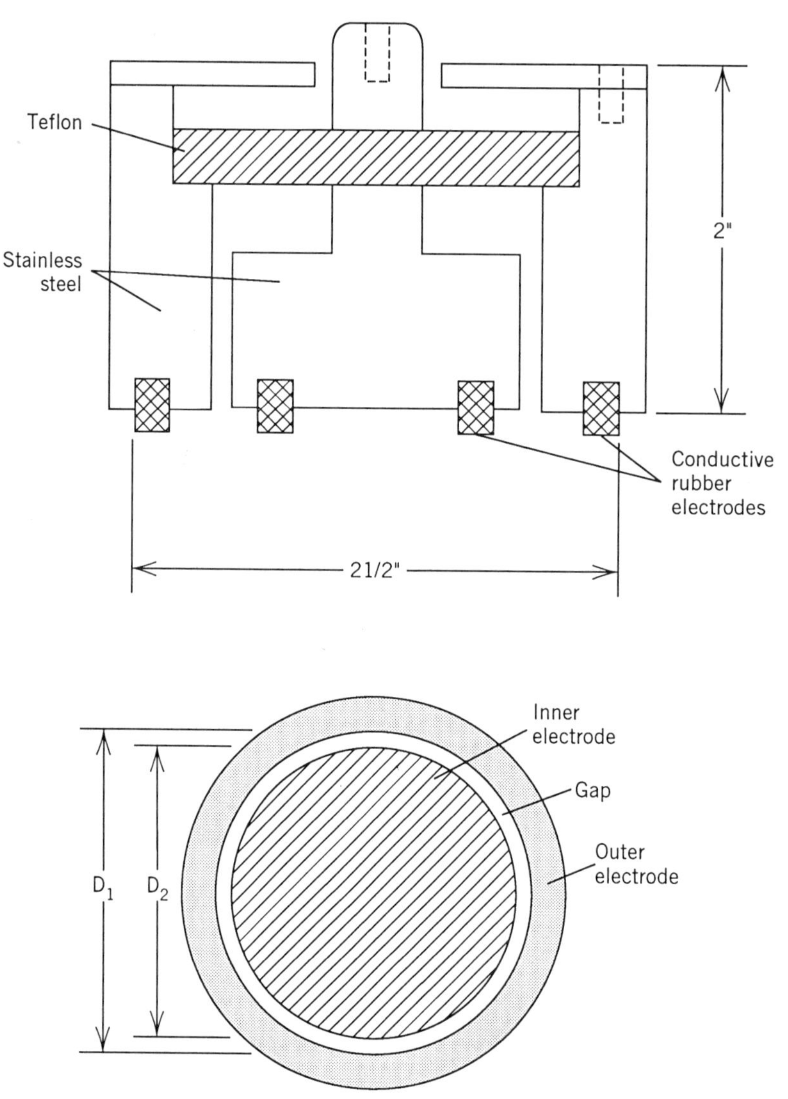

Figure 8-6. Cylindrical probe for surface resistivities. [From Kisler (1994), with permission of the author.]

8.4 MEASUREMENT OF STATIC CHARGES

An electrostatic voltmeter is used to measure electrostatic charges. It measures the total charge and does not distinguish between free and bound charges. One of the better meters (and one of the more expensive ones) recommended by Kisler (1993) is the Isoprobe Electrostatic Voltmeter, Model 244, made by Monroe. The probe of the meter does not touch the surface but is held 1 to 3 mm ($\frac{1}{32}$ to $\frac{1}{8}$ in.) above the surface of the object mounted on a grounded metal plate (Figure 8-7). For on-line measurements on the coating machine the probe is held above the web, traveling over a grounded roll. The outside metal of the square probe is always at a high voltage, so personnel should be warned that the probe should not be touched. Also, the probe can easily be damaged and is expensive—it costs many hundreds of dollars—so should be handled with great care. On plastic films the electrostatic charges can vary greatly within short distances. On films 75 μm (3 mils) with surface resistivities of $10^{14}\ \Omega/\square$ or more, charges can vary from +5000 to −5000 volts within a space of microns.

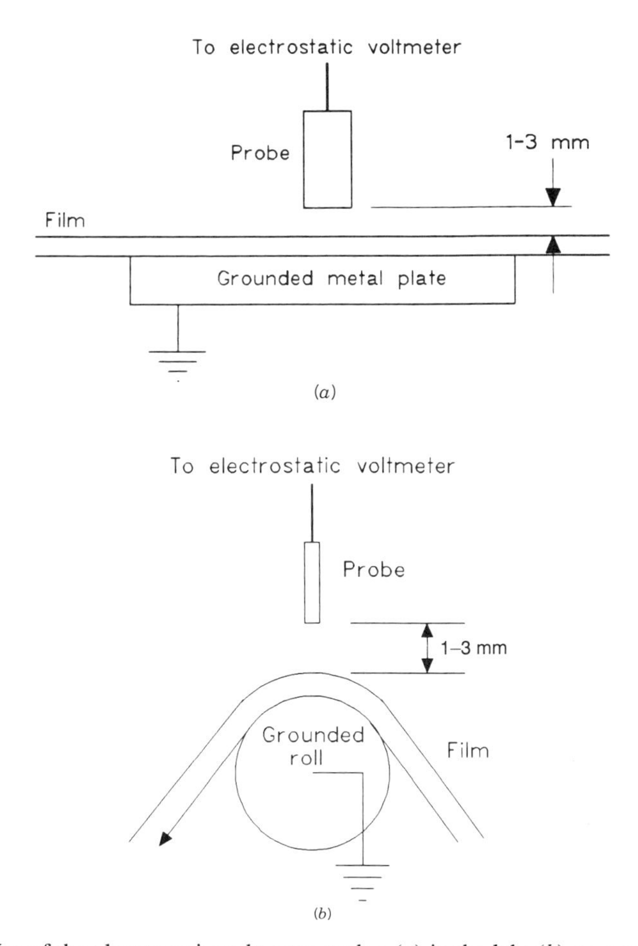

Figure 8-7. Use of the electrostatic voltmeter probe: (*a*) in the lab; (*b*) on a coating machine.

Figure 8-8. Toner test for charges on a web. [From Kisler (1994), with permission of the author.]

The electrostatic charge can be read from the electrostatic voltmeter, or it can be recorded on a strip chart, or it can be logged into computer memory. Most strip-chart recorders have too slow a response for accurate readings at line speeds over about 0.5 m/s (100 ft/min).

When an electrostatic voltmeter is not available one can get a qualitative picture of the charge and its distribution on a film by sprinkling a dry xerographic toner on a sample of the film and then shaking off the excess. Figure 8-8 shows the results of this test with a polyester film base that gave coating problems. These charges will not decay appreciably with time, as shown Figure 8-4.

8.5 REMOVAL OF CHARGES

Free charges are relatively easy to remove; it is only necessary to provide a conductive path to ground. Bound charges are difficult to remove; to do so it is necessary to disorient the dipoles. And as charges can build up in the moving web, the charge removal should take place as close as possible to the region where the static charges cause problems. This is often the coating stand.

One should try to remove all charges and one should also reduce the buildup of new charges in the system. All rolls should be aligned as well as possible, since friction between the web and a poorly aligned roll will generate charges on both the web and the roll. Metal rolls should be grounded so that these charges cannot build up; an ungrounded metal roll can act as a van de Graaff generator and build up very high charges, which will than arc to ground. To ground a metal roll through its metal shaft requires more than just grounding the shaft. The bearing lubricant should be conductive to ensure a good ground connection. Conductive greases are available.

The coverings on nonmetallic rolls should be chosen such that the buildup of charge by friction is minimal. This can be determined by actual test, or one can rely on published triboelectric series, where materials close together in the series develop the least charge. Unfortunately, the series order is not always reproducible, and cleanliness and humidity are very important for the order. A typical series is given in Table 8-3. Unfortunately, it is incomplete; it does not list poly(ethylene terphthalate), a polyester, also known as Mylar, and it does not list polypropylene.

8.5.1 Free Charges

To bleed away surface charges, a conductive path to ground can be provided either by grounding the surface, by making the surface conductive and grounding it, or by making the air conductive.

A grounded roller will bleed away surface charges from the surface it contacts. But it will not discharge the opposite surface. The second surface must also pass over a grounded roll.

As mentioned earlier, pointed electrodes are more effective than flat electrodes in creating strong electrostatic fields. Thus grounded tinsel has long been used to discharge a surface. The tinsel does not actually have to touch the surface, as long as it is close to it. The sharp tips of the grounded tinsel cause the electrostatic field induced by the charge on the web to be high, which increases the likelihood of ionizing the air and allowing the charge to bleed away across the small air gap. When placed on both sides of a polyester web it can also reduce the bound charges to under 1000 V. This, of course, is still a high charge. However, there are many gaps between the tips of the tinsel, and with a nonconducting surface there will be many areas that are not discharged. Tinsel has been largely replaced by fine metal brushes. The many rows of bristles ensure full coverage of the surface. The metal bristles should be no more than 10 μm in diameter to develop a high field strength when they do not contact the surface. The brushes can be used in contact with the surface or can be a short distance away. The fine bristles are so flexible that they will not scratch the surface. It is best for the brushes to contact the surface, unless there are other reasons, such as actual charge buildup by friction or possible surface contamination, for having a separation. Conductive cloth contacting the surface has also been used, but is less effective.

The surface of the web can be made more conductive by increasing the humidity of the air in the coating alley. Conditioning the air tends to be expensive and is

Table 8-3 Triboelectric Series

Rabbit's fur	Cotton
Lucite	Wood
Bakelite	Amber
Cellulose acetate	Resins
Glass	Metals
Quartz	Polystyrene
Mica	Polyethylene
Wool	Teflon
Cat's fur	Cellulose nitrate
Silk	

Source: Hendricks (1973), p. 67.

usually not done unless it is required for reasons other than static control. Surfaces can also be made conductive with special antistatic treatments, such as by a coating of quaternary ammonium salts or of cuprous chloride. These surface treatments may be unacceptable because they may alter the final properties of the coated web.

Ionizing radiation ionizes the air, making it conductive and allowing it to discharge the surface. Ionizing radiation can come from americium or from polonium bars, or, less often, from x-rays, beta rays, gamma rays, or ultraviolet rays. Americium-241 emits hazardous gamma rays as well as the desirable alpha particles, and suitable shielding must be provided. Polonium-210 gives off only soft alpha particles and is not hazardous. It does have a short half-life of 138 days, and therefore must be replaced at least once a year. Most radioactive bars have been banned from use because of their hazards. They also are not as effective as one might want, because their effect is often inadequate for high-speed coating operations.

Corona discharges will ionize the air and can also charge the surface; a corona discharge in a separate apparatus can ionize the air, which is then blown against the surface to discharge it. Most corona devices use alternating current. It takes about 3500 V to obtain a corona. Most of the power supplies operate at 4000 to 8000 V. Thus, as indicated in Figure 8-9, the alternating current generates a corona in only that part of the cycle when the voltage is above ±3500 V. Direct-current corona generators, on the other hand, are effective at all times. Two dc corona units, one giving a positive corona followed by one giving a negative corona (or vice versa), will be more effective than an ac unit giving both positive and negative charges. Such dc units are now becoming available and are recommended.

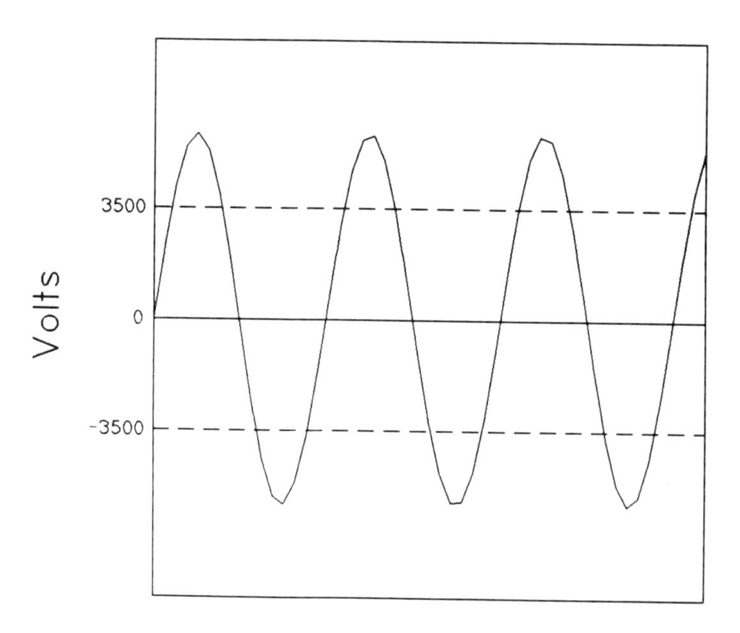

Figure 8-9. The ac corona unit is ineffective during that portion of the cycle when the voltage does not exceed the threshold needed for a corona.

The location of static eliminators is very important, for when the charged web comes close to or contacts a roll the electrostatic field strength becomes much lower, and the web will not attract the ionized air molecules that will neutralize the surface charge. For effective charge neutralization, Keers (1984) recommends that the eliminator bar be at least 15 cm away from any roller or guide and should be 2.5 cm above or below the web. Usually, it is necessary to have the bar on only one side of the web.

To remove static buildup from personnel, conductive floor mats or ionized air blowers are recommended. If the operators wear heavy insulated soles, a conductive shoe strap, running from the inside of the shoe to the bottom of the outer heel, should be provided for the conductive mat to be effective. The strap does not have to contact the skin; it will ground the person through the stocking. The ionized air blower would normally be used for discharging static on operators only if it is used for other purposes as well.

8.5.2 Bound Charges

To discharge the bound charges of a web by disorienting the dipoles is much more difficult, and no commercial equipment is available to do this. First, one has to charge the web to a uniform charge in an electrostatic field; that is, to charge the web to a charge higher than any existing in the web. Then the web is discharged by an opposite field (Gibbons et al., 1972), with the field strength set to bring the final charge to approximately zero, as indicated by an electrostatic voltmeter. Very fine stainless steel brushes held about 3 mm above the surface of the web on a grounded metal roller are suggested for charging and discharging. The power supply would be adjusted to give a uniform charge on the web. In many cases the voltage from the power supply should be in the range ±1000 to 2000 V. The discharging voltage, with opposite polarity, would be sent to a similar set of brushes but downstream of the first set. The voltage would be close to that of the first set or slightly higher. The setup could be similar to Figure 8-10 (Kisler, 1985). Here the two sets of brushes,

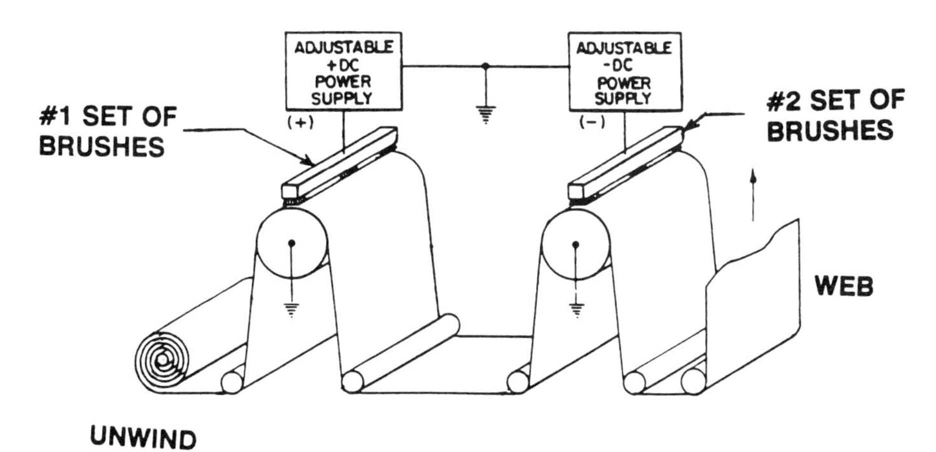

Figure 8-10. Discharging the free and bound charges on a web by first charging the web to a uniform high potential and then discharging it. [From Kisler (1985).]

one positively charged and one negatively charged, are against different grounded rolls. However, Kisler (1994) also shows a system where both sets of brushes are against the same grounded roll.

Voltages of several thousand volts can be extremely hazardous. The current needed to charge or to discharge a web is minuscule. The power supply should have a safety feature that does not allow a current flow above about 3 mA. This will not injure a person, but is it is still enough to give a good shock and make a person jump. Extreme caution should be taken to avoid contact with any object at a high voltage, including the surfaces of the electrostatic voltmeter probe.

REFERENCES

Gibbons, C. B., W. C. Kerr, and R. H. Maddocks, "Web treatment method," U.S. Patent 3,702,258, Nov. 7, 1972.

Hendricks, C. D., "Charging macroscopic particles," pp. 57–85 in *Electrostatics and its Applications*, A. D. Moore, ed., Wiley-Interscience, New York, 1973.

Keers, J. J., "Static electricity," in *Web Processing and Converting Technology and Equipment*, D. Satas, ed., Van Nostrand Reinhold, New York, 1984.

Kisler, S., "Method and apparatus for uniformly charging a moving web," U.S. Patent 4,517,143, May 14, 1985.

Kisler, S., personal communication, April 16, 1993.

Kisler, S., "Electrostatics," in *Technology of Thin Films Coatings*, AIChE Course Notes, AIChE, New York, 1994.

Miller, F. D., and J. J. Wheeler, "Coating high viscosity liquids," U.S. Patent 3,206,323, Sept. 14, 1965.

Nadeau, G. F., "Method of coating a liquid photographic emulsion on the surface of a support," U.S. Patent 2,952,559, Sept. 30, 1960.

CHAPTER IX

Problems Associated with Drying

In the drying operation energy is transferred into the wet coating, volatilizing and removing the solvents that have been added to solubilize and carry the coating formulation. In addition, chemical reactions such as cross-linking of the binder are initiated in some products. The solvents must be removed without adversely affecting the performance properties of the coating and without introducing defects into the coating. The drying step is important in defect formation because it is the last step in the process where the chemistry and the physical properties of the product can be affected. The only steps after drying are converting the product to the final size and shape needed by the customer and packaging it so that it is protected from damage before use. If the final coated product is not usable, most of the raw materials may not be reclaimable. The result will be an additional cost penalty for disposal besides the cost of the materials and of the operations. Some of the materials, such as polymers, precious metals such as silver, magnetic oxides, coating support, and solvents, may be recycled, but even so, this adds cost.

The most common way of introducing the heat needed for volatilizing the solvent is by blowing warm or hot air on the coating. The air also functions to remove the solvent from the coating surface. Convection is commonly used because it is the only mode of heating that increases the heat transfer coefficient and the mass transfer coefficient at the same time. Because of the high heat transfer coefficient of air impingement, the temperature of the air need not be too high and overheating the coating is avoidable. Hot-air convection dryers have been in use for a long time and there are a wide variety of nozzles, jets, and configurations in use. The two general types are single-sided dryers (Figure 9-1) and two-sided dryers (Figure 9-2). In single-sided dryers the air impinges on the coated side of the web from nozzles, jets, slots, or orifices, and rolls are used to support the web in the dryer. In two-sided dryers, the air impinges on both sides of the web and serves to float the web and aids its transport through the dryer. Two-sided drying increases the heat transfer rate compared with single-sided drying and results in a higher web temperature in the dryer.

Important components of the dryer include the control systems, process measurement instruments, the web transport system, the heating and ventilating hardware to move the air and remove the solvent, and the environmental control systems. All of these can contribute to the formation of defects in the coating.

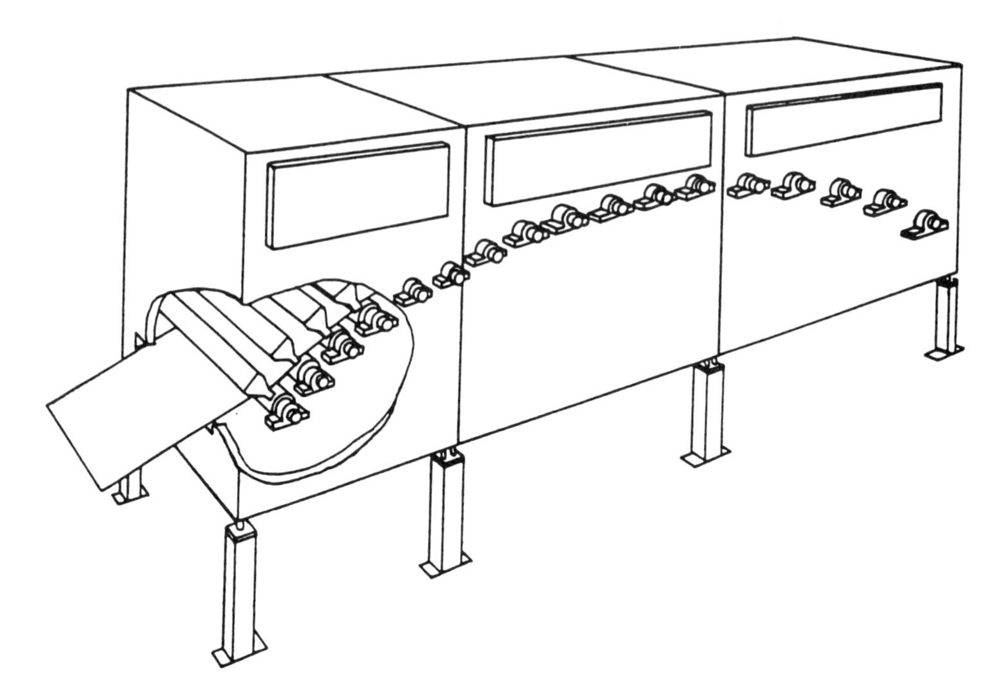

Figure 9-1. Single-sided arch dryer showing the air nozzles.

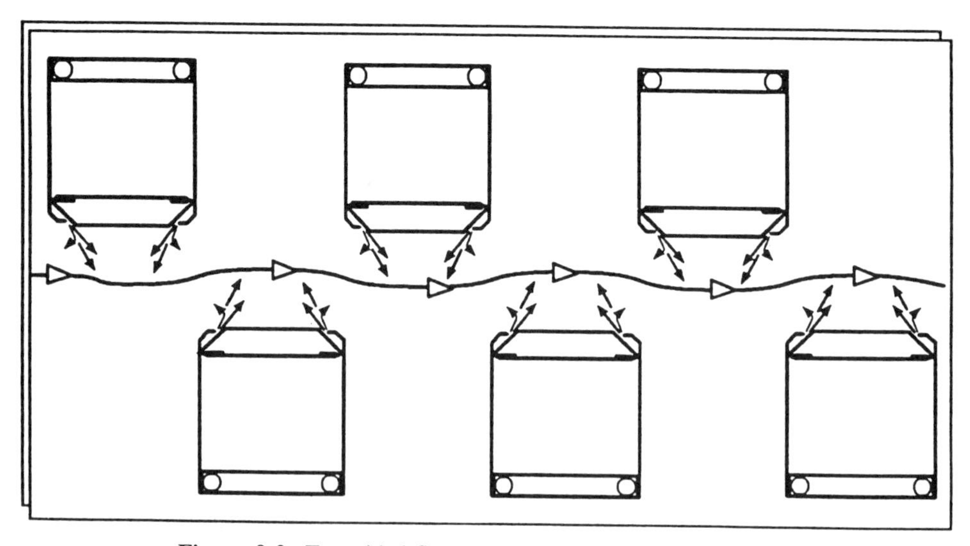

Figure 9-2. Two-sided floater dryer showing the air nozzles.

9.1 DRYER CONTROL AND SOLVENT REMOVAL

The primary function of the dryer is to remove the solvent from the coating by the end of the dryer without damaging the product. It is unacceptable for the coating to leave the dryer in a wet state. The consequences of a coating leaving the dryer wet are far more severe than overdrying. If the coating is wet, it will rub off on any face or wet-side rollers between the dryer and the unwind operation. This will lead to damage of subsequent product when the wet material on the roll transfers back to it. Winding a wet product often causes the wound roll to be fused together, making unwinding for further processing and packaging impossible. Such product will be scrapped, resulting in a cost penalty. When rolls become contaminated from wet coating the coater should be shut down, the rolls cleaned as quickly as possible, and corrective action taken to increase the drying rate or reduce the wet load.

Too high a drying rate leads to the possibility of overdrying the film by removing the solvent too early and having the final coating exposed to high drying temperatures for too long a time. This high-temperature exposure can lead to deterioration of the physical properties such as adhesion and coating strength, and to physical damage to the coating by remelting and flow of the coating.

The preferred situation is to control the drying rate precisely so that the residual solvent in the coating at the end of the dryer is at the desired level. Often this is done by controlling the location of the end of the constant-rate period to be at the same location in the dryer for all products. The end of the constant rate period is where the temperature of the coating starts to rise rapidly to approach the temperature of the drying air. In aqueous systems most of the drying occurs before this point, and at this location the coating appears dry. In solvent coatings the end of the constant-rate period occurs early in the dryer, and therefore its detection is not as useful for control purposes. In that case the solvent concentration can be measured at the end of the dryer and the drying controlled to give the desired residual solvent. Another approach is to control the temperature to an arbitrary value that is known to give the desired degree of dryness at the end of the coater.

The location of the end of the constant-rate period should be the same all the way across the coating. It should not fluctuate. However, variability in the coating weight profile in both the machine and transverse direction, as well as in the air velocities across the nozzles or slots, will cause variations in the location of the end of the constant-rate period both across the web and in the machine direction.

There will always be some level of residual solvent in the dry coating, as the solvent can never be removed completely. When the solvent is water, the coating will come to equilibrium with the moisture in the air. On some aqueous coaters air at the temperature and humidity of the storage areas is blown onto the coating in the last dryer zone, so that the residual moisture will not change in storage. Some binders will retain more solvent than others.

To provide the necessary information to troubleshoot dryer control problems, an explanation is needed of some of the important features of the system that provides drying air to the web. There is no single dryer system used for all applications, but the key elements, concepts, and definitions described here are applicable for many systems. For a more detailed explanation, see Cohen (1992) and Weiss (1977). The key elements are:

1. The air impingement nozzles to deliver the air over the coating.
2. The fan and motor assembly to move the air throughout the dryer.
3. Ducts to control flow of the air, carry it to nozzles, and return it to the conditioning system.
4. Coils to heat and or cool the air.
5. Filters to remove contamination from the air and provide the purity level needed.
6. A system to remove solvent from the air.
7. Environmental control devices.
8. Sensors to measure air temperature, air velocity, and solvent levels at several points in the dryer.
9. Dampers to control the flow of the air to the different dryer zones.
10. Sensors to measure web temperature and coating solvent level in the dryers.
11. A system to detect the location of the end of the constant-rate period.

A typical system that contains many of these elements is shown in Figure 9-3.

There are several methods of detecting the location of the end of the constant-rate period, that is, where the web temperature starts to rise rapidly. Infrared thermometers, thermocouples in the rolls contacting the film, and thermistors mounted close to the web will all work. Thermistors mounted close to the web do not read the web temperature but do read the air temperature, but in the constant-rate period this will

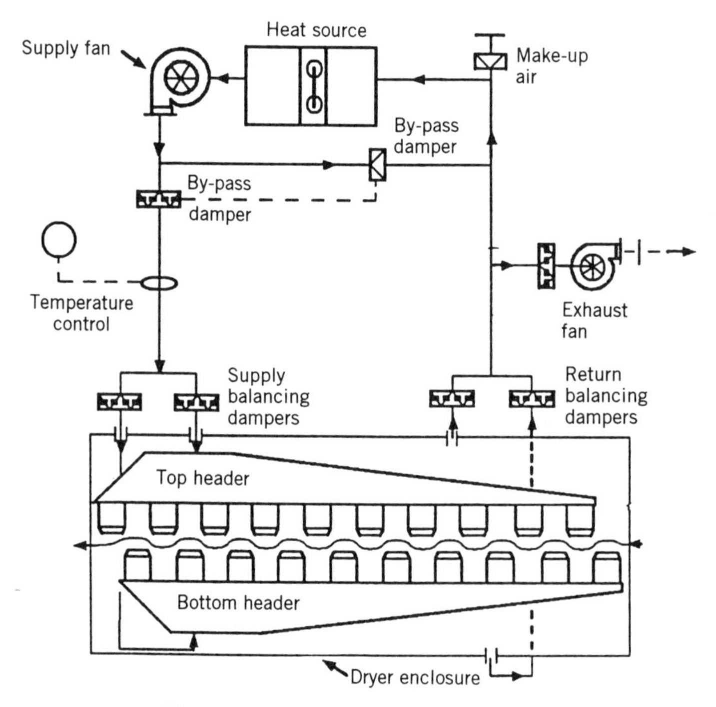

Figure 9-3. Dryer air handling system.

be very close to the web temperature, and in the falling-rate period will be reasonably close. It is a technique that has been used satisfactorily. In one case measurement of the return air temperatures gave good results. With all of these systems the drying conditions are set to give a specific location for the end of the constant-rate period. All these systems can be automated and can be controlled by a computer. The data can be continuously recorded. The solvent content of the coating can be monitored with an appropriate detection method, such as microwave absorption for water, and the coater controlled to give a specific solvent concentration at a specific dryer location.

In some cases one can also use the "hand" method, in which the back side of the coating is touched to determine where the temperature starts to rise rapidly, or, on the coating side, whether the coating is wet or dry. Obviously, safety is of great concern here, because a hand can easily get caught between the support and a coating roll or at other pinch points. The eye is also a good detector, since there are often different surface reflectivities on a wet and dry coating.

The overall dryer capacity is expressed as kilograms or pounds of solvent per hour that can be removed. This in turn is a function of the volume of air being delivered to the surface of the film, its temperature and solvent level, the effectiveness of the nozzle configuration in use, and the amount of any additional heat delivered to the coating, such as by infrared heating. For this discussion all the heat is assumed to come from the drying air. Then the rate of heat transfer to the coating is

$$q = hA(T_a - T) \quad (1)$$

where

A = area of a dryer section, width times length, m^2 or (ft^2)
h = mean heat transfer coefficient, $W/m^2 \cdot K$ or ($Btu/hr\text{-}ft^2\text{-}°F$)
q = rate of heat transfer to the coating, W or (Btu/hr)
T_a = temperature of air at the nozzle exit, °C or (°F)
T = temperature of coating, °C or (°F)

This equation holds for a section of the dryer where the conditions are constant. Using suitable mean values, it can be used where temperatures are changing, as over a dryer zone. For the full dryer, the total rate of heat transfer has to be summed over all dryer zones or sections.

The heat input by the drying air goes to heat the web and the coating, and to evaporate the solvent. As it usually takes relatively little heat to raise the temperature of the web and the coating, since both the web and the coating are usually very thin, one can estimate the dryer capacity by dividing the rate of heat input, found above, by the latent heat of vaporization of the solvent. The latent heat of vaporization is the amount of heat required to evaporate 1 kilogram or pound of solvent, in units of joules per kilogram or Btu per pound. The result is the weight of solvent that can be evaporated per unit time. This value is used to characterize the performance of a particular dryer since it can easily be compared to the drying load generated by the coater. It is also useful in determining the needs of the dryer solvent removal system.

The actual drying load is the weight of solvent coated per unit time minus the residual solvent in the dry coating. Assuming that the residual solvent is negligible, the drying load is the weight of solvent coated per unit area, times the width of the coating, times the coating speed.

The heat of vaporization of the solvent is very important. A solvent such as water with a high heat of vaporization of about 1060 Btu/lb requires six times as much heat to evaporate 1 pound as does methylene chloride with a low heat of vaporization of 170 Btu/lb. Organic solvents tend to have low heats of vaporization. For the same dryer conditions, aqueous coatings will require either a longer dryer, or operation at a lower coating speed, or higher air temperatures to remove the same amount of solvent. These relations are semiquantitative only. More precise methods of calculating drying are needed for precise control and optimization of a dryer. However, they can be used as a basis to define the important variables and concepts for dryer control in a troubleshooting process.

The dryer capacity should equal or exceed the required drying load. If the heat input is too low, the coating will be wet and a correction must be made. If the coating suddenly is no longer dry as it leaves the dryer, this can be due to a reduction of the heat input into the dryer or to an increase in the solvent loading. The troubleshooting approach is to identify the specific coater parameters that can influence each of these and determine what has changed.

Factors that reduce the heat input are:

1. Reduction in the temperature of the impinging air.
2. Reduction in the velocity of the air impinging on the film.
3. Increase in solvent concentration of the impinging air. This will increase the coating temperature in the constant-rate period, thus reducing the drying rate.
4. Incorrect settings of temperatures and air velocities.
5. Blockage of the return air, reducing the effective air velocity at the film surface.

Factors that can increase the solvent loading are:

1. An increase in coating weight.
2. A reduction in solids content of coating solution, giving an increase in solvent concentration.
3. A change in transverse direction coating weight profile, resulting in lanes of increased coating weight.
4. Defects, such as streaks or spots, which will cause localized areas of heavy coating weight.

The heat transfer coefficient is a function of the dryer design and the air velocity. The component of the heat transfer coefficient that is inherent to the design should not change; however, the heat transfer coefficient increases with air velocity.

To troubleshoot the reduction in heat input, the temperatures should first be checked to ensure that the instrument set points are correct and are the same as those listed in the standard operating procedures for the product being coated. The dryer air temperature records should also be checked to ensure that the desired set points are being obtained. The air temperature thermometers, air velocity probes (or plenum pressure gauges, if pressure is used for control), and air solvent level (such as humidity) instruments should be calibrated to ensure that the readings displayed are accurate and that the instruments are, in fact, operational. A good rule of troubleshooting is to have a questioning attitude in regard to the accuracy of all readings, particularly when there is a problem. The velocities of the air delivery nozzles, jets, slots, and so on, should be checked to ensure that they are at standard conditions and delivering the correct amount of air to the web surface. A good method of monitoring the consistency of air delivery to the surface is to identify key nozzles in each zone and to measure them periodically, say once a month. These data are then used to establish a baseline and normal variability so that deviations can be detected, if they should occur. These data can be maintained in a spreadsheet with the average and standard deviation being calculated continuously and made available when needed.

Possible sources of reduced airflow are leaks in the plenums and air handling equipment, open doors in the heating and ventilating units, contaminated filters that will cause a high resistance to flow, dampers coming loose and blocking flow, fan blades coming loose on the shaft of the motor, bearings seizing in the motor, and even obstacles in the ducts. When checking dampers, the position of the handle may not indicate the position of the damper. There have been many cases where a disconnected handle incorrectly indicated an open damper. One should look inside the duct to check the actual damper, if at all possible. The area of the dryer should not change; however, reduction of airflow to a zone or to a series of nozzles can have the same effect. One should check to ensure that air is flowing to the web surface.

A particularly useful calculation is to check the actual drying load for each zone using a mass balance calculation. For this, measurements of air volume, temperatures, and solvent content on both the inlet and return air are needed. The solvent loadings in the coating at the start and end of a zone could also be used, but these data are usually much more difficult to obtain. The actual drying load can be calculated easily as shown in Table 9-1. These types of calculations can also be used to verify that all

Table 9-1 Drying Load Mass Balance Calculation

Process Data

Aqueous coater

Air conditions to the dryer

Dry bulb	105°F
Dew point	30°F
Humidity	0.00346 lb water/lb dry air
Airflow	15,000 ft^3/min

Return air conditions from dryer

Dry bulb	85°F
Dew point	50°F
Humidity	0.00766 lb water/lb dry air

Calculations

$$\text{Mass flow of air} = \left(15{,}000\frac{\text{ft}^3}{\text{min}}\right)\left(60\frac{\text{min}}{\text{h}}\right)\left(0.075\frac{\text{lb dry air}}{\text{ft}^3}\right) = 67{,}500\frac{\text{lb dry air}}{\text{h}}$$

$$\text{Moisture pickup in dryer} = 0.00766\frac{\text{lb water}}{\text{lb dry air}} - 0.00346\frac{\text{lb water}}{\text{lb dry air}} = 0.00420\frac{\text{lb water}}{\text{lb dry air}}$$

$$\text{Total water removal} = \left(67{,}500\frac{\text{lb dry air}}{\text{h}}\right)\left(0.00420\frac{\text{lb water}}{\text{lb dry air}}\right) = 284\frac{\text{lb water}}{\text{h}}$$

the components of the air-handling system are functioning correctly. The coils used to heat or cool the air can be fouled from deposits from the heat transfer fluid inside these coils. This is similar to what occurs in a car radiator. As this fouling builds up, the efficiency decreases for heating or cooling. A quality control chart on the return part of the system will permit diagnosis of these effects.

9.2 DRYER CONDITION CASE HISTORY

In this section we describe a dryer control case history. The problem investigated was a high air temperature in a plenum in a drying zone, which led to overdrying of the product. When the problem was brought to the troubleshooter's attention, there was some confusion as to when the problem started, whether the standard procedure limits were realistic, and whether it was possible to obtain the specified conditions. Therefore, the first step was to gather some historical data and analyze it. For this aqueous system the data in Figure 9-4 show clearly that there was a process change at roll 75. The air temperature and dew point both increased significantly.

The next step was to characterize the process and define the reason for the change. The problem seemed to be associated with the air handling system. Therefore, small portable data loggers were placed before and after all coils in the heating and ven-

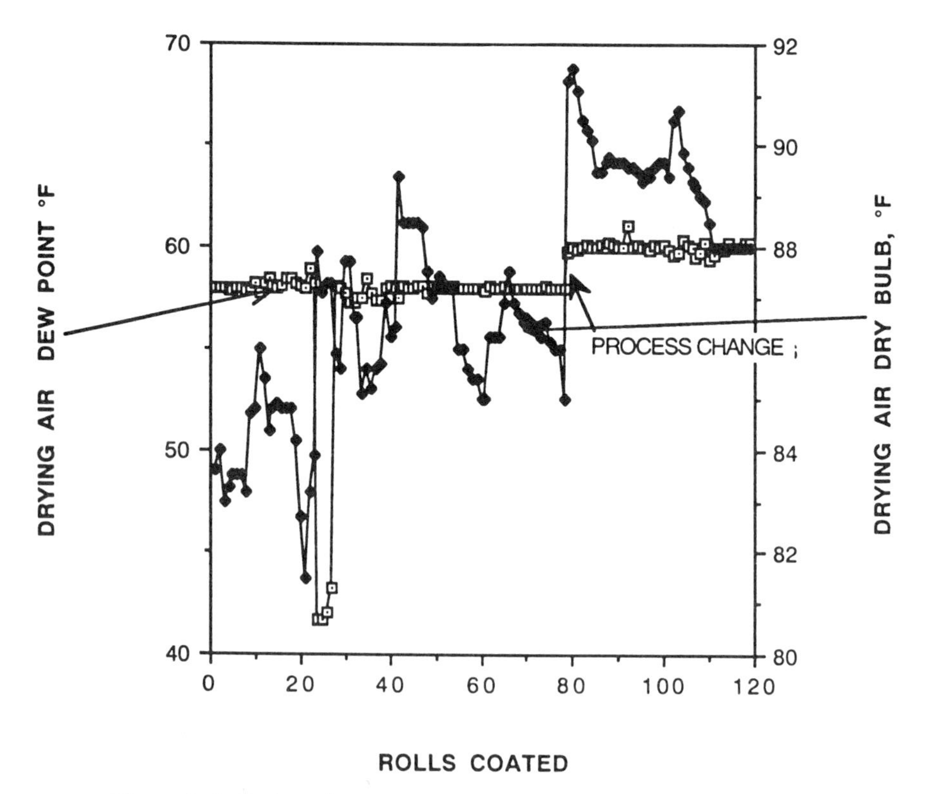

Figure 9-4. Drying air temperature and dew point versus roll number.

tilating system, to characterize their performance over an extended period. As this was an aqueous system, relative humidities and air temperatures can be used for solvent concentrations. Air velocity measurements established that the air velocity at the nozzle exit to the web and the total air volume were in the specified operating range. Figure 9-5 shows the temperatures of the air entering and leaving the cooling coil in the air-handling system. These show that there was a temperature drop of about 1.5°F across this coil, even though they were designed for at least a 5°F drop. This suggested that the reason for the higher temperature was the coil not functioning. Also, a check of the coater logs indicated that there was some routine maintenance performed on the coolant lines at the time of the observed change. It was postulated that this poor performance was the result of a restricted flow of the coolant in the coil. This could have been from scale since the coil had not been cleaned in a long time. It also could have been the result of loose scale in the upstream piping carried into the coil and blocking flow. The coil was cleaned and the unit functioned correctly, as can be seen in Figure 9-6, where the temperature drop in the coils was now in excess of 5°F. This problem can occur on any type of coil and can lead to either over- or underdrying. A nonfunctioning cooling coil will give higher-than-desired temperatures, leading to overdrying, and a nonfunctioning heating coil will give lower-than-desired temperatures, leading to underdrying.

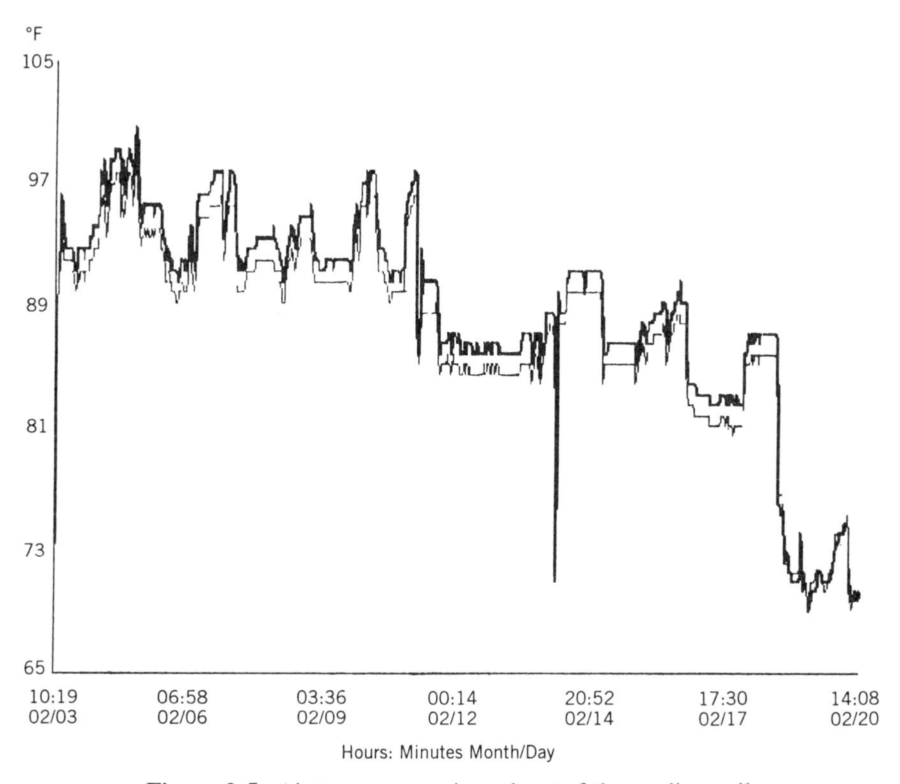

Figure 9-5. Air temperatures in and out of the cooling coil.

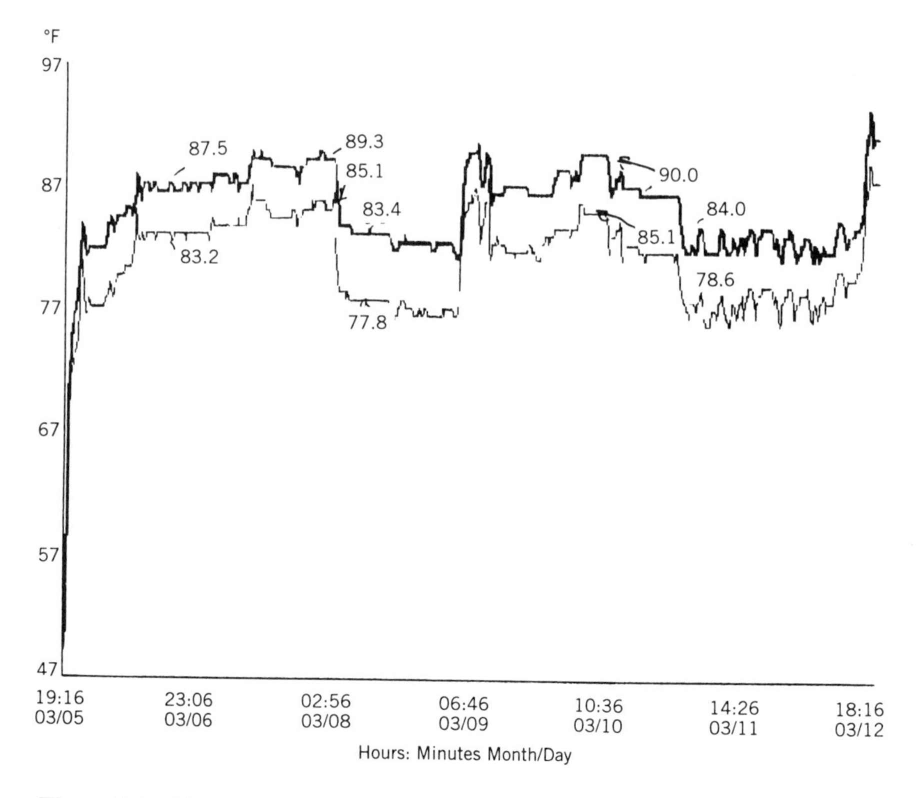

Figure 9-6. Air temperatures in and out of the cooling coil after cleaning the coil.

Problems with the coils can also be diagnosed by monitoring the temperatures and flows of the cooling or heating medium into and out of the coils, as well as the air temperatures and flows. The heat transfer rate and heat transfer coefficient can then be calculated. The measurement and calculation is more complicated with steam than with a liquid medium but it is still possible. If the flow monitoring sensors are not in place, it can be time consuming and expensive to install them. The measurement of the temperature drop or rise of the drying process air across the coil is an easier method to use and is preferred. The data loggers can be installed easily in any location and all of the desired data obtained and the calculations done rapidly.

To troubleshoot dryer problems from solvent loadings, the coating weight of the finished product and the solids concentration of the coating solutions should be measured on all batches and monitored with a quality control chart. Typically, a couple of points at the start and end of the rolls are sufficient for coating weight determination. This will give an indication of the coating weight, which can be a contributing factor. A complete transverse direction profile should be run to isolate its effect. The drying load in every zone should be calculated from the recorded data. An increase in drying load is also an indication that the coating weight is high.

One strategy sometimes used to avoid dryer control problems with a new product is to keep the line speed slightly below the nominal steady-state speed, about 5% less, until the coating process produces the desired coating weight across the sheet and drying is complete before the end of the dryer. The economic incentive for starting at full line speed does not warrant the risks associated with starting too fast and not drying completely. Therefore, the process is started at a conservative speed and then the speed is increased to the desired level as quickly as possible, while checking that the residual solvent in the dry product is within specifications.

For aqueous coating processes the residual moisture level in the product is usually relatively high. It is often detrimental to product quality to dry the coating much beyond the moisture level in equilibrium with normal humidity air. In solvent coating, on the other hand, the level of residual solvent in the final product must be kept below some maximum level for environmental and quality reasons. This is usually a low residual solvent level.

9.3 DRYING DEFECTS

The maximum solvent removal capacity of a typical dryer can be determined using the approach described in Section 9.2. However, in actual practice the dryer may not be operating at this capacity, and drying itself may not be the rate-limiting step or the bottleneck of the coating process. This can occur when the drying process is operating on the verge of causing some defect in the coated film. There are dryers that operate at their maximum solvent removal capacity; but often a close look at the drying limitations shows some unspoken reason for not operating at higher line speeds. Usually, this unspoken constraint corresponds to the onset of a defect.

A variety of factors can lead to this situation. In addition to the heat transfer mechanism, mass transfer and diffusion effects in the drying process are important. Once the energy is transferred into the coating, the solvent has to migrate to the surface of the coating for it to evaporate and transfer into the drying air. If this migration and transfer can occur easily, it will not be the rate-limiting step; the rate of heat transfer will be the controlling process. When the migration and transfer is the limiting step (and the process is mass-transfer limited), the introduction of more energy into the system does not increase the drying rate. This heat energy, instead of being used to evaporate the solvent, heats the wet coating and its temperature rises. This effect occurs in the falling rate period. In aqueous systems the coating is semisolid and appears dry. In solvent systems this can occur when the coating is still quite wet, and can even occur throughout the entire dryer.

The variation in the controlling mechanism is most apparent in the behavior of the web temperature in the dryer and can result in four distinct regions of drying, as shown in Figure 9-7. The predryer is the section between the coating applicator and the dryer entrance. With normal airflow the coating temperature can rise or fall, depending on the volatility of the solvent. Then comes the constant-drying-rate period, where the rate of evaporation is constant and, for single-sided drying, a function only of the air temperature and the solvent level in the air. This is characterized by a constant web temperature. In the falling-rate period the drying rate falls as mass transfer becomes limiting. With constant heat input the temperature rises and the drying rate decreases. This is characterized by the web temperatures rising to approach that of the drying air. The final or equilibration regime is where room-temperature air is used to bring the coating temperatures down to desired levels before winding up the product. (For a more detailed discussion of drying, see Cohen, 1992.)

During drying, stresses can build up in the coating. As the film dries the coating shrinks. It tries to shrink in all directions, but it cannot shrink in the plane of the coating because of adhesion to the base. As a result, the coating is stressed. Tensile stresses remain in the plane of the coating. The magnitude of these stresses will depend on the material, on the amount of shrinkage, on the properties of the binder system, and on the time-temperature relationships. The stress will decrease somewhat with time due to relaxation mechanisms occurring. If not relieved, the residual stresses can lead to a variety of defects.

There is a wide range of drying defects. One of the problems in discussing defects is the naming problem and the use of poetic names, as discussed in Chapter 1. Since there is no agreed-upon naming convention, the most common names that have been encountered in our experience and in the general literature will be used in this discussion. To simplify the discussion, defects will be classified by their typical length scale, going from large to small. Within a given size range they will be grouped by appearance. This scheme is designed to assist in looking for the causes of the defects.

The first step is to separate the defects caused by coating from those caused by drying. There is no rigid or guaranteed formula to do this. One good guideline is that typically the coater problems can easily be seen at the applicator. Chatter, streaks, ribbing, bubbles, spot defects, and heavy edges are all obvious. If it cannot readily be seen at the applicator, consider drying as a possibility. In addition, the problem-solving procedure should point to the proper point in the process. Describing the defect instead of giving it a name will also help. Knowledge of the type of defects that can be caused by drying will help define the source. Designed experiments can isolate the source. As mentioned previously, stopping the web and drying in place can help locate the cause. For this discussion, we define drying defects as those that are caused in the drying regimes described above. This means that the web path from the dryer entrance area through the dryer to the windup are included. The problem-solving process should always begin by finding the characteristic dimension of the defect and proceeding through a systematic investigation of the physical phenomena that can be expected to dominate over that scale.

The approach in the next sections will be to use this classification scheme and discuss some typical defects. The specific defects will be characterized, the mechanism reviewed, and possible methods to eliminate the defects presented. Table 9-2 is a summary of the classification scheme being used and the defects that will be discussed.

9.3.1 Large Defects

Large defects are defined as those large enough to be measured with a ruler and seen easily with the unaided eye.

Curl

Curl is the tendency of a coated film to bend toward either the coated or uncoated side instead of lying flat. It results from residual tensile stresses, normally in the coating as explained above. If in the coating, the curl will be toward the coated side. Curl is resisted by the stiffness of the base. Curl can be seen in both the dryer and the finished product. In a dryer the edges of the film can curl toward the coated side

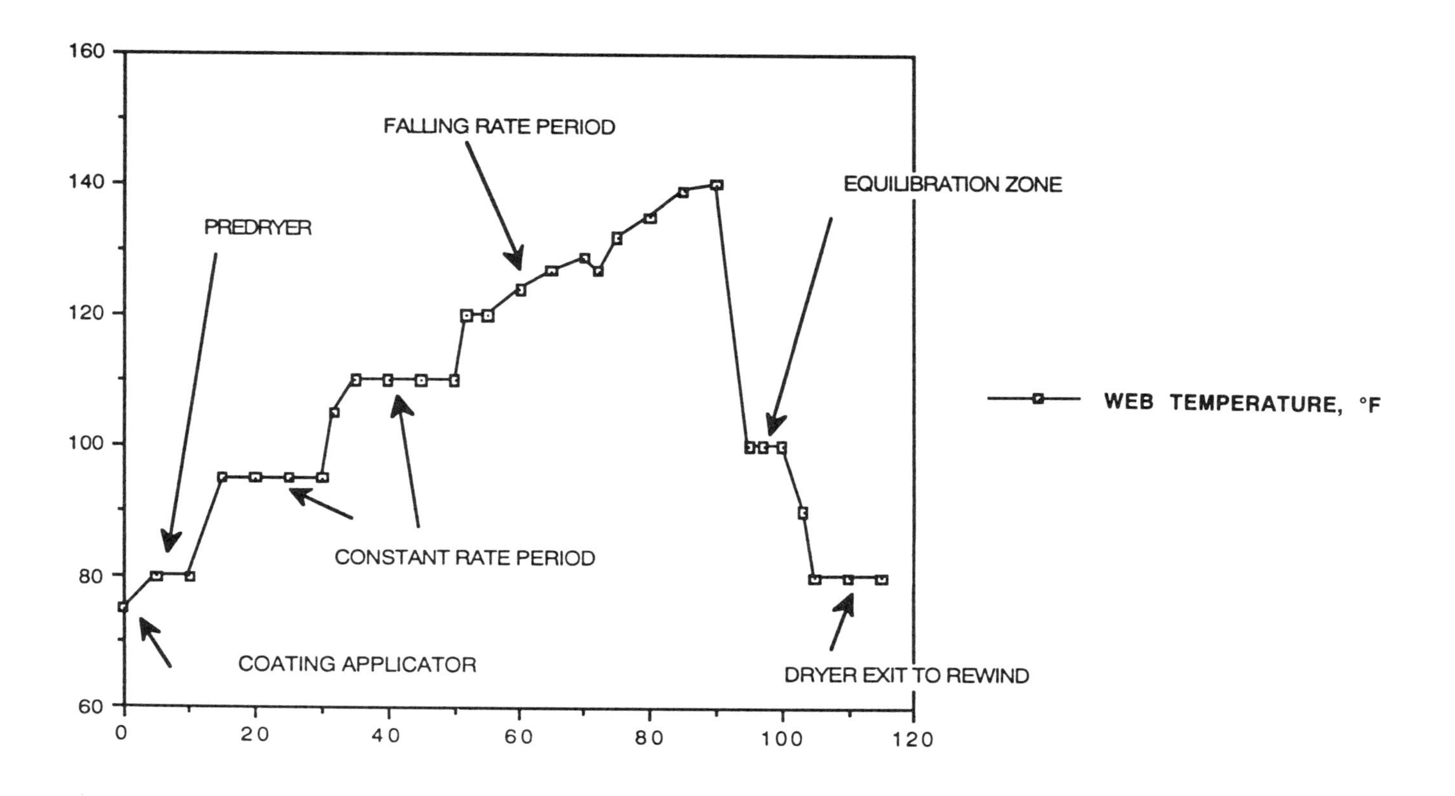

Figure 9-7. Web temperature profile in a dryer.

Table 9-2 Dryer Defects Classification

Large Defects	
Curl	Delamination
Cockle	Wrinkling
Air bar rubs	Dryer bands
Remelt	
Macroscopic Defects	
Spots	Bubbles
Blisters	Mottle
Surface blow-around	Orange peel
Crinkling	Reticulation
Mud cracking	Craters
Blushing	Overspray
Microscopic Defects	
Haze	Starry night

to the extent that it will rub against the air nozzles, leading to film defects. It can also occur after the coating is dried and the film is converted into the final form. Occasionally, the entire sheet of film will "curl up like a cigar," although typically only the edges turn up. There are two components of the curl mechanism that need to be discussed. The first is the mechanism of the generation of tensile stress in the coating, and the second is the effect of the stress on the overall product structure.

The effect of stress can be understood by comparing the curl of the film to that of a bending beam. This effect is used to measure temperature by fabricating a strip out of two metals that expand or shrink differently as the temperature is changed. The differential change in length between the two metals puts one in compression and the other in tension, exerting a bending moment toward the one in tension, which causes the strip to curl. Only if the metals are the same, and thus the contraction or expansion are the same, will there be no bending moments. This analogy can be extended to a coated product structure, which in its simplest form is composed of a support and a coated layer (Umberger, 1957). The tensile stress generated in the coating layer puts the support in compression, and depending on the modulus of elasticity and the thickness of the support, the structure will curl as a response to the force. A stiff support, with a high modulus such as of poly(ethylene terephthalate), will curl less than paper or cellulose acetate. Curl is illustrated in Figure 9-8.

As explained earlier, the residual in-plane tensile stress in the coating is due to the shrinkage of the coating as it dries, while it still covers the same area on the web. The general understanding of curl is due in large measure to the work of Croll (1979). Studying simple systems, Croll reasoned that it is the deformation of the coating after solidification that leads to curl. While still soft and rubbery, that is, below the glass transition temperature, the solid structure allows enough molecular motion to relieve the tensile stresses that tend to form. Only after the coating loses enough solvent to reach the "glassy" state do the residual stresses remain. As the coating dries the glass transition temperature increases until, with many materials, it is well above room temperature in the dried state. Croll neglected the stresses generated before the

a

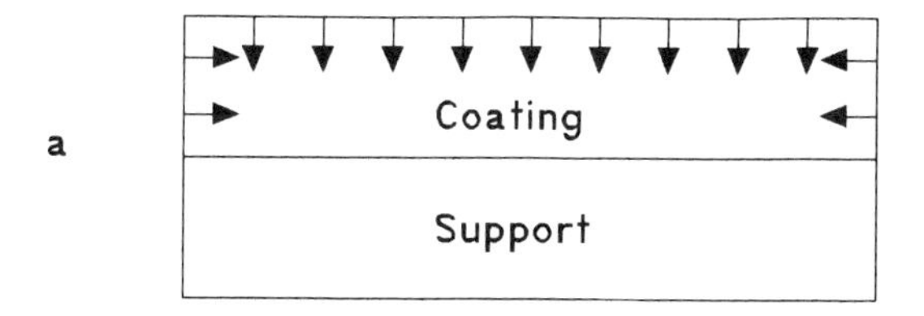

b

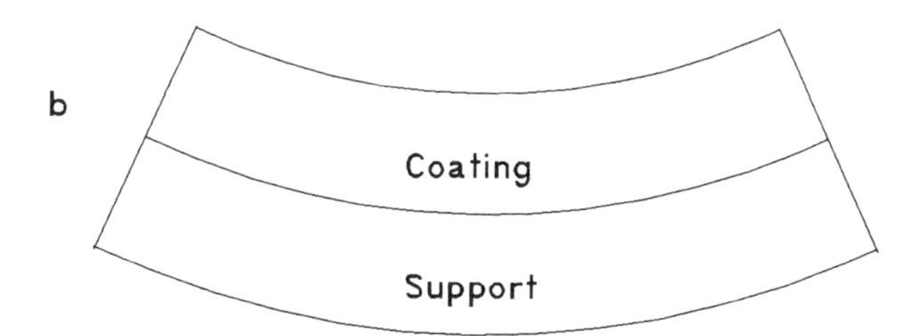

Figure 9-8. Curl illustrated. The coating shrinks during drying (*a*) and goes into tension. The support is therefore put into compression, and the system curls toward the coating side (*b*).

coating reached the glass transition point and treated the material as a linear elastic solid from that point on. This led to the general picture of curl being caused by residual stress development.

In terms of solving curl problems, the basic structure of the coated product determines the inherent tendency toward curl both in the dryer and in the final application. Good examples are photographic products. These are very sensitive to curl because of the inherent sensitivity of gelatin layers to the moisture content of the air. As the relative humidity of the air decreases, the photographic emulsion and other gelatin-containing layers lose moisture, contract, and generate stresses leading to the curl. Thus photographic prints tend to curl more in the dry air of wintertime and less on the humid days of summer. To counteract this curling tendency, a gelatin layer is often coated on the opposite side of the support. This provides a counteracting tensile force on the other side. Even though the layers on both sides are contracting, there is no net effect on the product, except to put the support in compression. With these materials curl can also be controlled by rehumidifying the coating. This tends to swell the coating and also tends to soften it. Both of these effects tend to reduce residual stresses in the coating. Stresses can also be relieved by conditioning the material at a high temperature. This should be above the glass transition temperature to permit appreciable molecular motion. Adding a plasticizer to the formulation, such as moisture to a gelatin coating, softens it, allowing increased molecular motion. Theoretically, curl can be minimized by minimizing the volume change of the coating after solidification, but it can never be truly eliminated in a single-sided coating.

As already mentioned, the shrinkage forces that are generated during drying can lead to curl both on the coater and in the final use. Mild curl may not be seen in the dryer or after it, because the web tension needed to transport the web acts as a counteracting force to keep the web flat. However, at times severe curl may be seen in the coater and can be a problem. Moisture is a plasticizer for a wide range of materials, and variations in the air humidity can lead to shrinkage or swelling and can cause curl. This can occur both in and after the dryer. Again the photographic system is a good example. A polyester support is coated with gelatin-coating layers. These contract at low relative humidities due to loss of moisture and can curl. When curl is seen outside the dryer, check the relative humidity. Winter can be a problem for certain products. It may be necessary to use humidifiers in the coating rooms. In the coater an uneven drying profile can lead to additional stresses and curl. Check the coating weight uniformity to determine if this is a problem.

Rapid drying causes more residual tensile stress than does slow drying, because less time is allowed for stresses to be relieved before being locked in. Thus, if curl is a problem, gentler drying in the zones where the structure forms and is set, around the end of the constant-rate period, would be desirable. Thus in these zones lower temperatures and a higher solvent concentration in the drying air can minimize curl. Check the location of the end of the constant-rate period, and confirm that the coating weight is on target. As explained above, curl is generally toward the coated side.

The solvent content of the air on the dry side of the web can also affect curl, even though it may have no effect on the drying process itself. This occurs when the support, or a coating on the back of the support, absorbs solvent. This is particularly noticeable with aqueous coatings on paper. If the dryer air has less moisture in it than in equilibrium with the paper, the support will lose moisture, tend to shrink, and the system will tend to curl toward the dry side.

An increase in web tension on the coating line can "pull out" some of the curl. As the base properties strongly influence curl, these properties should be checked when curl is encountered. Those include the thickness uniformity, the modulus of elasticity (Young's modulus), the tensile strength, and the elongation to break. These properties should be checked in both the machine and transverse directions.

Delamination

Delamination is closely related to curl. If the tensile force in the coating is greater than the strength of the adhesive bond between the coating and the support, the coating can separate from the support, resulting in delamination. Delamination is often seen at the edge of a coating, for the residual stress rises sharply at the edge. It is sometimes observed that a small strip of the coating at the edge will "delaminate" from the base. Usually, the strip will remain cohesively attached to the coating. The same considerations for reducing curl by reducing the stresses will also apply to delamination problems.

Delamination can also occur between coated layers, not only between a coated layer and the support.

There are some unique factors that can contribute to the delamination and adhesion failure problems. The surface properties of the support are very important. The surface must not only have a high surface energy, but a proper balance of surface forces. Surface treatments such as by flame, plasma, or corona are often used to improve wettability and adhesion. If adhesion problems occur, these processes should be subjected to a troubleshooting process to determine if any changes in them correspond with changes in adhesion. The effects of corona treatment are often not permanent, and corona treatment should be done on-line, just before the coating station. Tests on the wettablity of the surface should give some indication if this is a factor. Surfactants are often used to improve adhesion.

At times a thin, high surface energy adhesive or subbing layer is coated on the support to improve adhesion. This layer helps bond the coating to the substrate. For products that have this adhesive layer, if failure occurs, verify that the layer is indeed on the coating and that it is uniform and of the correct coverage. Also, check that the correct side of the support has been coated. There are instances when the primary test is that the adhesive layer is always wound on the outside of the roll. When a roll with the adhesive layer on the inside is used occasionally, it can lead to trouble. There may also be problems with the adhesive properties of the coating. Simple peel tests with tape can give a clue to this. If the coating comes off when a piece of tape is placed on the coating and pulled off rapidly, and this normally does not occur, the source of the problem has been found. If the support is suspected, try a different lot of the support. Also check the wettability of the surface. However, an acceptable support is needed to use as a standard of comparison to determine if chemical or physical differences have occurred. Age since application of the adhesive layer or the treatment process can be important, since adhesion can decrease with time.

Changes in the coating solution can also lead to these types of defects. One of the factors in good adhesion is good wettability of the support by the coating solution. If the surface tension is too high, it will not wet and can lead to adhesion and coating problems. Similarly, if the surface of the support is contaminated, it can prevent wetting and reduce adhesion. Contaminants in a coating solution do not always show up as coating quality problems; they can lead to wetting and adhesion problems. Check both the surface tension and the viscosity of the coating solution and make up a new coating solution batch with different lots of raw materials.

Cockle or Wrinkling

Another large class of drying defects results from permanent deformation of the web caused by distortions while held at high temperatures. Such defects are only seen in applications where the web is held at a sufficiently high temperature that it is softened. Common examples would be deformation set into the web by tension wrinkles (troughs) that occur in spans of film between rollers (often due to misaligned rollers), and deformations set into the film when it passes over a roller while hot. A particularly intriguing example is a defect sometimes called *Venetian blinds*. Film affected by Venetian blinds exhibits alternating bands of tension wrinkles and lateral smoothness as the film moves through a span. These defects are always associated with a physical problem in the web transport system. Usually, the source of the defect can be pinpointed by measuring the spacing between the defects and finding where in the process such a spacing can occur.

Air Bar Rubs

Air bar rubs are marks caused by the wet coating coming in contact with the air nozzles in the drying oven. They are a particular concern with air flotation ovens because the distance from the web to nozzle is small, on the order of 0.5 in. (12 mm), and a small flutter will lead to contact and wet spots on nozzles. Typically, air bar rubs are irregular diffuse "scuffs," although the appearance is somewhat variable. They tend to be discrete with finite edges, since they are caused by the nozzle pressed against the wet coating. They may also tend to be in the machine direction, because the web instability causing the contact is in the same place. Characteristically, these may tend to get worse with time. The deposit on the air nozzle will build up with each contact, grow larger, and eventually form a stalactite which actually touches the web and causes a streak. The easiest way to verify air bar rubs is to shut down the coater, open the dryer, and look for the contact point in the same locations as the defects observed.

There are several possible causes of air bar rubs. A poor web transport system can cause a vibrating web. Oscillating tension control and low tension can lead to an unstable web, resulting in contact. In floater dryers, the slot velocities of the top and bottom air bars need to be balanced to ensure a stable web. If the balance is changed in a couple of the air bars, the web may become unstable in that area and hit a nozzle. It is good practice to monitor the air velocities of representative nozzles so that changes can be detected. If an air bar rub occurs, check the air velocity in that region to ensure that it is correct. The alignment of the nozzles is also important. The nozzles should all be on the same centerline within an accuracy specified by the manufacturer. Usually, this is not a problem after the dryer has been started up, but it is a concern for a new dryer. The quality of the base can also be a factor in air bar rubs. If the base is not uniform and does not lie flat, problems can occur in the dryer, since the out-of-level spots will react differently to air pressure. Try a different lot of base.

The factors that lead to curl in the dryer can also lead to air bar rubs. Typically, the curl is at the edges of the film and can be intermittent. If the rubs are at the edge, consider the steps outlined above for eliminating curl.

Dryer Bands

Dryer bands are areas of nonuniformities that appear as lanes running in the machine direction. Depending on the system, they may have a rough surface, a difference in reflectivity, or may appear simply as hazy lanes. It is also possible that they will not be visible but will appear in use as lanes of different properties, for products that are sensitive to drying conditions. Dryer bands are most common when round holes, rather than full-width slots, are used for the drying air. The distinguishing characteristic of dryer bands is their uniform spacing across the web, and also the correlation of their spacing with that of a physical characteristic of the dryer. When circular jets are used for the air nozzles the dryer bands appear with a thickness approximately equal to the diameter of the air nozzles. The band itself can be caused by a difference in drying rate, or by an actual disturbance in the surface of the coating caused by the air jet. Dryer bands are less common with full-width slots, although they can still occur. They are caused by nonuniformities in the air impinging on the web and the reinforcement of that nonuniformity over a wide range of nozzles. If the airflow disturbs the surface of the wet layer, dryer bands can be caused by a single row of nozzles; however, if the dryer bands are caused by differences in the rate of drying, several rows of nozzles are usually needed to form the bands.

Dryer bands are the direct result of the maldistribution of airflow in the dryer, either in the delivery system or in the exhaust system that removes the air. Depending on the nature of the defect, the airflows, air temperatures, or coating composition may need to be adjusted. However, they are often an inherent characteristic of the dryer hardware design, and modification in the hardware to improve airflow uniformity should be considered. The air velocity may need to be reduced to eliminate the dryer bands.

The flow of the wet coating caused by the force of the impinging air increases with the air velocity and also with the thickness of the coating, and decreases with the liquid viscosity. To reduce dryer bands, then, one should reduce the air velocity. This reduces the drying rate in the constant-rate period at the beginning of the dryer. Also, the coating viscosity should be increased to make the coating more resistant to flow, and the coating solution should have a higher solids concentration to give a thinner coating, which is less susceptible to deformation. Increasing the solids concentration simultaneously increases the viscosity, both of which are desirable. Farther down the dryer, as drying continues the solvent evaporates, thus the coating becomes thinner and more viscous and therefore is less likely to be disturbed. Air velocities can be higher in succeeding zones. Dryer bands usually appear at the beginning of the first drying zone, where the viscosity is lowest and the wet layer is thickest. However, if the air velocity is high enough in the later zones, dryer bands can occur there.

Dryer bands have also been seen at the start of the falling-rate period. In this period the coating temperature approaches the air temperature. Because of this higher temperature it is possible for the viscosity to be low enough for dryer bands to occur, even though the solvent concentration is much less than at the start of drying. In the falling-rate period the drying rate is controlled by the rate of diffusion of solvent through the coating to the surface. The diffusion rate of the solvent is a strong function of the solvent concentration. The solvent concentration in the layer at the surface will be in equilibrium with the solvent concentration of the drying air. When this is zero, the solvent concentration at the surface will be zero, the diffusion rate through this surface layer will be very low, and drying will be very slow. Also, the surface layer will act as a surface skin over a soft coating, and this skin can easily be deformed to give dryer bands. Increasing the solvent level in the air will increase the surface concentration of solvent, which will increase the diffusion rate and thus increase the rate of drying. There will no longer be a surface skin, with a fairly uniform, relatively high solvent concentration throughout the rest of the layer. The solvent will now have a "high" concentration only near the base, and the coating will not be as soft. It will therefore be more resistant to dryer bands. Thus, strange as it may seem, increasing the solvent concentration in the drying air increases the drying rate at the start of the falling-rate period and reduces the tendency to form dryer bands.

In the final stages of drying, dry air has to be used. Obviously, the solvent level in the coating cannot be reduced below that in equilibrium with the drying air. To get a completely dry coating, there has to be no solvent in the drying air in the last drying zone.

Remelt

Remelt occurs in those coatings that are chilled to solidify the liquid coating into a gel before entering the dryer, to allow higher air velocities in the early stages of drying. Remelt occurs when the temperature of the gelled coating exceeds its melting point. An example of chill setting is in the coating of photographic systems. The gelatin binder dissolved in water forms a solid gel when chilled below the setting temperature. The system is coated at about 40°C, which is above the transition temperature, and the coating is in the liquid state. Chilling the coating before it enters the dryer converts the liquid to a semirigid solid gel. If the coating heats up in the dryer to above the transition temperature the coating will convert back to a liquid, and then dryer bands can easily form from the high-velocity drying air. Fortunately, the temperature of the gelled coating is well below the temperature of the drying air, due to the heat needed to vaporize the water. In fact, in single-sided drying, the coating temperature is the wet-bulb temperature of the air. For defects to appear, the entire coating does not have to be melted. The surface need only soften slightly, and high-velocity air impingement can distort it.

Remelt is seen either as dryer bands or as cross-web bars that may appear to be similar to chatter. It can occur at any part of the process, even at the end of the dryer. Here the coating temperature approaches the air temperature. If this temperature exceeds a transition temperature at that particular solvent level, the coating can flow and distort. To eliminate these defects the phase transitions of the coating solution should be known and compared with the temperatures in the dryer. If there is an overlap in temperatures, the dryer temperature should be lowered. When the phase transitions are not known, a scan of the coating solution in a differential thermal analyzer will give this information.

9.3.2 Macroscopic Defects

By the term *macroscopic* we refer to defects that can be seen with the naked eye but are too small to measure with a ruler. This size range is in many ways the most interesting, for it is at this scale that the widest range of physical phenomena can become important.

Spots

Spots and discontinuities are defects that are typically found in coatings. Their size is usually within the "macroscopic" size, but occasionally they can be large enough to be measured with a ruler. Drying spots arise from a contaminant in the airstream impinging on the wet coating and disturbing it. Airborne dirt is a typical contaminant and examination of the defect will often show the particle in the center. This is a difficult area to diagnose precisely, since contaminants in the original coating solution cause similar defects. The contaminant itself can be small and covered by the thick wet coating. However, as the coating dries, its thickness decreases and the defect becomes visible. Highly filtered air should be used to minimize spots by keeping particles out of the airstream. Also, the dryer should be under a slight positive pressure so that air blowing out of the dryer will keep unfiltered room air off the coating and thus will help reduce spots. If the dryer cannot be at a positive pressure, it should at least be kept as close to neutral as possible, to avoid sucking room air into the dryer. It is easy to check this by hanging a thread over the dryer openings; the thread is light and will quickly respond to the slightest air currents, showing whether air is flowing out of or into the dryer.

It is also essential to keep the dryer clean. The wet coating can rub off on the nozzles, dry, and then drop back on the coating, causing spots. The dried coating in the dryer can also cause streaks if the nozzles are close together and the contamination is large. Keeping the dryer clean will eliminate this defect. Also, be aware that the entire heating and ventilating system can be a source of spot-causing defects. Ducts can get dirty, and since they are usually hard to access, they are not often checked. Routine cleaning should include the ducts wherever possible. Filters are effective but they are not foolproof. They can get damaged and the seals can leak, permitting dirty air to enter the system. When spot defects are found, the filter pressure drop is checked. If it has increased, change the filters. Also check all the seals and door gaskets for their integrity. Replace them if there is any doubt.

Solvent condensation in the dryer can also cause spots. If solvent-laden air is chilled below its saturation temperature, the solvent will condense and drop onto the coating. This can occur if the outside walls of the duct, at or near room temperature, are below the saturation temperature of the air. This is more likely to occur in relatively stagnant regions of the air return system. It can also occur at the wall separating the dryer from the chiller, if a chiller is used. These defects are hard to determine, since in most cases by the time the process is shut down and dryer doors opened, the solvent has a chance to evaporate and can no longer be detected. This defect is characterized by the appearance of a clear center surrounded by splash marks.

The key to solving spot problems is to characterize all the sample and be aware of all the contaminant-causing spots. This information can then be used to compare with the current defect. When cleaning hard-to-access hardware such as air delivery ducts and nozzles, take samples of any dirt or contaminant and characterize them. It may not be possible to obtain samples in the future. Having these data in the files will help to solve future problems.

Bubbles and Blisters

Bubbles are spot defects where the contaminant is a gas such as air in the coating solution, instead of a solid or an incompatible liquid. Usually, the gas is introduced into the coating in the solution preparation step or in the feed system to the coating applicator. Bubbles usually appear as such, but bubbles can remain in the coating bead (of a slide coater) to cause streaks. Bubbles can also be introduced in the coating process by entraining a thin film of air under the coating solution. In addition, it is possible for the drying process to introduce bubbles into the coating. With a high temperature in the dryer and a volatile solvent, it is possible for the temperature of the coating to approach the boiling point of the coating solution and for bubbles to form from the solvent. These bubbles will grow as the gas expands and then will break through the surface. An example of this behavior is shown in Figure 9-9. Initially, there is the full bubble, with the coating solution forming a dome over it. As drying progresses the gas expands and the dome collapses. Another possible source for air can be the coating solution. If it is saturated at the coating station, then at the higher temperature in the coating, bubbles can form. They too will expand at the higher drying temperatures and will appear in the dryer. Bubbles can also form in the coating lines where the pressure is reduced, such as in valves where the liquid velocity increases, and these bubbles may not redissolve.

One method of diagnosing the sources of bubbles is to stop the coater, dry the coating in place, and mark the various drying zones. If the bubbles or blisters are present before the dryer entrance, they were introduced prior to the dryer. If they are

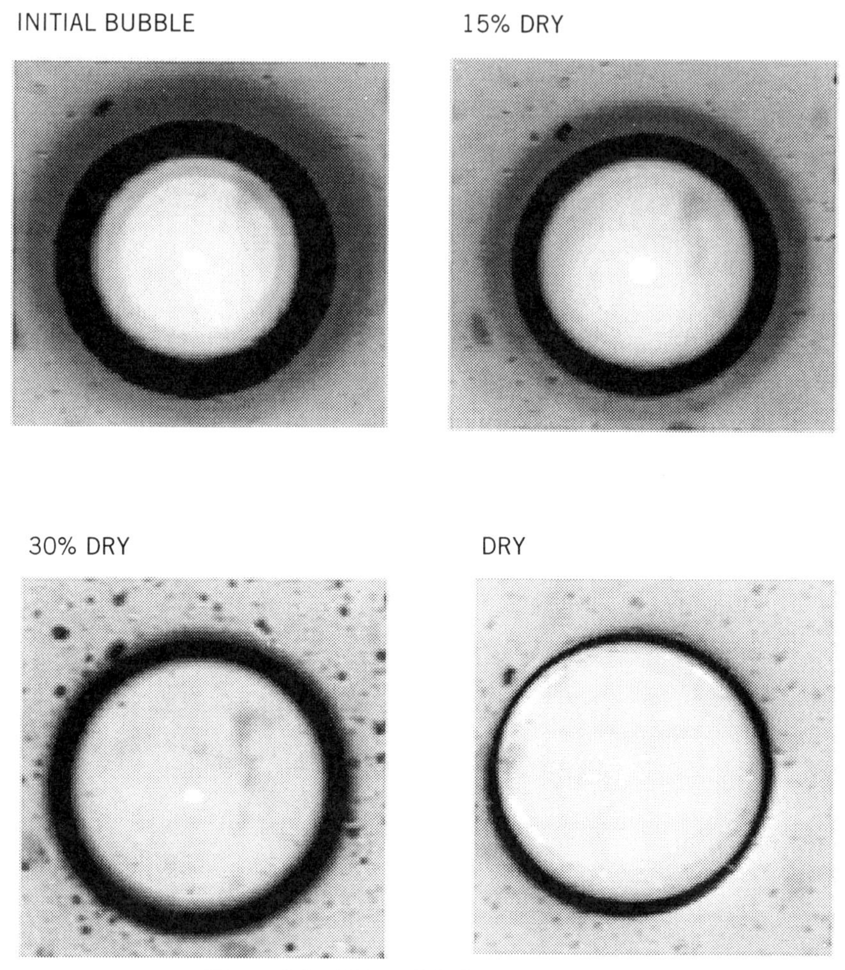

Figure 9-9. Effect of drying on a bubble.

not seen at the dryer entrance but are seen at the end of the dryer, they are likely to be caused by boiling of the solvent. This technique will help give some insight into the fundamental cause.

To eliminate the dryer-induced bubbles, the solvent composition and the dryer temperature must be controlled so that the boiling does not occur. If boiling is suspected, lower the dryer temperature and slow down the coater. Also, keep the temperatures low in initial stages of the dryer, so that any bubbles present will slowly diffuse to the surface while the viscosity is low and the surface can flow out and fill in the spot. If bubbles do occur, degas the coating solution to remove any bubbles. This can be done by pulling a vacuum on the coating solution for some period of time. It can also be done using ultrasonic energy to move the bubbles out of the flow path (Holhlfeld and Cohen, 1993). If neither of these are available, it is possible to let the solution stand in a relatively shallow kettle for several hours, to allow the bubbles to rise to the surface. Drawing off the solution from the bottom of the kettle will give a coating solution relatively free of bubbles.

Mottle

Mottle is an irregularly patterned defect that can be gross or quite subtle, and at times it can have an iridescent pattern, depending on its severity. The scale can be quite small to quite large, on the order of centimeters. It is believed to be caused by a nonuniform airflow blowing the coating around in early stages of the drying process when the coating is still quite fluid. The nonuniformity of the air motion causes a slight motion in limited areas of the web which appear as the mottle. The surface difference is what appears to be the mottle. Sometimes mottle is not visible in the dried coating but is seen later when the product is used. A uniform airstream would affect the entire surface and would not appear as a defect. One should first investigate the transition area between the coating applicator and the initial drying zone. Air motion can occur in this area, and air can blow out of, or into, the dryer entrance door. This can be checked by holding a thread at the entrance door.

If the transition area is open to the room, consideration should be given to enclosing this area to minimize air motion. This does not have to be elaborate and can be as simple as plastic sheeting attached to the framework between the coater and the dryer. This will also protect the web from airborne dirt. If air blowing into or out of the dryer entrance door is causing the mottle, the air pressure should be carefully balanced so that no air motion occurs. A slight motion into the dryer is preferable to flow out of the dryer, as such a flow is with the motion of the web. If mottle occurs in the first dryer zone, the velocity of the drying air should be reduced.

As with dryer bands, mottle is reduced by making the coating more resistant to motion. This can be done by concentrating the coating liquid so as to have a thinner layer, and by raising the viscosity of the coating liquid. Concentrating the solution also increases the liquid viscosity, so this method is highly recommended.

Surface Blow-Around

In most coating systems the coating does not gel at any point in drying; rather, the coating becomes progressively more viscous as drying proceeds. In such systems, disturbances of the surface by airflows can be much worse than those called *mottle*. This defect usually occurs at the beginning of the dryer where the coating is most dilute and is minimized by minimizing the airflow to the surface in the initial stages of drying and increasing the airflow gradually throughout the dryer. The scale of surface blow-around can vary from quite small to quite large.

Orange Peel and Crinkling

Orange peel is a rough finish in the dried coating that is generally associated with surface-tension-driven flows, although use of the term is not well controlled. Orange peel is not a regular hexagonal pattern. It tends to be exaggerated at high drying rates and, in particular, with higher air velocities in the dryer. Orange peel may occur in bands. Crinkling is a phenomenon that occurs when a thin skin forms on the surface of the wet coating. Crinkling is produced intentionally to produce a rough surface for painted metals, such as on typewriters. Crinkling is normally not seen in thin-film coatings.

Reticulation

Reticulation is a somewhat irregular pattern of distortions in the coating that occurs not in drying, but in uneven swelling of a previously dried layer, such as in processing an exposed photographic film or paper. The pattern appears as sharp wrinkles or cracks in the coating. When heavy it has the appearance of chicken tracks. These tracks will lead to an objectionable haze. Reticulation is a concern with photographic systems and was first studied by Sheppard and Elliot (1918). A typical example of reticulation is shown in Figure 9-10. Reticulation results from the surface of the coating buckling as a result of the stress in swelling of a dried film. The reason that reticulation is discussed under drying defects is that in the photographic field, the susceptibility of the coating to reticulation is controlled in the dryer. That is, the conditions in the dryer control the mechanical properties of the coating and the level of residual stress in the coating. Both of these determine whether or not reticulation will occur.

Reticulation is particularly hard to troubleshoot because it is not seen until the product is rewet and swells. When reticulation occurs, the rate of drying has to be reduced to reduce the residual stresses. This can be done by lowering the drying temperatures or by increasing the solvent concentrations in the air. Also, hardening or cross-linking can be a factor. The use of cross-linking agents should be checked to ensure that the proper amounts and levels are used. A simple statistically designed experiment can help define the correct drying conditions and levels of cross-linking

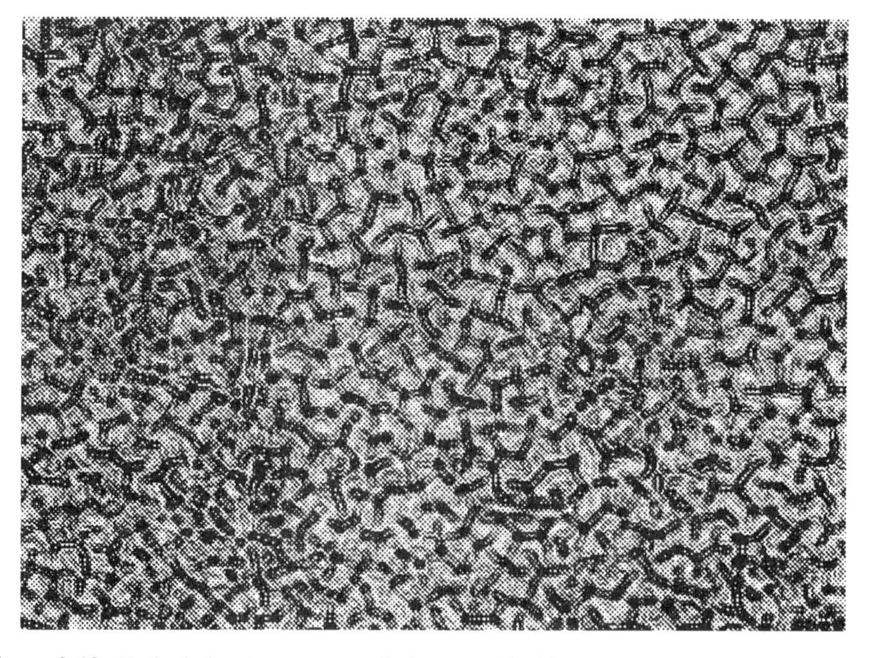

Figure 9-10. Reticulation in a processed photographic film. [From E. D. Cohen and R. Grotovsky, *J. Imag. Sci. Technol.*, **37**, 133–148 (1993), with permission of the Society of Imaging Science and Technology.]

agent. Coating formulation modifications to increase the modulus of the coating to resist the deforming forces can be tried.

There is also a mild interaction between reticulation and delamination. The driving force for both of these is the stress generated by drying. There are two types of adhesive bonds, those between the support and the coating, and those between the coating components. The specific failure from the stress will determine the defects that results. Both types of bonds are stressed during drying and also during subsequent swelling. Failure of the adhesive bond between the coating and the support results in delamination. Failure of the adhesive bond between the coating components can result in reticulation.

Mud Cracking

Mud cracking refers to an irregular breakup of the coating into small plates still attached to the substrate. It is similar to the appearance of a field of mud after it dries out in the sun. The cracks can be very close togther, or they can be more than 5 mm apart. The tensile stresses that occur in drying were discussed under curl. When these stresses cannot be relieved by curling because the substrate is too rigid, they may exceed the cohesive strength of the coating (or of one layer in the coating) and the coating may crack. Mud cracking most commonly occurs in weak coatings on rigid supports. It is often seen in coatings were there is not enough binder to fill the spaces between the solid particles and air is drawn into the coating during drying. A common occurrence of mud cracking is in the clear coat/top coat of multilayer automotive paint systems. In this case, the top coat is both damaged and thinned by ultraviolet light, breaking the backbone of the polymer chain. At some point, the top coat can no longer support the stress and it develops cracks.

Mud cracking can be reduced by the same factors that reduce curl. In addition, it is sometimes helpful to add particles to the coating liquid. The type and amount of particles have to be determined by experiment.

Blushing

Blushing refers to a "frosted" or hazy surface that results from water condensing on the coating in solvent coating. In the initial stages of drying, perhaps even before the entrance to the dryer, the coating will cool down to supply the heat necessary to evaporate the solvent. In single-sided aqueous coating, the coating temperature reaches the wet-bulb temperature of the air. In solvent coating it reaches the wet-bulb temperature of the air with respect to that particular solvent. As the air normally has no solvent, the temperature of the coating can be quite low, even considerably below the freezing point of water. On humid summer days this can be well below the dew point of the air, and water will condense out of the air onto the coating. This water will evaporate in the dryer, but will often leave a hazy surface.

To avoid blushing a number of companies do not coat certain products in summer. The other solution is to dehumidify the air.

Overspray

Overspray is a drying defect so named because affected film looks like a painted surface that has been inadvertently exposed to an incomplete second coat of paint. The term is commonly used in the paint industry. As a drying defect, overspray results from a phase separation where some additives become insoluble in the coating as drying proceeds. The insoluble materials exude out of the film. Overspray may appear similar to blushing, but the cause is completely different. Overspray is usually controlled by adjusting the formulation of the coating, although it may be possible to avoid it by judicious choice of drying conditions. Overspray may be considered a milder form of a defect that has been referred to as "scrambled eggs," in which the coating phase separates on a gross scale to produce a curdled appearance.

9.3.3 Microscopic Defects

Microscopic defects are those too small to be seen with the naked eye. Depending on the application, defects in this very small size range may not present a functional problem. However, contrary to that old adage, what you cannot see still can hurt you. Applications such as photographic film and magnetic tape are very sensitive to microscopic defects because the size of the stored information is the same size or smaller than the defect and the defect will lead to errors in the product use. The examples given here range from common to very specific.

Haze

Haze here refers to an internal haze in the coating rather than a surface haze which results from blushing or from reticulation. Haze can have at least three distinct causes, depending on the nature of the coating. In coatings that have a latex binder, the latex particles must coalesce into a continuous phase in the early stages of drying. Incomplete coalescence manifests itself in haze. Haze can also result from a phase separation, as in overspray. Additives that are more soluble in the coating solvent than they are in the polymer binder may separate out as drying proceeds. In addition, haze can result from microscopic voids that occur when the volume fraction of dispersed solids exceeds a critical value, which may be far higher than 50%, depending on the shape and size distribution of the solids. There is simply not enough binder to fill the space between the solid particles, so that air penetrates into the voids, resulting in haze that can be detected with the naked eye. However, the physical defect that causes this haze is typically too small to see.

Starry Night

Starry night refers to apparent clear spots left in photographic film as a result of "matte" particles from an overcoat layer being pushed down by drying stresses (Christodoulou and Lightfoot, 1992) into the silver halide layer, thus pushing aside the silver halide crystals. The matte particles are used to give a rough surface, allowing uniform contact during film exposure in a vacuum frame by providing channels for the air to escape from the center of the film. The defect is a series of random white holes of 1 to 10 μm diameter which are seen against the black exposed film background. They look very much like the night sky, hence, the name *starry night* or *galaxies*. They appear white because the silver halide crystals that were present had been pushed aside. Thus, after a uniform exposure and processing, that area has no silver present and therefore appears white. While starry night is not visible to the naked eye and can be seen only by a microscope, the effect is to reduce the optical density of the exposed and processed film, and this has an adverse effect on its performance.

The term *windowing* is commonly applied to a similar defect in the paint industry. It is a typical drying defect for most coatings that have particulates in them. Reducing the drying rate, which reduces the stresses in curl and in reticulation, also reduces the severity of starry night. In particular, reducing the drying rate toward the end of the drying process is particularly important. If the solvent concentration, here water, in the drying air is low and the air temperature high, the surface will dry out and will force the particles to migrate along the wet–dry boundary. Increasing the solvent concentration in the air or reducing the air temperature keeps the solvent concentration in the surface higher. The matte particles themselves also play a role. If starry night is encountered, the particle-size distribution of the matte must be checked to ensure that it is within quality control limits. Large particles are a particular problem. A small increase in the fraction of large particles leads to a large increase in the severity of starry night. If possible, check the particle size of the matte in the coating solution. It is possible for an ineffective dispersion process to give large agglomerated particles, even though the raw material is acceptable.

Pinholes

There are also a wide variety of pinhole defects which, in photographic films, can appear to be similar to starry night but obviously have different causes. Correct identification is essential. It is important to analyze the defect and to ensure that it is indeed what you think it is. An example of this is shown in Figure 9-11. A white spot defect

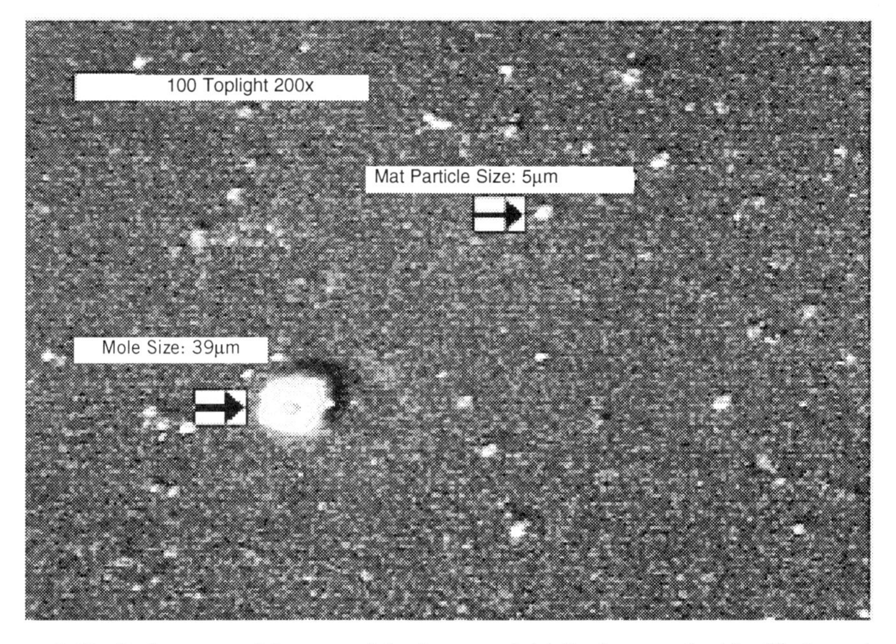

Figure 9-11. Defect caused by a particle that was initially incorrectly identified as starry night. [From E. D. Cohen and R. Grotovsky, *J. Imag. Sci. Technol.*, **37**, 133–148 (1993), with permission of the Society of Imaging Science and Technology.]

in a photographic film was called starry night and was submitted for analysis as a standard part of the troubleshooting process. The digital image analysis showed that the defect was larger than starry night, had a particle in the center and was not starry night. The problem was not dryer induced but was traced to incomplete dispersion and poor filtration. Trying to correct the problem by reducing drying stresses would never have cured it. The cure would have been delayed, perhaps, until it would have gone away on its own.

9.4 THE ROLE OF DRYER SIMULATION IN TROUBLESHOOTING

Since all of the defects discussed require modification of dryer conditions for their elimination, it is very helpful to be able to simulate the drying process on a computer and to try out the proposed changes before the actual experiment. It is far more effective to learn that a proposed reduction in air velocity will give a wet film at the end of the dryer in a simulation than it is to do it on line. The advent of the personal computer has also placed extensive computing capability in the hands of most industrial engineers and has given them the ability to do more realistic calculations. It has also created the need for readily available process simulation tools.

The objective of this section is to give an overview of the computational tools available for the analysis of drying of the thin films found in industrial precision coating processes. The required complexity of a mathematical model that simulates a drying process depends on both the physics of the process, mainly the rate-controlling mechanism, and on the phenomenon that needs to be analyzed. There are two basic types of models available for use in the industrial drying processes to predict the drying rate, to ensure that the coating is completely dry in the specified time, or equivalently in the available dryer length. The first type is an approximate model that looks only at the constant-rate behavior. The second type includes the falling-rate period and therefore is more accurate. These models are described next.

9.4.1 Models for Prediction of Drying Rate

In starting up and running a production or pilot coating process, it is important to be able to predict the rate of drying with a moderate accuracy of about ±3 to 5%. The line speed may be kept slightly below the nominal steady-state speed by about 5%, until the coating process produces the desired coating weight across the sheet and the drying is complete before the end of the dryer. In both aqueous and solvent coating, adequately accurate calculations of the drying rate can be accomplished by one of two types of mathematical models. The first type calculates only the drying during the constant-rate period, using a simple lumped-parameter model for the coating and for the temperature. The temperature will be uniform across the coating and the support in most thin coatings on thin webs. In aqueous systems most of the drying takes place in this constant-rate period, where the limiting factor is the heat transfer from the air and solvent, and what takes place within the wet coating has no effect. Given a specific solvent and coverage of solvent, the controllable variables are the air temperatures, and solvent levels in the air, and the air velocities, in each dryer zone. It is easy to calculate the drying rate, and thus the residual solvent level at the end of each zone, and the location of the end of the constant-rate period as long as it can be specified by its solvent level per unit of total solids or of dissolved solids. Computations can be done by hand or on a personal computer using spreadsheets (Cary and Gutoff, 1992).

A more complicated drying model takes into account the falling-rate period, where, in most cases, solvent diffuses through the coating to the surface. In solvent coatings most of the drying takes place in the falling-rate period. The main governing equation in such models is a partial differential equation for the solvent concentration in time and in one space dimension, film thickness. Added complications are those needed to account for the shrinking of the film during drying (Okazaki et al., 1974; Gehrmann and Kast, 1980) and the often unknown dependence of the diffusion coefficient on solvent concentration. In their coarsest forms such models can be solved on personal computers (Gutoff, 1994), but they may require the more powerful workstations.

References

Cary, J. D., and E. B. Gutoff, "Constant rate drying of aqueous coatings," *Chem. Eng. Prog.* **87**, 73–77 (1991).

Christodoulou, K. N., and E. J. Lightfoot, "Stress induced defects in the drying of thinfilms," Paper 46c, presented at *AIChE Spring National Meeting*, New Orleans, LA, 1992.

Cohen, E. D., "Thin film drying," pp. 267–298 in *Modern Coating and Drying Technology*, E. D. Cohen and E. B. Gutoff, eds., VCH Publishers, New York, 1992.

Croll, S. G., "The origin of residual internal stress in solvent-cast thermoplastic coatings," *J. Appl. Polym. Sci.* **23**, 847–858 (1979).

Gehrmann, D., and W. Kast, "Drying gels, as exemplified by thin, flat gelatin layers," *VDI Ber.*, **391**, 590–615 (1980).

Gutoff, E. B., "Modeling the drying of solvent coatings on continuous webs," *J. Imag. Sci. Tech.*, **38**, 184–192 (1994).

Hohlfeld, R. G., and E. D. Cohen, "Ultrasonic debubbling using acoustic radiation pressure," presented at *46th Annual Conference, Society for Imaging Science and Technology*, Cambridge, MA, May 1993.

Okazaki, M., K. Shioda, K. Masuda, and R. Toei, "Drying mechanism of coated film of polymer solution," *J. Chem. Eng. Jpn.* **7**, 99–105 (1974).

Sheppard, S. E., and F. A. Elliot, "The reticulation of gelatine," *Ind. Eng. Chem.*, **10**, 727 (1918).

Umberger, J. Q., "The fundamental nature of curl and shrinkage in photographic films," *Photogr. Sci. Eng.* **1**, 69–73 (1957).

Weiss, H. L., *Coating and Laminating Machines*, Converting Technology Company, Milwaukee, WI, 1977.

CHAPTER X

Problems Associated with Web Handling

by Gerald I. Kheboian

Web-handling problems can be caused by any part of the web transport system. The solution to web-handling problems is frequently a shotgun approach where a "posse" of vigilantes rides off to find the perpetrator without knowing its identity. Frequently, the posse will "do something" so that the machine will run. The action taken often does not solve the actual problem but does relieve the symptom. If enough such "solutions" are implemented, the machine may eventually require a complete startup and tuneup of all the elements of the web transport system.

The solution of web-handling problems must begin with a knowledge of the web-handling system. The web-handling system consists of the web and its physical characteristics, machine alignment, dryer and coating parameters, tension and speed control systems, drive system, and coating formulation. Anything that can cause web wrinkling, delaminating, scratches, edge cracks and web breaks, uneven slit edges, dusty slit edges, or damaged coated web, ballooning of the web in the dryer, or mistracking are part of the web-handling system.

It is important for an efficient operation that the operator, machine supervision, process engineering, plant engineering trades, and technical support group know and understand the machine design parameters as well as the acceptable operating parameters.

10.1 WEB

Web (Figure 10-1) is any paper, plastic, or metal that is presented to the machine in a continuous flexible strip form for coating, laminating, plating, or covering or any other process. The film coating industry, the textile, the metals, and the wire industries share common problems although the terms for similar processes are different.

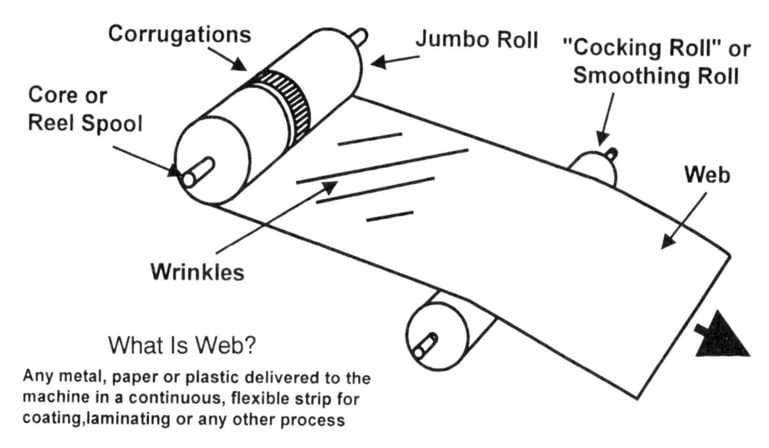

Figure 10-1. Definition of web.

10.1.1 Web Characteristics and Problems Associated with the Web

Wrinkling, creasing, web breaks, or mistracking are a few problems that can be caused by web characteristics. Wrinkles in the base web can be caused by web camber, nonuniform basis weight, uneven cross-machine web thickness profile, and an uneven moisture profile. Web breaks can be caused by poor vendor splices and vendor-produced edge cracks in the web. Mistracking can be caused by web camber or by the frictional characteristics of the driven surface of the web.

Base Web Defects That Cause Wrinkling

If the web in the base roll has camber, nonuniform thickness profile, basis weight profile, or moisture profile, a wrinkle problem may exist. One can feel "corrugations" or hard and soft spots in the "jumbo" roll if camber or uneven profiles exist. One will also see wrinkles form as the web is transported through the machine. If the wrinkles form before a nip point or coater applicator, they will form a crease and will cause a web break.

Wrinkles can be caused by the machine or the web itself. A machine out of alignment will cause the same wrinkles as a web that is cambered or one that has an unacceptable web caliper, web moisture, or web basis weight profile.

The difficulty of identifying the cause of wrinkling comes from the idea that the web is not the problem. The web is almost never identified as a cause of wrinkling because there are specifications that are written for the web and it is assumed that the vendor-supplied web is in specification and that it is perfect. In actual fact, the incoming materials quality control is frequently ineffective or nonexistent.

Specifications for the incoming base are not policed, tested, or updated to ensure that the incoming material is appropriate for the needs of an ever-changing product mix. Nor are web problems communicated to the incoming materials quality control department to allow them to work with vendors to produce better web.

The web specifications are frequently developed by an inexperienced incoming materials inspector from specifications supplied by the vendor of the web. It does not follow that if the vendor stays within his or her own specifications, the web is acceptable for all coating processes.

If the coating operation "purchases" its web from an in-plant machine, it may be difficult to reject web that is unsuitable for processing. If the web is produced within the coating mill, a neutral quality control function must exist between the web producer and the coating process that follows. If the "neutral source" is responsible for the production of the web, they may overrule the specifications and ask the coating plant to process web that is out of specification. Losses normally chargeable to the web operation fall on the coating process.

Each coating plant must understand the parameters of acceptable web and set up "practical" web specifications for the process and the vendor. Practical specifications represent a commitment by the vendor and client to establish web parameters that can be processed satisfactorily without making the production of the web impossible or economically unfeasible.

The responsibility for determining web specifications falls on the user of the web. If specs cannot be met by one vendor, try new vendors. Paper webs are being made on more modern machines that can limit defects delivered in the base roll to the user. Similar improvements are being made in the manufacture of plastic webs. Find vendors who can make the webs that increase productivity, increase yields, and reduce your cost of goods sold.

Web problems are often masked because the operators will do what they need to do to make salable product. They may misalign the idler rolls, put the rolls out of level, and change process speeds, tensions, and air handling in order to produce a product. Such techniques will be used until the machine is so out of specification that production comes to a stop! At that time a task force will be formed, the machine will be blamed ("it never ran right"), all rolls will be realigned and leveled—and the wrinkle problem will still exist.

The real causes for wrinkling may be:

1. Web specifications are not complete or realistic.
2. Web specifications are not policed.
3. Web is not in compliance with specifications.
4. Some machine rolls are misaligned.
5. Both the web and the machine are at fault.

Causes of Web Defects

Web defects such as camber, cross-machine thickness and moisture variations, and cross-machine basis weight variations occur during the manufacturing process. The web manufacturer has a limited ability to keep the web within the specified parameters with existing equipment. Modern equipment has a better ability to monitor the nonuniform quality of the web and then to make the necessary adjustments to restore the web's desired uniform characteristics.

For instance, a paper web is manufactured by mixing fibers with water to a consistency of $\frac{1}{2}\%$ fiber and $99\frac{1}{2}\%$ water. That solution is pumped onto a flat traveling screen where some of water is drained from the solution. The paper web is transferred to a press section where additional water is pressed from the web before it is passed into a steam-heated dryer section where the paper is heated to a point where additional water is removed from the web. The web may then be passed to a precoating press, to a calender to establish a surface finish, and then wound into a jumbo roll before being delivered to the coating process. Each of these steps can cause a defect that is described below.

10.1.2 Web Attributes

Web Camber

Camber (Figure 10-2) occurs in the web manufacturing process. Older machines tend to produce more cambered web than new machines. For instance, the processing of paper web in a paper machine requires the web to run through a dryer section. The web is run at speeds in excess of 1500 ft/min through the dryers. On older machines, the web tends to float through the dryers as the water in the web is heated to an evaporating temperature. On older machines the temperature on the drive side of the dryer is higher than the temperature on the operator's side of the machine. Thus the drive side of the web dries faster than the web on the operator's side of the machine. Paper web tends to shrink laterally (sidewise) and longitudinally (lengthwise) as it dries. Since the paper on the drive side of the machine dries faster than the paper on the operator's side, there is more shrinkage on the drive side of the web. This difference in shrinkage produces a web that is shorter on the drive side than it is on the tending side. That difference is called *camber*. If the camber is severe, it may be difficult or impossible to process the web into a useful product.

Modern machines have more uniform temperatures across the surface of the dryers, and the web is not allowed to float through the dryer section. This combination of better temperature control and the containment of the web on the surface of the dryer creates more uniform drying while allowing less web shrinkage than was experienced in the older machines. This is a simplistic explanation of the drying phenomena. Volumes have been written on drying and drying techniques that explain the process in much better detail than is presented here.

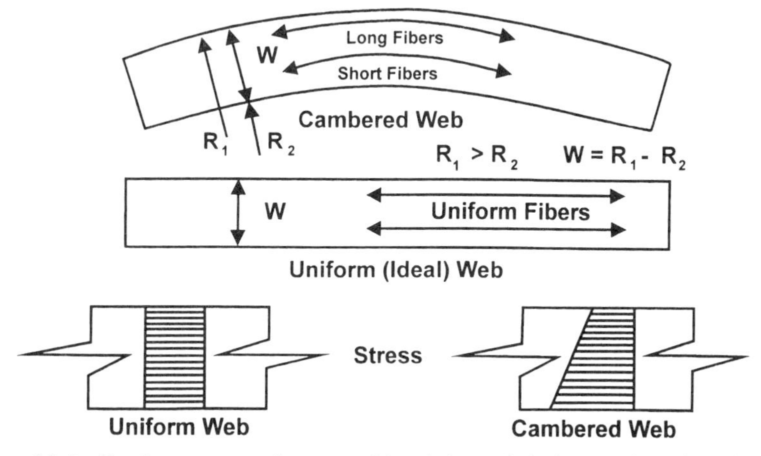

Figure 10-2. Camber occurs when one side of the web is longer than the other side.

Web camber is an attribute of the web that makes the web want to track to one side of the machine. Camber can be identified by laying a 25-ft-long straight line on the floor. Cut a 25-ft length of web from the base roll and lay one edge against the straight line. If the web forms an arc around the line, the web is said to have camber. If the web forms an arc that runs to the left, the left inside of the arc is said to be the *tight side* and the right outside of the arc is called the *loose side*. If you measure the inside and outside of the 25-ft sample, you will find that the inside dimension is shorter than the outside dimension. Thus the outside edge is said to be longer than the inside. If you examine the fully wound jumbo roll, you will find that a cambered roll frequently will have a "spongy" edge while the other edge is "hard."

If we try to run this cambered web on a coating machine that is perfectly aligned and level, the web will tend to run toward the left side of the machine. If we are pulling the web through the machine, the web will want to track left while we are trying to pull the web straight. This will produce diagonal wrinkles. If these wrinkles encounter a nip, the wrinkles will be pressed into a crease that can cause a web break or a defective product. Devices such as "cocking" rolls (Figure 10-3), are designed to remove wrinkles caused by cambered web. They do this by distributing the forces uniformly across the web. This action makes the tension across the web more uniform. Without the cocking roll, the tension on the tight side of the web would be higher than the tension on the loose side. These devices are limited as to how much camber they can remove, depending on the web and the location of the cocking roller within the machine. The camber in a web can vary between web manufacturers and between base lot numbers of a given web manufacturer. This means that the operator will continually adjust the "cocking" roll to compensate for varying camber. Some cocking rollers are mechanical control systems that automatically adjust themselves to control the wrinkling of the web. I have seen the manually operated cocking rolls remove wrinkles but have not seen an automatic system in operation.

It is better to specify acceptable limits of camber than to depend on the cocking roller to remove an infinite amount of camber. Establish practical specs for the web and then use the cocking rolls to trim up the web. Some webs are cambered to the left, and others are cambered to the right. It will be necessary to readjust the "cocking" rollers to compensate for the change in direction of the camber.

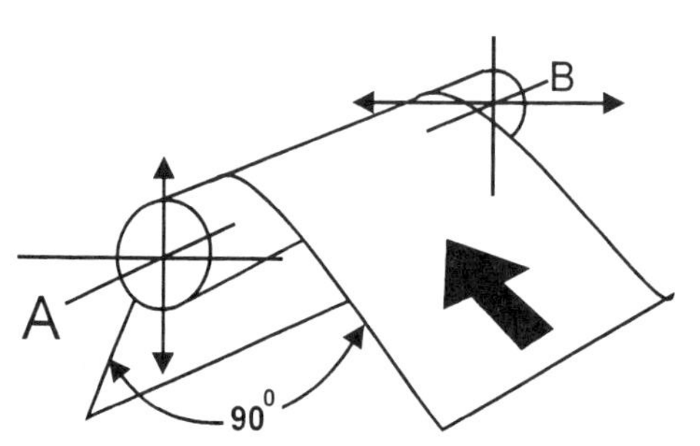

Vertical motion of side A effects the slack edge support (on either side)

Horizontal motion of side B compensates for the steering effect generated by the motion of side A

The roll in the neutral position (parallel) when the roll is located at 0-0

Note: A slack edge is a permanent deformation of the web and can't be eliminated. The "cocking" roler offers a means to remove the effect at crucial points of the web transport. This roller has been used sucessfully at Draw Nips, Coaters, Laminators, Rewinders, and Printing heads.

Manufacturers of slack edge removers are Eberlin (located in Cheltenham, England), Yorkshire Industries, Andover, Ma., and Mt. Hope of Taunton, Ma..
System designs will vary between vendors

Figure 10-3. Cocking rools are one way to eliminate a slack edge.

Some problems caused by cambered web are:

1. Diagonal wrinkles and creases that result in web breaks at nips, blade coaters, or extrusion-type coaters.
2. Diagonal wrinkles and creases which result in coating voids that prevent effective processing of the web. For instance, a coating void on release coatings prevents labels from releasing from the coating void. Such a defect will result in returns from the label manufacturer.
3. The loose edge of the web will fracture when it hits a web guide nozzle as the web passes through the machine.
4. The difference in length between drive side and the tending side makes it difficult to get an accurate registration of multicolor printing processes on a web. Camber combined with poor tension control can lead to poor printing results. Printing processes differ from coating processes in that coating processes need smooth perturbation-free web transport. Printing processes have the added need for exact position control of the printed feature at each print head.

To deal with the effects of camber:

1. Identify and group incoming rolls by process characteristics such as vendor, the date made, and the machine which made the roll. If the rolls were cut from a wider parent roll, keep track of the individual cut numbers from that roll.
2. Run all rolls with a common vendor, date, machine and cut position at one time. When all rolls with a common characteristic are run, run the next set of rolls to completion. Many coating and printing plants have spliced rolls into the machine in random fashion and then found that the roll characteristics between vendors and lot numbers vary so much that it is impossible to guarantee trouble free transportation of the web.
3. Test each new batch of vendor-supplied material to determine if it meets your specifications. Do this testing in your incoming materials testing area or before the material is delivered to the machine for processing.
4. Do not begin modifying the machine before you know what are the parameters of your incoming material.
5. You may find that the material that you have is off-specification and is the only material available which can be converted into product. Your vendors may have a delivery schedule that will cause you to shut down unless the defective material can be processed. In this case, you may be forced to try some temporary machine changes to continue to produce product. The cost of lost sales cannot be underestimated. If changes are made to the machine, document them so that the machine can be returned to its ideal condition when qualified vendor material is received.

Web Cross-Machine Thickness Profile Variation

Web cross-machine thickness profile variation (Figure 10-4) can be measured by taking a full width sample of the web from the roll and measuring the thickness of the

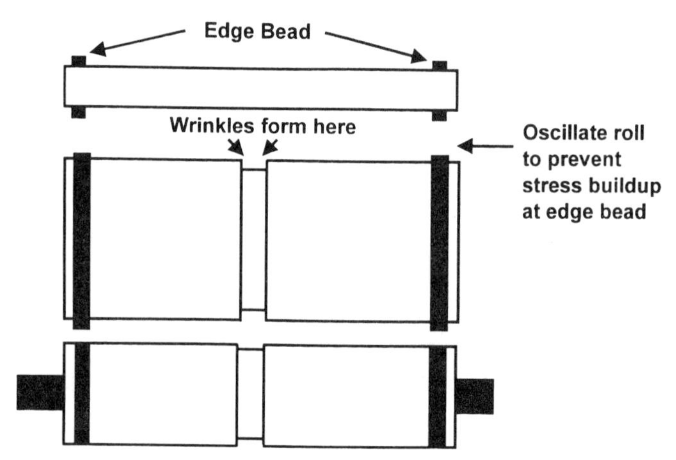

Figure 10-4. When the caliper varies across the web, wrinkles can form in the soft areas. Oscillating the role can reduce the effect of edge beads.

web at frequent locations across the web. A device as simple as a micrometer is effective, as is a computerized thickness profile instrument which produces a thickness profile chart. It is important to identify significant thickness variations in a short span. For instance, if a web is supposed to have a nominal thickness of 2.5 mils (0.0025 in.) and it is 2.8 mils thick (0.0028 in.) at one spot and 2.4 mils thick next to that very spot, that is a significant difference. The 2.8-mil thickness will show up as a "hard spot" on a wound roll. As a matter of fact, a 10% variation is significant enough to cause a "hard spot" on the surface of the wound roll.

Web thickness profile variations of paper web can be caused by uneven forming of the web or by the pressing and calendering process that is part of the paper machine process. It can also be caused by variations in the moisture content. If the press or calender rolls are not ground properly, the web thickness may vary in a cross-machine direction. All rolls are supported on the ends in bearings. When the roll is suspended by the ends, the roll develops a *sag*. Each roll must be ground to present a straight face to the web for each operating nip pressure. If the roll is not ground properly, the roll may be larger in the center than on the ends or can be smaller in the center than on the ends. Each condition can create thickness variations. A roll with a large center will crush the web in the center. A roll with large diameters on the edges will crush the edges of the web. When the roll is ground, it is said to have a *crown*.

Examples of problems caused by improper crown or nip pressure are:

1. If the press or calender is operated at too high a nip pressure, the edges of the web will be crushed and be thinner than the center of the web.
2. If the press is operated at too low a nip pressure, the center of the web may be crushed, causing a thin spot in the center of the web. If the web is run through a coater, straight wrinkles will form in the center of the web, which can form creases at a nip. If the web is bisected, we will also bisect the thin section of web in the center. The left side of the bisected web will have a thin right edge.

The right side of the bisected web will have a thin left edge. The thin edges will feel soft in a fully wound roll.

3. Improper crowning and nip pressure can cause the web to crush at the quarter points, causing thin spots in the web at the quarter points. If the web is run through a coater, straight wrinkles will form that can become creases at a nip. If the web is bisected, the thin quarter points will become the center of the two resulting rolls. If the web is slit into three or four cuts, the thin section will appear at the edges of the web. Each of these cuts should be identified by lot number and whether it was slit from the drive side, middle, or operator's side of the web. Run each roll with identical characteristics as a group. That will reduce the web transport problems that are caused by randomly introducing to the machine rolls with different characteristics. Thickness profile variations can also be introduced at the calender if the calender rolls are crowned improperly or if improper nips are used for a given calender roll crown.
4. Straight wrinkles will form on each edge of the hard spot that will form creases at nips that will cause web breaks or coating voids. Operators will try to "pull" the wrinkles out by increasing tensions between sections. Higher tensions tend to reduce the amplitude of the wrinkles but may increase the chance of web breaks and may affect registration if the process includes a print operation. The reduction in wrinkle amplitude may prevent creasing at a coater nip or other nips within the machine. Devices such as spreader rolls are used routinely at the inlet to a coating section to spread out the wrinkles before the web enters the nip.

Basis Weight Profile Variation

In making a paper web the forming process delivers a very liquid formulation to a moving screen. The process may not deliver a uniform density of fiber to the traveling screen. It is possible that the fiber distribution across the web is heavy in some places and light in other places. This variation in fiber density is called cross-web basis weight variation. One can feel these dense places as hard spots in the circumference of the wound roll. The less dense spots will feel very soft in the circumference of the wound roll. The hard spot in the roll will often look like the cable stitch in a sweater. The result of the hard and soft spots across the web can result in wrinkles in the longitudinal direction. These wrinkles can form into hard creases at a nip or process point, causing coating voids or web breaks. It is certain that the variations in basis weight and moisture profile variations can cause variations in the absorption of water in an aqueous coating process. Paper web will expand when wet. The paper expands as it absorbs water but is restrained from moving laterally. As a result, the excess material due to expansion must go somewhere. The excess material goes into the formation of wrinkles.

These wrinkles can break at a following coater unless web spreading devices remove the wrinkles at the following section. The operators will often increase tensions between sections to "pull out" the wrinkles. Variations in the thickness of the web can also cause coating problems. In slot and slide coating (see Chapter 5), there is a narrow gap between the coating die and the web traveling around the backing roll. Thickness variations in the web will cause equivalent gap fluctuations, and this can cause chatter.

Surface of the Web

A basic tenet of coating defect-free film is that the web itself should be defect free. In most coating systems, defects in the web will cause defects in the coating above them. Rarely are defects in the web eliminated or reduced by the subsequent coating and drying process.

It is a good practice to take representative samples of the uncoated support routinely and have them available for analysis, so that when web is suspected there is a set of samples available to establish the normal characteristics of the support. A small number of samples should be taken for each base lot in routine production and more when support problems are suspected. It is a good practice to take routine samples of the uncoated web and have them available for analysis, so that when the web is suspected, there is a set of samples available to establish its normal characteristics. A small number of samples should be taken from each lot in routine production, and more should be taken when web problems are suspected.

Variability and changes in the surface layer can lead to the following types of defects:

1. Adhesion failure between the substrate surface and the applied coating
2. Coating nonuniformities because of nonuniform wettability
3. Coating nonuniformities because of nonuniform static charges on the web
4. Spot defects from contamination on the substrate surface
5. Coating nonuniformities because of variability of any coated barrier and/or adhesive layers that may be present

All coatings usually require a specific range of adhesion between coatings and the support to meet their end-use properties. When this varies outside specified limits there is a defect, and the web should be considered as a possible factor, along with all the coating and drying processes. The defect can be poor adhesion in a system that requires high adhesion, such as paint to steel in automobiles or photographic emulsion to film base in photographic systems. It can be too high an adhesion in systems that require low adhesion, such as with release layers. The first step is to check the adhesion to the representative support and to the retained samples from previous production runs using basic lab tests. These tests can be as simple as coating on a small sample, drying it, and then checking the adhesion by trying to pull off the coating with a piece of tape. More complex tests in which the actual force required to remove the coating can also be run. The thickness of the support layer, the quality and composition of the raw materials, changes in composition, age of the support and curing conditions are all obvious factors that need to be considered. In addition, the support needs to be checked to ensure that the correct side of the base was coated on, and that the surface is indeed what it should be.

Variability of the wetting properties of the surface can lead to coating instabilities. Small differences will show up as subtle defects such as unstable edges and mottle. Wetting variability can also lead to gross failures such as the inability of the applicator to form a uniform coating. In case there is gross coating failure and the inability to form a uniform coating bead, consider the possibility of changes in the surface layer wettability as well as changes in the properties of the coating solution and the coating process conditions. Using the retained samples and the suspected web the wettability can be measured as described in Chapter 2.

It is also possible for the surface to be contaminated. This can lead to a wide variety of spot defects. Ideally, the surface should be clean and have no foreign material. However, the coating can pick up a wide variety of contaminants in manufacture, during slitting into different widths, and during transfer to the unwind stand. These contaminants can be dirt from the air, lubricants used in the process, lint from the packing material, dust from abraded coating material that builds up on rolls and is transferred back to the web surface, and so on. It is useful to understand the method of manufacture so as to be aware of possible contaminants. This contamination of the surface with foreign material can affect the coating in several ways. It can cause a physical effect in which the applied coating is disturbed by the particle and leads to an absence of coating surrounding the particle, as in Figure 10-5. It can lead to pinholes. A foreign organic material on the surface can result in high surface tension, leading to a repellency spot, as seen in Figure 10-6.

When troubleshooting localized spot defects, the support should be considered as a possible contributing factor. The uncoated support should be examined visually and with a microscope to determine if there are defects on the support which are

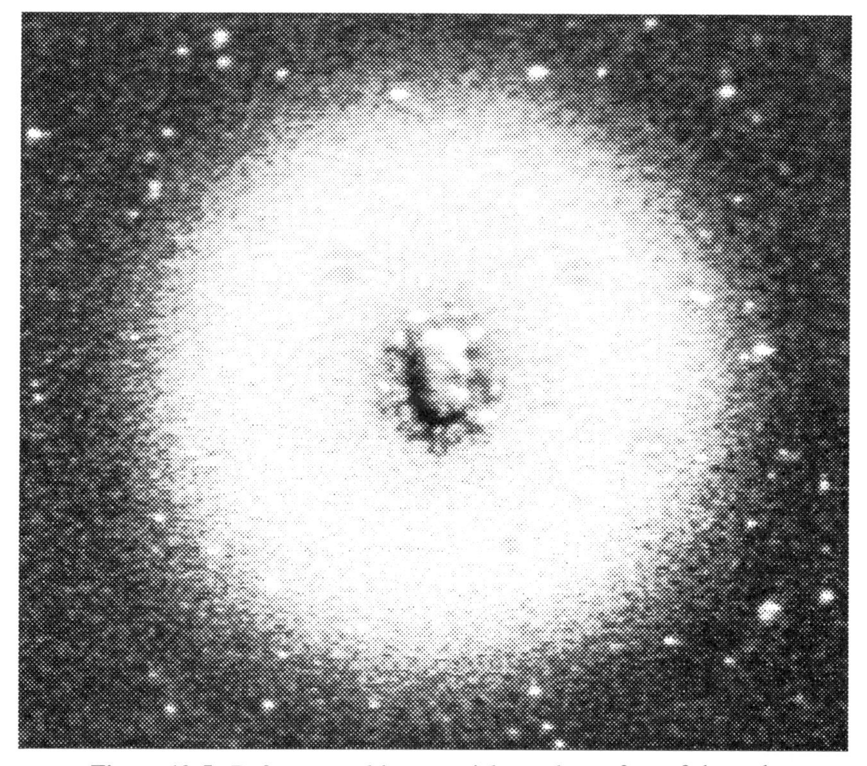

Figure 10-5. Defect caused by a particle on the surface of the web.

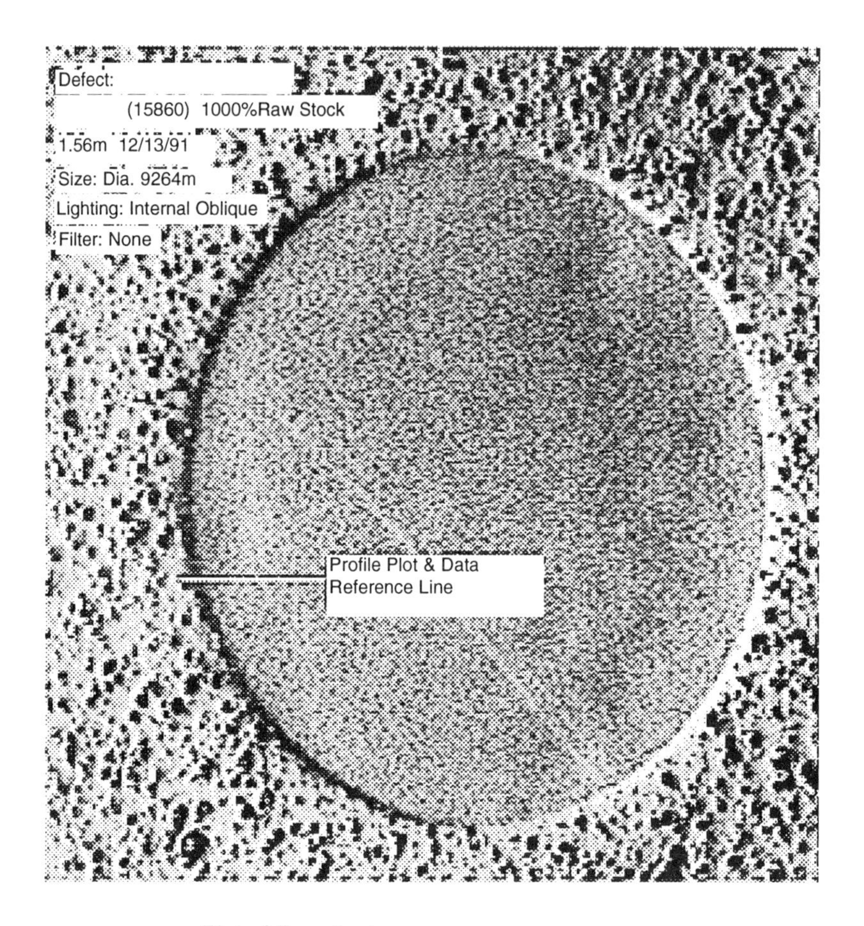

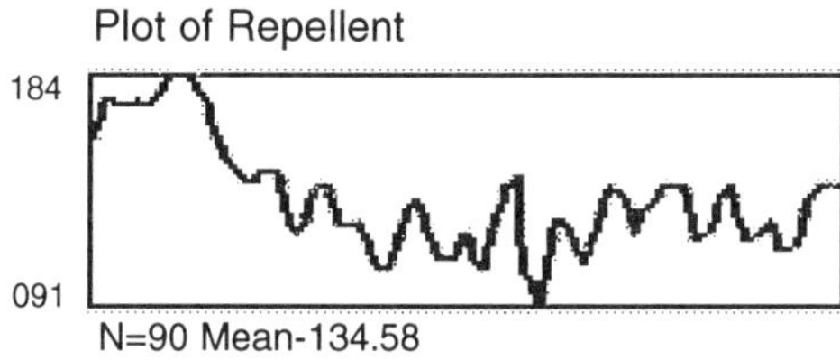

Figure 10-6. Repellency spot caused by contamination of the web.

similar to those seen on the final coated product, as in Figure 10-7. The same analytical techniques that are applied to the coating defect should also be used to characterize the support. This is easy when a sample of support is available and the same defects are seen in the coating and on the support. It is more complicated when there is no uncoated sample. However, it is still possible to analyze the base under the defect by removing the coating using abrasion techniques, chemicals, or selective enzymes. Also, microtome and freeze-fracture techniques can be used to get a cross section of the defect and give a sample in which all layers can be studied.

UNCOATED SUPPORT

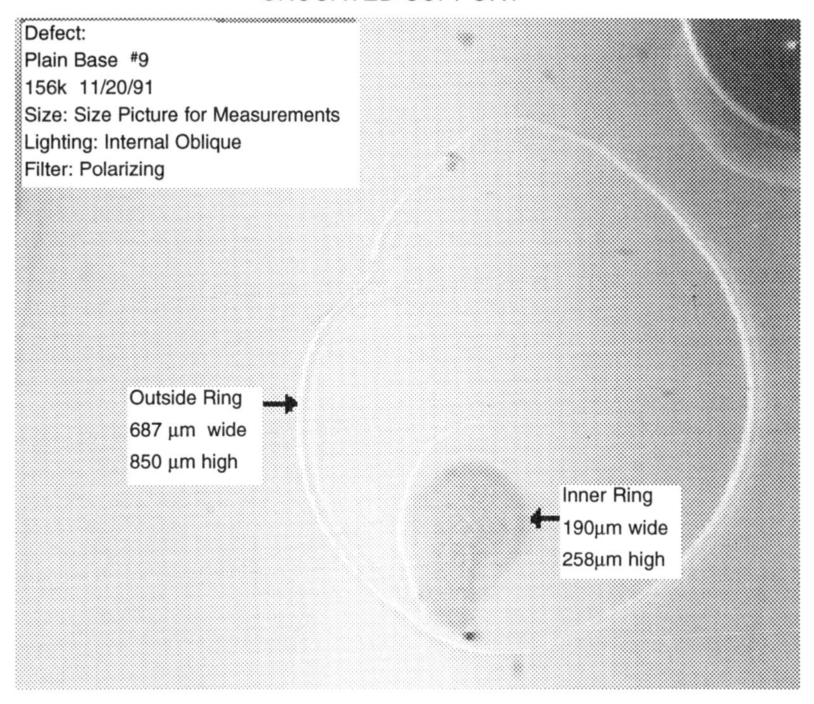

DEFECT IN COATED FILM

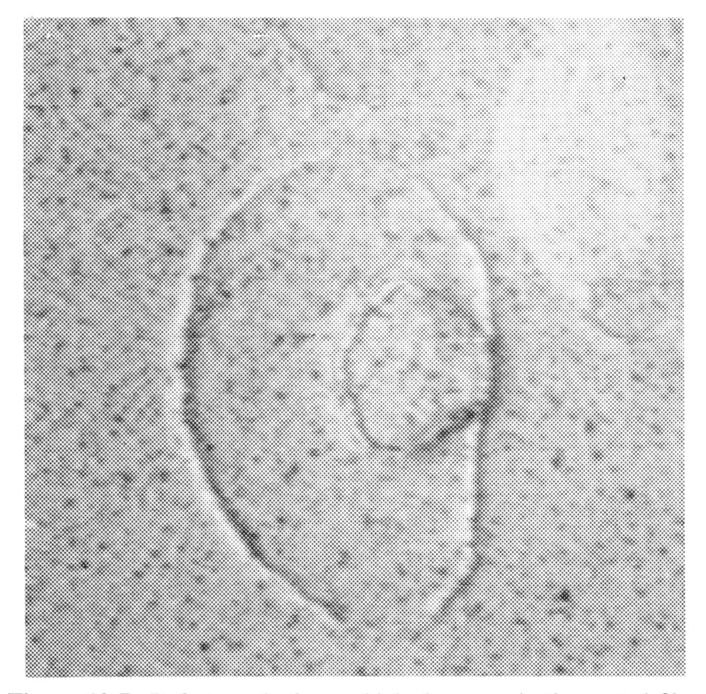

Figure 10-7. Defect on the base which shows up in the coated film.

Plastic Web

Plastic web may be produced by extruding material onto a drum and then stretching it to the desired width in a "tenter" frame. The tenter frame has grippers that clamp the edges of the extruded web and then stretch it laterally as the web moves toward the rewinder. Plastic web is considered to be perfectly flat. But it can have thickness variations and camber due to the nature of extrusion, casting, and stretching process. Plastic web will tend to shrink laterally if the process heats the web to a point where the web becomes soft and compliant. That point is called the *glass transition temperature* of the plastic. Many processes require high temperatures for the desired chemical reaction of the coating structure. Those temperatures can exceed the glass transition temperature of the material and require us to apply tension control (or speed control) techniques carefully if we are to avoid deforming the material, causing web breaks, or causing excessive lateral shrinkage. Later in the chapter we will apply tension control, speed control, torque control and combinations of these techniques to various web transport processes using different materials.

10.2 WEB TRANSPORT SYSTEMS

Web transport systems consist of the web; the rolls, which deliver power to the web; the motors, which deliver power to the rolls; and speed and tension sensors to provide feedback to the controllers, which control the motors.

1. The web transport systems include the web itself. How easy is it to transport the web through the machine? What is its coefficient of friction? What effect does a layer of coating have on its "driveability"? Is the coated surface less driveable? Can we increase tensions to help transport the slippery web, or does the process limit us to the average tension values for such a material?
2. The web transport system also includes the power transmission machinery, such as unwinds, vacuum rolls, vacuum tables, pull rolls, coaters, and rewinds, which transport the web through the machine. What characteristics provide reliable, consistent, defect-free, and cost-effective motivation of the web through the machine? What type of drive section will encourage the operator to include the section in the web path consistently instead of bypassing it because it causes web breaks or defects in the finished product?
3. The motors, gears, and couplings that drive the machinery must be selected to provide the best results for the end product. Motors and gears must be matched to requirements of the process.
4. The selection of dancer rolls, force transducers, tachometers, encoders, torque sensors, or load-measuring devices is important to the performance of the power supplies that are part of the control of the motors and drive rolls.

10.2.1 Tension

Tension through a machine is the force that is applied to the web between adjacent machine sections. This force tends to stretch the web and is usually expressed as some unit of web width.

Most web is transported through the machine using tension values that are high enough to allow the web to track properly but are low enough to minimize web breaks. Figure 10-8 shows some typical tensions for a wide variety of materials. These tensions are average values and may differ depending on the temperatures within the process, the length of span between driven rolls, or condition of the web when it comes to the coater. The units are defined as (1) pounds per linear inch, (2) ounces per linear inch, (3) grams per linear inch, (4) kilograms per meter, and (5) newtons per meter.

It is important to record the operating tension values for each product and to determine if the machine and web are performing normally. Tension information can be used to design new processes, equipment, and in problem-solving machine problems. A knowledge of how each product runs may help determine if a machine process can be speeded up with a minimum capital outlay for drive equipment.

Tension can also be expressed as total tension, which is the product of tension per unit width times the total web width. For example, a tension of 3 lb/lin. in. exerted on a web width of 100 in. = 300 lb total tension. Total tension represents the amount of work that the drive motor or braking system must do. Total tension is used to calculate brake, clutch, or motor size.

Reasons Why Operators Must Change Tensions to Transport Web Successfully

1. High temperatures in the process may cause the operators to run at low tensions to prevent deformation of the plastic web.
2. Laminating a thin elastic web to a thicker inelastic web may require lower tensions if delaminating is to be prevented.
3. If paper web is too dry, it will be brittle and weak. Edge cracks become a potential for web breaks. Operators will reduce tensions on dry paper web in the hope that edge cracks and web breaks can be avoided.
4. Operators will increase tension if wrinkles are present in the web, hoping that the extra tension will eliminate the wrinkles.

Suggested Rewind Tension Levels			Suggested Unwind Tension Levels		
Product		Tension pli	Product		Tension pli
Paper	15#	0.5	Paper	15#	0.25
	20#	0.75		20#	0.5
	30#	1.0		30#	0.75
	40#	1.5		40#	1.0
	50#	2.0		50#	1.25
	60#	2.5		60#	1.75
	80#	3.0		80#	2.0
Board	8pt.	3	Board	8pt.	2
	12pt.	4		12pt.	2.75
	15pt.	5		15pt.	3.25
	20pt.	7		20pt.	4.75
	25pt.	9		25pt.	6.0
	30pt.	11		30pt.	7.25

Product	Lb/mil/in
Acetate	0.5
Aluminum foil	1.0
Cellophane	0.75
Cryovac	0.1
Glassine	1.5
Mylar (polyester)	0.75
Nylon	0.25
Polyethylene	0.25
Polypropylene	0.25
Polystryene	1.0
Pliofilm	0.1
Saran	0.1
Vinyl	0.1

Copper Wire*	
AWG	Tension
8	30 lb
10	20
12	12
14	9
16	6
20	5
24	4.5
30	1.25
36	0.25
40	0.1

* Run Aluminum wire at 1/2 to 1/3 of copper tension levels. Multi-strand cables are run at the sum of the tension of all strands. Example: 6 conductor, all 20 AWG copper, 6 x 5 = 30 pounds.

Figure 10-8. Typical tensions for a number of different webs.

Let us look next at the sections that transport the web through the machine. For purposes of this discussion, we assume that a coating line will consist of an unwind stand, pull rolls, a coater, dryers, a laminator, and a rewind section.

10.2.2 Unwind Stand

The most important function of the unwind is to introduce the web to the rest of the process. The unwind may have only one arbor if the process is to be stopped at the end of each roll. That is the way that many research or test coaters are operated. In the research mode, one roll of web may be coated with many short samples of prototype coatings. The roll is removed from the machine and the coatings are evaluated to determine which coating has the most desirable attributes.

Many *production* operations are continuous processes and cannot be shut down at the end of each roll. It is too costly to dump pans of coating, reduce air temperatures, and pay the operating crew and support departments while the machine is idle. Unwinds on continuous processes must provide continuous web to the machine by allowing splicing and transferring from a spent core to a new incoming "jumbo" roll. Depending on the product's value, the unwind will have two, three, or even four arbors on which new rolls are held in preparation for a spice. The extra positions ensure that if a poor transfer is made from a core to a new roll, a splice can be made immediately, from a third position, to minimize waste and losses. Modern unwind splicing systems have improved so much that two arbor unwinds are most frequently used. Triple unwinds may be used if experience with the web indicates splicing problems due to the quality of the wound web. Typical problems may be that the new roll telescopes or has "blocking defects" that make it necessary to remove the roll from further processing. A third position on the unwind makes it possible to splice in a new roll quickly.

Splicing adhesives should be selected carefully. Many of the early liquid adhesives had a shelf life that was affected adversely by time and temperature. Solvent-based adhesives were stored in air-conditioned rooms because the summer heat caused rapid deterioration in the holding power of the splice adhesive. Frequent web breaks occurred when splice patterns pulled apart immediately after a splice was made. Modern splicing tapes have improved splicing consistency.

One of the most common faults of unwinds is the method selected for tension control. The unwind is supplied with a manually adjusted brake that requires constant attention from the operator. The brake may consist of a leather strap around a part of the core, or a pneumatic or electric brake that is adjusted by the operator. Invariably, the operator must also fill out a log, check on coating quality, and spend time in the inspection booth at the rewind end of the machine. It is not difficult to see that the unwind will receive little attention from the operator unless the web breaks. If the operator ignores the unwind, the tension will increase in inverse proportion to the roll diameter.

The braking torque, which is constant, is equal to the total web tension times the lever arm, which is the roll radius (half the roll diameter). Thus

$$\text{braking torque} = \text{tension} \times \frac{\text{roll diameter}}{2} = \text{constant}$$

and therefore

$$\text{tension} = \frac{2\times \text{ braking torque}}{\text{roll diameter}}$$

For instance, if the jumbo roll is 50 in. in diameter and is wound on a 5-in. core, that represents a diameter change of 10 : 1. If the operator does not adjust the manual control, tension will increase by a factor of 10. That can introduce the following problems:

1. The web is extensible and shrinks laterally at increased tensions. As tension increases the web gets narrower until it is too narrow to make product. Consider the case of a foam laminator that makes a blanket out of two layers of foam laminated to a center skrim. If the tension is increased to objectionable levels, the blankets will be rejected for twin, full-size, or oversized beds. Ideally, the unwind should have an automatic tension control system that senses and controls tension within acceptable limits. If the material is too weak and compliant, it may be necessary to control the speed between sections very precisely to guarantee the exact width required.
2. The web is part of a multicolor printing process. As the unwind tension increases, the web stretches and begins to affect the registration of the print on each print station. Printing will be out of registration, causing defective product.
3. The web is part of a medical test membrane manufacturing process, which to have repeatable test results, must have a predictable pore size. If tension increases with diameter change, the pore size will change and the test results will be variable. In fact, in the case of membranes, it is doubtful whether the operator can maintain the tension constant enough to guarantee the accuracy that is required of this critical process.
4. The web is part of a laminating process. As the tension increases with diameter change, delaminations can occur because the webs will revert to their original state at different rates when the tension is released. If the structure does not delaminate, it may provide an objectionable level of "edge curl."
5. The web being slit shrinks laterally as the tension increases. Lateral shrinkage will ensure that the edge trim will be lost and the product will be out of spec. If the material has "memory," the lateral dimension will expand, causing the material to be out of spec. For instance, photographic film is designed to fit into a plastic cartridge with dimensions within certain tolerances. Those tolerances may be exceeded if the film stretches laterally after it has been slit. If the tolerances are exceeded, each image may jam as it is ejected from the camera.
6. Most materials will not sustain a 10 : 1 change in tension without an increase in web breaks. Certainly, weaknesses in the web will be exploited by high values of tension.

Why do we persist in using manually operated unwinds when our new processes and products demand more precise tension control? Because we have equipment that satisfied the needs of the older products and processes. Rather than replace the older equipment with new equipment, we constantly tend to force them somehow to run the new product. We then accept the losses that come from using inappropriate equipment as part of the cost of being in a high-quality market. When we speak of old equipment we are talking about equipment that is 40 to 50 years old. Many of the materials that exist now were not in existence 40 to 50 years ago.The problem is compounded by the desire to "make do" with what we have because we want to keep capital expenditures to a minimum. On the other hand, management can justifiably be suspicious if we demand a new machine for every new process. The answer frequently is to make gradual changes to machinery as there are gradual changes in processes or products.

When to Use Drive Motors on an Unwind Stand

Should the unwind use a motor or a brake for tension control? A brake can be used as long as the product is not damaged. The product can be damaged during acceleration or if the braking system cannot reduce its braking torque to zero. How can a brake cause defects in the product?

Let us consider the roll with a core that is one-tenth of the diameter of the full roll. The brake torque, at full roll, is set to yield the proper tension in the web. As the roll unwinds and becomes smaller, the brake torque must be reduced to keep the tension constant. Frequently, the brake is oversized to handle a wide tension range (say $\frac{1}{2}$ to 3 lb/lin. in.) and the web is run at low tensions. If the brake is run at $\frac{1}{2}$ lb/lin. in., it is possible that the frictional drag torque of the brake and power transmission system will be fairly high even if the pressure (or power) to the brake is turned off. There have been cases where the frictional drag torque was so high that it exceeded the value of tension that was desired. The web began to shrink laterally and caused web defects. Registration problems can develop if tension rises at the end of the roll. A brake is designed to hold back against the rest of the line. Its ability to hold back allows web tension to be developed. However, it is frequently necessary to reduce braking to zero and even to apply a motivating force on many light-tension applications. That means a tension control system must be capable of applying power to the web in a braking mode (regeneration) or a positive mode (motoring). Motors with a regenerative controller can provide braking or motoring.

The same is true during acceleration. When the roll is accelerated at the unwind, power must be applied to overcome the inertia of the unwinding roll in addition to the power that is required for tension control. If a brake is used at the unwind, it cannot supply the motoring effort that is required to accelerate the roll. The machine must supply that effort through the web that connects the unwind to the rest of the machine. Rapid acceleration of the unwinding roll may cause the tension in the web to increase. Tension may increase to a point where the web is damaged or even broken. Printing presses frequently see out-of-registration defects during acceleration, because web tension increases throughout the machine. The tension increase causes the web to stretch temporarily. The web stretches and shrinks as a function of tension changes and cause the printed pattern to be out of registration. Motor-driven unwinds minimize tension transients due to acceleration. Further benefits are derived by using properly designed dancers and draw roll systems before the main coating or printing stations. Thus brakes can be used very successfully on slow-speed machines, where the web, splicing transients, and acceleration are not a problem. Brakes are also used on high-speed machines where the web has lateral stability and where the machine can be accelerated slowly. We have seen 6000-ft/min winders using braking systems, but the web is strong and the acceleration rate is reasonable. These intermittent duty winders process the web into shipping diameters and widths.

Motors are being used on more unwinds when the material has little lateral stability and where the "memory" of the web can cause it to introduce registration defects in printing processes. Many production facilities are rejecting the notion that splicing losses are part of the process. Productivity increases are possible if more attention is paid to the materials that are run and the methods that are used to process those materials.

Motors are also required on high-speed coaters, where flying splices require the incoming roll to be accelerated to line speed prior to a splice attempt. High-speed coaters can run from 2000 to 5000 ft/min, but unwind motors are being used more frequently on coating lines that are designed for operating speeds of 300 to 500 ft/min.

10.2.3 Web Control

The unwind and rewind are the "bosses" on the coating line. The transients that are generated by splicing or transferring at the unwind and rewind will affect the entire machine. Production engineers continue to be surprised to see how an unwind splice or rewind transfer can influence the coater or print station. A multipen recording oscillograph can show how each section is affected as the transient passes through the machine from the unwind to the rewind. It is necessary to provide properly designed dancer rollers or load cell systems which will isolate the machine from the unwind/rewind transients that can produce product defects. If no product defects are produced, isolation may not be a necessity. The new computer-controlled drives can measure the effect of the splice transient on the rest of the process and can be programmed to cascade the tension transient corrections to all other sections on the machine. This technique has been used successfully to minimize both the tension transient amplitude and the tension interaction between adjacent sections. Tension interaction depends on the speed of the machine, the material being processed, and the length of web between driven sections. Knowledgeable computer control can compensate for all these factors and definitely contributes to more stable machine performance.

10.2.4 Core Selection and Build-Down Ratio

The jumbo roll comes to the coater wound on a core. These cores can be a source of dirt and product defects. Cardboard, metal, and plastic cores can all produce fibers or chips that will fall onto the web and cause dirt defects in the final product. Web cleaning devices have been installed which remove dust and dirt from the web. The cleaning devices are located before the coating head. The best experience with cleaners has come from web cleaners that incorporate a method of neutralizing static electricity, and following that, with a system that removes the dirt from the web. To keep away from potential dust and dirt sources, the web threading path is designed to keep the web about 1 to 2 ft above the floor. That also allows the operator to keep the areas under the web clean and free of dust.

The ratio between the maximum wound roll diameter and the core diameter is called the build-down ratio. Equipment manufacturers and drive suppliers prefer to use large-diameter cores to keep the build-down ratio small. Larger core sizes do not need a proportional increase in maximum wound roll diameter to store the equivalent length of web. A small build-down ratio allows selection of standard drives or brakes without the need for special control to accommodate large build-down ratios.

10.2.5 Draw Rolls, Pull Rolls, Capstans

The terms *draw rolls*, *pull rolls*, and *capstans* are used interchangeably. Draw rolls are used in a coater to allow differential tension to set up tension zones or to motivate the web when there are long sheet leads between driven sections. Draw rolls can be nipped, S-wrapped, single rolls, vacuum rolls, or vacuum tables. Nip rolls, vacuum rolls, and vacuum tables are used to overcome the boundary layer of air that accompanies a moving web. The trapped layer of air frequently contributes to wrinkling, creasing, and web slippage. If these problems are to be avoided, the draw rolls must be designed to handle the different tensions that the process demands of that section and may need web spreader rolls at the section to eliminate wrinkles.

Web Slippage and What Slip Affects

We consider web to be slipping when the web is running at a speed that is different from the driven section. Slip obviously affects the deposition of coating, the drying of the coated web, and the tracking of the web through the machine. Slip can also occur when the web slips over undriven rolls. We can expect to see scratches on the web when undriven rolls slip.

Devices such as web guides and web spreader rolls cannot function properly if the web slips over them. We have seen slipping web run to one side of the machine and break even as the web guides were trying to correct web position within the machine. Thus slippage not only affects our ability to coat and dry, but also affects our ability to deliver wrinkle-free web that is aligned properly with the coating process.

Nip Rolls

Nip rolls (Figure 10-9), are especially effective on high-speed machines where the web drags in a boundary layer of air that allows the web to float and make it impossible for *unnipped* rolls to control the web. The paper industry, converting industry, and laminating industry will use nip rolls to help transport the web and to set up tension zones within the machine. Nip rolls are *not* used where the machine applies a cosmetic or optical coating that cannot be touched once the coating is applied. Nip rolls consist of a driven chrome-plated roll in partnership with an undriven rubber-covered roll. In cases where the coating quality can be affected by a nipped section, S-wrap rolls, suction rolls, and suction aprons are used to provide isolation of tension zones. Cosmetic quality is important to photographic papers, sunglass lens material, filter papers, labels, and multistation printing presses.

Nip rolls are usually designed for a maximum nip pressure of 15 lb/lin. in. That pressure will help isolate the tension on the inlet side of the nip from the tension on the outlet side of the nip. Be aware, *every* pull roll has a limit as to how much differential tension it can develop before the sheet slips in the nip. Slip is a real

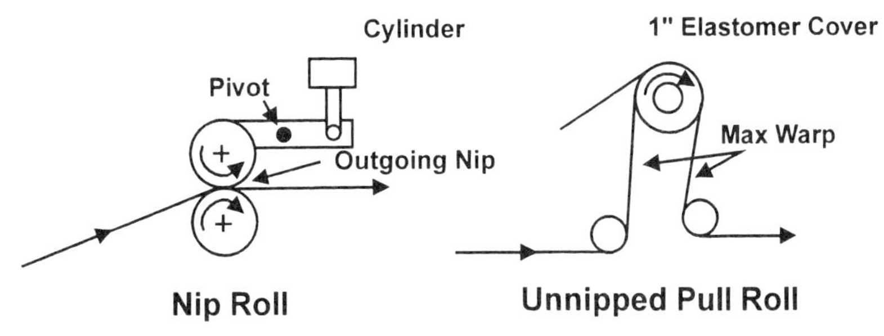

Figure 10-9. Nip rolls and elastomer-covered unnipped rolls with a high wrap are used to drive the web.

problem that has been experienced by anyone in the web-handling field. The most effective web motivators are the unwind and rewind. The rewind in particular can overwhelm the rest of the machine if the tension is set significantly higher than the rest of the machine. Slippage is a factor at paper mills running paper web at 4000 ft/min or at converters running lightweight web at 10 ft/min. It is common to see the operator at either type of plant run high tension at the winder in order to process a good wound roll. Winding tension is usually higher than process tensions. The high winder tension can cause the sheet to slip through the nips and accelerate the web to speeds that are limited only by the limits of control that are imposed on the winder. If the web slips through the nips, we can be sure that there will be defects such as scratches, wrinkles, creases, poor web guiding, and web breaks.

Machinery builders generally try not to have a long straight run from the section preceding the draw roll to the draw roll nip. Long straight paths drag air into the nip, especially at high speeds. The air will "boil" as it enters the nip, causing wrinkles, creases, and web breaks. That is why short runs (draw) are used between sections and why the roll is wrapped at the entry point of the pull roll to minimize the boundary layer of air. Machinery manufacturers often will direct the web over an entry roll and then have an entry wrap angle before the nip. That extra roll and redirection minimizes the air that is dragged into the nip. The roll at the nip entry is frequently a "bowed roll," which can lay the web flat in the nip, thus reducing the possibility of wrinkles and creases. It is a good practice to place spreader rolls before each process point or drive roll to ensure the most wrinkle-free web transport system.

Nip rolls can cause the following defects:

1. Wrinkles or creases due to poor web characteristics, due to web slippage, boundary air layer, or wrinkling due to machine transients such as unwind splices.
2. Wrinkling can also occur within the nip if the nip rolls are ground with improper "crowns." The crown is determined by the operating nip pressure and natural deflection of the roll. Operating experience will determine what types of nip pressures are best for the process. The operator can determine if the rolls are not crowned properly by using special paper to take a nip impression. This paper is called *nip impression paper*.
3. Wrinkling can occur in the nip if one roll is skewed (not parallel) relative to

the other roll. In this case the faulty roll must be realigned. Nip impression paper can also expose a skewed nip.

4. Wrinkling can be formed when an unwind splice passes through a machine. Operators tend to open the nip or bypass it if the wrinkles lead to web breaks. If the nip section is the machine lead section or is tension controlled, opening the nips or bypassing the section can disturb the transporting of the web. At the very least, the web will slip through an open nip at low levels of differential tension.
5. The web will be marked by lumps of coating or dirt that accumulate on the roll surface or by cuts that occur to roll covers when operators splice a web on a roll surface. The roll cover is damaged when the operator cuts the web with a razor or utility knife. A splice table should be provided at convenient locations in the machine to discourage splicing on machine rolls. Similar defects also come from unnipped rolls that collect dirt, coating, or are gouged when they are used as splice tables.

Unnipped Rolls (Elastomer Rolls)

Unnipped rolls (Figure 10-9) are used frequently on low-speed machines and where coating quality can be affected by anything that contacts the coated side of the web. Low speed is considered to be speeds under 200 ft/min. In this case the coated side of the web is not touched by any roller once the coating is applied. These rollers use a wide variety of elastomer coverings and a high degree of wrap to provide the grip that is required to control the web effectively. As in the case of the nip rolls, bowed rolls and short draws are used before the S-wrap rolls to ensure that wrinkles and creases are not formed. The S-wrap drive roll is inserted into the web path where a large natural wrap angle exists. This may be at the exit of a dryer or within a dryer that is arranged in clamshell fashion. The web will wrap a drive roll 180° at the end of a clamshell dryer. The temperature limits of the elastomer cover should be considered. Vacuum rolls might be needed if the dryer temperature exceeds the elastomer temperature limits. Soft elastomers are limited to temperatures of 180 to 200°F (82 to 93°C).

The use of elastomer covers such as those listed in Tables 10-1 and 10-2 are used to provide a good grip on the web in order to transport the web through the machine. Polyurethane roll covers have excellent properties for drive rolls but are limited to low temperatures. Other elastomers are satisfactory in a high-temperature environment but may lose their ability to grip the web quickly. Some plants have had success by spiral grooving the elastomer rolls to give the boundary layer of air an exit path without causing the web to float over the roll. Some plants have also placed a vacuum box under the unnipped roll to evacuate the air from grooves. Speeds up to 500 ft/min have been experienced with grooving and a vacuum system.

When elastomer rolls become smooth and ineffective drive sections, many plants replace them with suction rolls and/or suction aprons. If elastomer rolls continue to be used, a preventive maintenance program must be established that examines the rolls to determine if they are becoming "shiny" and are slipping. Operators will sand the covers in place while rotating the roll at very slow speed.

If the process can tolerate chrome-plated rolls, two chrome-plated idler rolls should

Table 10-1 Chemical Resistance of Basic Roller Coverings [a,b,c]

	Gasoline/Naptha/Kerosene	Toluene	Methyl Ethyl Ketone	Perchloroethylene	Trichloroethylene	Mineral Oil	Vegetable Oil	Glycerine	Ethyl Alchohol	Sulfuric Acid	Phosphoric Acid	Hydrochloric Acid	Acetic Acid	Chromic Acid (12%)
Buna N[d] (nitrile)	Exc.	Good	N.R.	Fair to poor	N.R.	Good	Good	Exc.	Good	Fair	Good	Fair	Good	N.R.
Neoprene[e]	Good	N.R.	N.R.	N.R.	N.R.	Fair	Fair	Good	Fair	Fair	Good	Fair	Good	Fair to poor
EPT[f]	N.R.	N.R.	Good	N.R.	N.R.	N.R.	Good	Good	N.R.	Fair to poor	Good	Fair	Fair	Fair to good
Hypalon[g]	Good	Fair to poor	N.R.	N.R.	N.R.	Good	Good	Good	Good	Good	Good	Fair	Fair to good	Fair
Urethane[h]	Exc.	Good	Fair to poor	Fair to poor	N.R.	Fair	Fair	Good	N.R.	Fair to poor	Fair to poor	Fair to poor	Fair to poor	Fair to poor
Silicone[i]	Fair to poor	N.R.	N.R.	N.R.	N.R.	Fair	Fair	Fair	N.R.	Fair to poor	N.R.	Fair	Fair	N.R.

[a]Among the factors to be considered in determining a roll cover for use with the chemicals listed below are:
1. Primary chemical 2. Presence of auxiliary chemicals 3. Temperature of operation 4. Presence of abrasive materials 5. Cycle of operation 6. Product contamination (Would cover affect customer's product?) 7. Are rolls to be used interchangeably (with any two or more chemicals)?
[b]Buna N is used at the standard spreader roller cover. All other covers MUST be special order ONLY. Standard durometer for spreader roller covers is 55–60 Shore A.
[c]N.R., not recommended.
[d]Buna N: Temperature range, 250°F max. Abrasion, fair to good; Color, off-white.
[e]Neoprene: Temperature range, 250°F max. Abrasion, good; Color, gray-black.
[f]EPT: Temperature range, 300°F max. Abrasion, good; Color, blue-green.
[g]Urethane: Temperature range, 220°F max. Abrasion, Excellent; Color, amber.
[h]Hypalon: Temperature range, 300—350°F max. Abrasion, Excellent; Color, brownish-tan.
[i]Silicone: Temperature range 500°F max. Abrasion, poor; Color, dull red.

be used to provide the maximum wrap angle on the drive roll. This should work well before the coating stand. In many cases a two- or three-roll S-wrap configuration will be used before the coater. Each roll will be covered with elastomer. The dominant roll cover is considered to be "sacrificial" and will pass the web transport load onto the next elastomer-covered roll as it becomes worn.

If the process cannot tolerate any contact with the coated surface of the web, one should use a large-diameter vacuum roll or a vacuum table to transport the web. These drive rolls must be kept clean and free from coating lumps and must not be used as splice tables. They can cause product defects as easily as nip rolls.

Table 10-1 (*Continued*)

	Potassium Hydroxide	Amonium Hydroxide	Sodium Hypochlorite	Sodium Carbonate	Sodium Cholride	Aluminum Chloride	Aluminum Sulfate	Triethanolamine	Formaldehyde	Natural Wax	Petroleum Wax	Oxidized Starch	Chlorinated Starch	Ozone	Water at 212°F
Buna N[d] (nitrile)	Good	Good	Fair	Fair	Good	Good	Good	Fair to good	Good	Good	Good	Good	Good	Poor	Fair
Neoprene[e]	Good	Good	Good	Good	Fair	Fair	N.R.	Good	Fair	Good	Good	Fair	Fair	Fair	Fair to good
EPT[f]	Good	Good	Good	Good	Good	Good	Fair	Good	Fair	Fair to poor	Fair to poor	Good	Good	Exc.	Exc.
Hypalon[g]	Good	Fair	Good	Fair	Good	Good	Good	Good	Good	Good	Good	Good	Good	Good	Good
Urethane[h]	Fair to poor	Fair to poor	Fair to poor	Fair	N.R.	N.R.	Fair to poor	Good	N.R.	—	—	—	—	Fair	N.R.
Silicone[i]	Fair to poor	Fair	—	Fair	—	—	—	Fair to poor	—	—	—	—	—	Fair	Good

Suction Rolls

Suction rolls (Figure 10-10) are used where elastomer rolls are not effective. Suction rolls consist of a drilled shell that is covered by a screen or by a sintered bronze cover. Suction rolls have been effective up to 3000 ft/min. Suction rolls can cause the following defects:

1. Scratches if the roll slips against the web. This usually happens when a vacuum roll is applied to a position in the machine where the web has a minimum wrap around the vacuum roll and is then subjected to more differential tension than the section can control.

   All drive rolls require a sufficient wrap for good web control. Adding vacuum to a well-designed drive section enhances its control of the web. Adding vacuum to a poorly designed drive section can only lead to poor drive performance when the roll begins to slip. Resist the temptation to use small-diameter suction rolls because they are less costly than large-diameter suction rolls. Large suction rolls (12 in. or larger) with 180° wrap will have the ability to handle large differential tensions while treating the web gently. Small rolls with

Table 10-2 Material Comparison Chart

Property	Material[a]							
	Neoprene	Nitrile	Hypalon	Thiokol	EPDM	Silicone	Polyurethane	Hydrin
Tensile strength	G–E	F	G	P	F	P[b]	E	G
Hardness range	10–95	20–100	45–85	15–80	25–90	35–95	10–95	50–90
Maximum service temperature (°F)	250	250	300	212	350	500+	212	275+
Ozone resistance	F–G	P	E	E	E	E	E	F–G
Cut resistance	G	F	F–G	P	F	P[b]	E	F–G
Tear strength	G	F	G	P	F	P[b]	E	F
Resistance to compression set	G	F–G	F	P	G	E	F–G	F
Abrasion resistance	G–E	F	G	P	F	P[b]	E	F
Resistance to heat buildup	E	P	P–F	P	F–G	G–E	E	F–G
Swell resistance to:								
ASTM No. 1 oil	F–G	E	F–G	E	P	G	E	E
ASTM No. 3 oil	F	E	F	E	P	F	E	E
Reference fuel B	F	G–E	P	E	P	P	E	G–E
Ketones: Methyl ethyl ketone	F	P	F	E	E	F	P	P
Aromatics: Toluene	P–F	G	P	G	P	P	E	G
Aliphatics: Hexane	G	E	G	E	P	P	E	E
Esters								
Ethyl acetate	G	P–F	F	E	E	G	P–F	P
Cellosolve	E	F	G	E	E	E	P–F	P
Chlorinated solvents								
Methyl chloride	P	P	P	G	E	F	P	P
Trichloroethylene	P	P	P	P	P	P	P	P
Glycols: Diethylene glycol	E	E	E	E	E	E	G	E
Alcohols: Isopropyl alcohol	G–E	G–E	E	E	E	G	G	G–E
Water: Distilled (75°F)	G	G	E	E	E	G	G	G
Caustics: 10% NaOH	G–E	G–E	E	G	E	E	P	G
Acids: 10% H_2SO_4	G–E	G–E	E	G	E	E	P	G

[a]Ratings: E, excellent; G, good; F, fair; P, poor.
[b]At elevated temperatures, these properties equal or exceed those of most other elastomers.

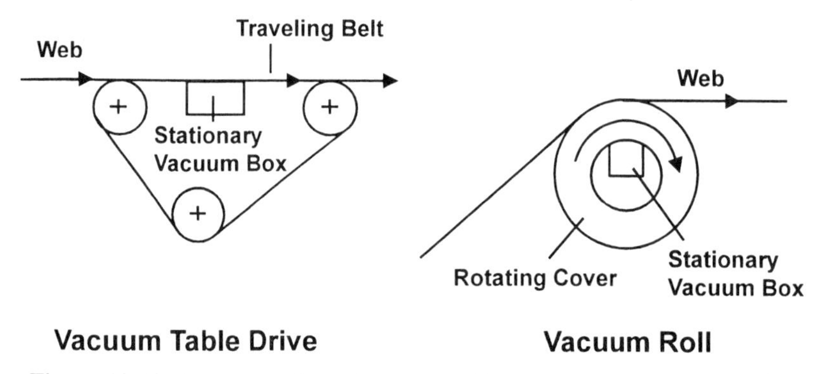

Figure 10-10. Vacuum rolls and vacuum tables are used to drive the web.

a small wrap angle are only capable of supporting small levels of differential tension and tend to cause scratches on the web when they slip.

Involve your supplier of vacuum tables and vacuum rolls in the application. He will need to know what material you are processing, how fast the web will run, what tension levels you will use, and the web's coefficient of friction. Your equipment supplier should be able to help you select the best roll diameter or table length for your vacuum drive sections.

2. If the vacuum roll is located directly after a coater, the pattern of the screen cover may show up on the coating surface. It appears as though the smooth sintered bronze cover, with its small pore size, has less tendency to affect freshly applied coating.

Suction Aprons

Suction aprons (Figure 10-10) are frequently used whenever very small degrees of wrap are available to nip rolls, unnipped rolls, S-wrap rolls, or vacuum rolls. The flat vacuum apron presents a large flat surface to the web and motivates the web whenever there is little or no wrap angle available. Examples of such a condition is the horizontal space between two consecutive drying zones or a vertical drop from a top drying zone to a bottom drying zone. Another location can be at the dryer exit just before a laminator or rewind. Vacuum tables have been used routinely at speeds up to 2,000 ft/min. Some vendors have recently claimed higher speeds for their vacuum tables.

Paper machines run vacuum rolls and vacuum tables at speeds over 6000 ft/min, so it seems certain that coating lines should expect elevated speeds from these devices. Material selection for the screens and vacuum boxes are important considerations to the reasonable life of the screen materials.

Examples of defects caused by vacuum tables are similar to the vacuum roll. Scratches and coating deformation can occur if the web slips or the section is located close to the coater where the freshly applied coating can be distorted by the combination of vacuum and a coarse screen pattern.

The vacuum roll or vacuum apron both have the ability to vary the width of the vacuum box to compensate for a range of web widths that are run on typical converting operations. The operator must match the width of the vacuum box area to the sheet width. If the vacuum box opening is wider than the sheet, vacuum will be short circuited and the roll will slip, causing mistracking or scratches on the web.

A special caution is in order for those who want to use a web cleaner to clean the web before it is coated. Do not tie the web cleaner into the vacuum roll vacuum fan. The volume requirements of the web cleaner are much higher than the volume requirements of the vacuum rolls. When web cleaners are connected to a vacuum roll vacuum fan, the vacuum may be reduced to a point where the vacuum roll becomes ineffective.

10.2.6 Increasing the Reliability of Draw Roll Sections

If a draw roll is needed to assist the web through the process, it is important to decide what type of draw roll is best for the application. Too often, a vacuum roll or vacuum table is rejected because it costs too much. Alternative types of rolls are used, cause problems such as scratching or wrinkling, and are then bypassed. Evaluate how a draw roll configuration will perform with your process. Mylar webs may slip over chrome or elastomer rolls at low speeds because the boundary layer of air cannot escape through the nonporous web. A vacuum draw roll or apron may be a better application—unless it is located directly after a coating stand. A vacuum apron with a sintered cover may work well after the coater. Try a test run at the vendor's plant to confirm it. Nip rolls after unwinds may crush the splice pattern as it proceeds into the machine. An S-wrap or vacuum roll may be a better selection.

Manufacturers of paper coaters design nipped draw rolls into them for speeds up to 4000 ft/min. However, the same manufacturers are trying unnipped rolls using extremely tacky elastomer covers. Most manufacturers apply spreader rolls, such as bowed rolls, before the draw roll section.

No matter what the configuration of the draw roll, the condition of the rolls must be policed. If the draw roll is being bypassed, was the bowed roll removed, which allowed wrinkle formation at a nip? Are roll surfaces dirty or damaged? Are suction roll covers cut and puncturing the web? Are elastomer covers aging and becoming "polished"? Are they losing their ability to motivate the web?

Machine operators are rewarded for meeting production goals. They frequently will take whatever steps are necessary to ensure that goals are met—even if it means bypassing the draw rolls when they begin to slip or if they cause problems. An engineer or maintenance technician must "walk" the assigned area and determine if any problems have occurred or will occur. Part of the job is to read the operating log and to talk to the operators. The aim is to find out if the operator has taken any action that makes the machine vulnerable to defects in the future. Action can be planned and implemented by the support staff, which will reduce the chances of defects.

10.2.7 Tension Isolation

Draw rolls are used to separate a coating operation into identifiable tension zones. This capability is called *isolation*. Isolation is valuable only when there is a need for zoning. Many processes that run a constant level of tension throughout the machine do not require the extra complexity of draw rolls. However, if we are laminating webs of different characteristics, heating a plastic web until it is soft and compliant, or coating a paper web that may stretch or become weak when it is wet, isolation can be useful. Draw rolls may also be helpful when low tensions are used in the machine. This is especially true when thin films are being processed at low tensions. Here, extra draw rolls may provide web transport assistance when drive points are far apart. The paper industry used a rule of thumb that required a drive section every 20 ft to ensure proper tension control and web transport. We are seeing that to be true on machines that are designed for speeds above 3,000 ft/min. Drive sections on these coaters must be designed specifically to transport web without slipping. In particular, steam-heated drum dryer sections, with felts, are receiving special treatment to prevent slippage. The dryer is separated into two sections, the dryer drums and the dryer felts. The drive system is arranged to allow one section to be the master and the second to be a follower. The dryer drive control is arranged to allow the operator to split the total section load between the dryer drum motor and the felt drive motor so as to ensure that the dryer section operates as a slip-free drive section which is capable of providing effective isolation at a reasonable differential tension. We have noted significant improvement in web tracking through the machine when the dryer drive consists of a drum drive and felt drive.

Examples of different tension zones are:

1. Unwind versus the rewind tension.
2. A loosely wound roll may unwind at a lower tension than the entry side of the first coater.
3. The entry side of a coater versus the exit side where the wet paper may stretch.
4. The dryer exit versus the entry point of the rewind.
5. The entry side of a laminator versus the exit side of the laminator. Differential tension across the laminator will depend on the characteristics of each material being laminated and the tension needed to make the web track properly. If different materials are being laminated together one web may require a lower tension than the other if delamination is to be avoided. The resulting laminated structure may or may not require a higher tension at the laminator exit, depending on the laminating process and the tension needed to track the heavy laminated structure.
6. A draw roll before the rewind, or reel, on the coater will allow the operator to have one tension in the coating process while a different tension is being run to wind the roll. The draw roll before the rewind (or reel) also serves to buffer the machine from a failed rewind transfer or a web break between the rewind and draw roll. The draw roll isolates the web break to the rewind itself and eliminates the time that it takes to clean and rethread the entire machine.

Differential Tension

We have mentioned tension isolation and where it is used. If a drive section is designed to "isolate" one part of a process from another, the section must be able to maintain a tension on the inlet side that is different from the tension on the outlet side of the section. That ability is called *differential tension*. Differential tension is the difference in the web tension between the ingoing side of a section and the outgoing side of the same section (Figure 10-11). Every section is limited to the differential tension that it can control. When that differential tension limit is exceeded,

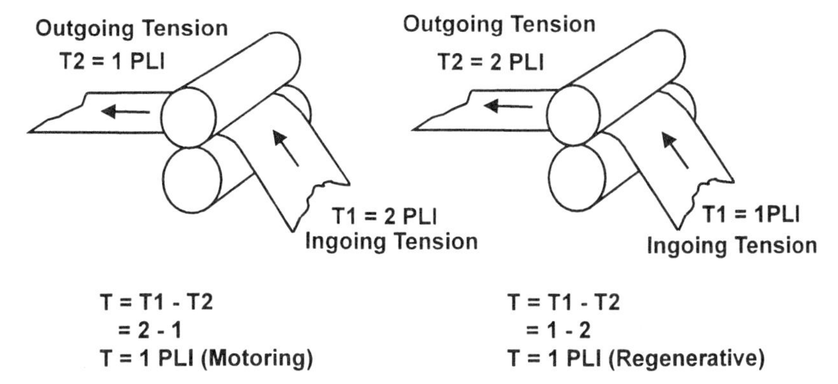

Figure 10-11. Differential tension is the difference in the web tension between the ingoing side of a section and the outgoing side of the same section.

the web will "break away," the section will slip against the web, and the section will no longer control web speed or web tension.

Drive sections that will isolate parts of the process must be designed to accommodate the levels of tension differential that the process demands. How much differential tension can a section handle? We need to know the following:

1. Is the drive roll a nipped section, or is it a suction roll, an S-wrap, or a suction apron?
2. What is the coefficient of friction of the web?
3. What is the coefficient of friction of the drive roll?
4. What is the diameter of the drive roll?
5. Has the roll become old and "polished" and suffered a reduction in coefficient of friction?
6. How fast is the machine running, and what is the boundary air effect?

Most coating machines tend to run best when we set the tensions throughout the machine at the same level. For instance, if 1 lb/lin. in. is run at the unwind, then 1 lb/lin. in. is used at the coaters, the draw rolls, and the winder. Running the same tension throughout a machine is called a *flat tension profile*. Using the same tension throughout the machine reduces the differential tension that any section must handle during steady state and increases its chance of handling tension transients during splicing, transferring, acceleration, and deceleration of the web. If the section is handling a large differential tension during steady-state conditions, any transient can cause the section differential to increase beyond its capability to control the web and the section will slip.

We know that elastomer roll covers eventually age and become polished while they process web. A roll cover inspection program must be established to identify rolls that may begin to slip or are slipping. Roll slippage will occur if the roll surface appears shiny or exhibits surface scratches. Surface scratches indicate that the roll is already slipping and a shiny cover indicates that the roll is probably losing its ability to handle the desired differential tension.

10.2.8 Driving Web-Carrying Rolls Within a Dryer

Impingement dryers may require tendency drives on the web-carrying rolls if very low tensions are run in the dryer zones. Tendency drives drive the shafts of idler rolls, with separate bearings supporting the rolls on the shafts. Thus the drives bring the rollers to approximately the line speed, and the web has only to supply a very low torque to adjust the roll speed to the web speed. Tendency drives eliminate slippage of the web on the idler rolls.

Low tensions and high impingement pressures will tend to cause the web to accumulate between web-carrying rolls and will cause mistracking. The addition of a drive to the web-carrying rolls seems to improve web transport through the dryer. We have seen machines where the idler rolls were driven within very long impingement drying sections. That drive helped transport the web, scratch free, through the dryers with excellent tracking at low tension levels. Attempts to eliminate the idler roll drive have resulted in web scratching, lateral web wander, and an increase in web breaks each time the machines were accelerated.

Consider the case where a machine was specified for a tension range of $\frac{1}{4}$ to 3 lb/lin. in. Attempts to run below $\frac{3}{4}$ lb/lin. in. resulted in the web forming deep catenaries in the dryer section. The dryer web-carrying rolls were undriven. The weight of the hanging web exceeded tension setpoints below $\frac{3}{4}$ lb/lin. in. We noted that the web entered the dryer section but did not exit it. We were fortunate that this machines' normal tension was 1 lb/lin. in. or higher. If tension below I lb/lin. in. had been required, the web-carrying rolls would have required a tendency drive. In any event, the machine could not accommodate the tensions that were specified by the client and vendor.

Undriven rolls frequently stop turning and cause scratches in the product. A tendency drive reduces the potential for scratching.

10.2.9 The Rewinder

The rewinder is the exit point for the wound rolls. If the process is intermittent in nature, such as a pilot coater, a single set of arms is enough to wind the jumbo roll. If the process is continuous, the rewind will need at least a pair of arms to ensure that the process will be uninterrupted. If the product is costly and easily damaged as a result of a poor start, the rewind may have three or four pairs of arms.

Use of Accumulators at the Rewind and Unwind

Some materials are easily damaged if a wrinkled start occurs on the core. Some of these materials also avoid the use of tape on the core. In those cases the rewind is stopped while the rest of the machine continues to run. An accumulator is located between the rewind and the machine. That accumulator is kept in a collapsed condition until a rewind transfer sequence is initiated and the rewind comes to a stop. When the rewind stops, the accumulator simultaneously begins to store the web that approaches the rewind. When a wrinkle-free tapeless transfer has been completed successfully, the rewind is restarted. The accumulator stops storing web and returns to a collapsed position in preparation for the next transfer.

The unwind frequently will utilize an accumulator to produce a zero-speed butt splice. The accumulator in this case is kept in the "filled" position and is emptied when the unwind is stopped. Thus the machine keeps running as the unwind is stopped and the operator splices the incoming roll to the expiring web. Accumulators require much space and are limited to the equivalent of 1 minute of running time at the fastest operating speed. If the transfer is not completed before the accumulator fills, an "accumulator full" switch signals the drive to stop. A high-speed manual transfer is sometimes used to initiate new operators on the coating line. Large audiences tend to congregate to see the first attempts of the new operator, and a great deal of gleeful backslapping occurs if a successful transfer is made.

It is necessary to design the accumulator so that it does not adversely influence the tension range of the machine. We have seen cases where the accumulators had so much friction in the moving roll carriages that it was impossible to run the tension levels that were specified in the machine. In those cases the tension specifications were $\frac{1}{4}$ to 3 lb/lin. in., but the accumulator friction could not accommodate tensions below 1 lb/lin. in. Fortunately, the operating tension was above 1 lb/lin. in. and the accumulator did not limit the existing process. However, as lightweight webs of thinner caliper are run on this machine, the accumulator will need to be modified to accommodate lighter tensions.

Surface/Center Rewinds

Surface rewinds typically cause wrinkling at the core and are not used for winding coated webs that need to be wrinkle-free. Most coaters that coat web for publication-grade papers use the surface rewind.

Surface/center rewinders start winding the web in the surface mode and then finish the roll in the centerwind mode. Some surface/centerwinds use both surface and center winding while the roll is building. The surface-driven roll is typically controlled by a dancer roll and the centerwind is torque controlled. This combination tends to build a good wound roll structure for materials that are otherwise difficult to wind. Surface/center rewinds are sometimes used to get a firm start for pressure-sensitive coatings. After the firm start, the operator has the option to retract the winding roll from the surface wind and to complete the roll in the centerwind mode.

The newer rewinders contain a "pack roll" that functions as a surface-driven rewind while the centerwind drive functions as a typical centerwind. It is necessary to determine the proper percent of power that each drive will contribute to the winding roll. Much of that depends on the material being wound and how much taper tension will be required.

Center-Driven Winders

Center-driven winders are tension controlled by force transducers or dancer rolls. These sensors can sense tension changes during transient periods such as transfer, acceleration of the machine, or when the rewind turret is rotated. A taper tension system maintains tension under steady-state conditions as the roll diameter builds.

The taper tension system (Figure 10-12) compares line speed with the speed of the winding roll and calculates diameter. It reduces tension in conformance with a taper tension controller that is controlled by the operator. Sometimes, the taper tension system responds incorrectly to temporary transients such as acceleration or turret

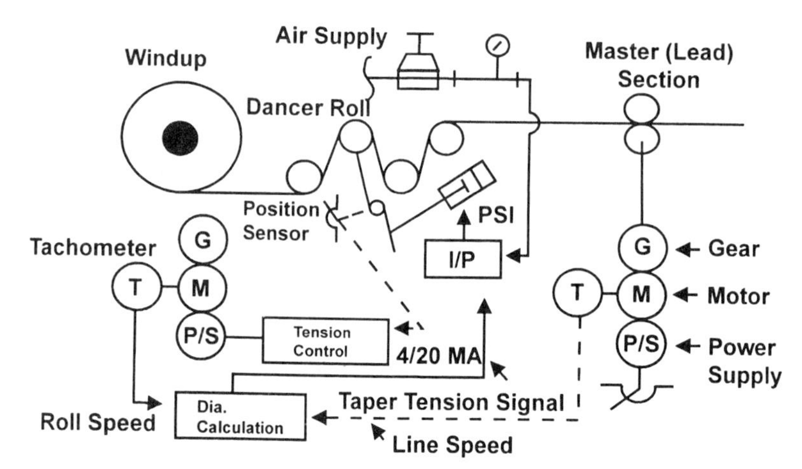

Figure 10-12. The taper tension system compares the line speed with the speed of the winding roll, calculates the diameter, and reduces tension in conformance with a taper tension controller.

rotation. Drive vendors have provided controls that lock out the taper tension system during transient periods. Techniques have been developed to minimize the influence of turret rotation by placing a variable-speed drive on the turret and to control it with the tension sensors.

Additional special control is required at the centerwind when a transfer is made. When transfer is initiated, the paste roll pushes the web against the core, causing the tension to increase. When the web touches the core, it isolates the tension sensor from the winding roll. The tension sensor senses an increase in tension and causes the tension controller to reduce tension. The web will go slack between the winding roll and the paste roll. As a result, the web will be slack when the cutoff knife attempts to complete the transfer cycle. The knife will not sever the slack web and the transfer will be missed. The machine will need to be stopped and rethreaded.

Modern drive systems transfer from tension control by dancer or force transducer to a current memory circuit that is enabled when the transfer is initiated. The current memory circuit will hold the winding motor current constant while the transfer is completed. The rewinder is returned to its normal tension control mode when the knife severs the web.

10.2.10 Drives for Laminators

The laminating process consists of two webs that are fed through two rolls that press the two webs together. The process can be "wet" or "dry." If an adhesive is added to the nip between the two webs, it is a wet process. If the adhesive is applied to the web at a coating station and then delivered to a laminator, it is a dry process. In some cases one web is laminated to a "carrier web," is coated, and later separated from the carrier web at the end of the process. The carrier web gives the product some strength and provides stability to a weaker web while it is being processed.

When to Drive Both Laminator Rolls

If the existing process provides good laminations that are free of delaminated sections, the existing drive is fine for the product. However, if you are experiencing delaminations, it is time to examine your laminated structure and the laminator drive system. Most laminators have a single drive to motivate the section. Let us evaluate what is happening at the laminator. First, a drive is connected to one of the laminator rolls. Then the second undriven roll is pressed onto the first roll to form a nip. The undriven top roll is driven by the nip of the driven roll. That means that the undriven roll represents a load (or drag) to the driven roll. If we introduce the webs to the nip that is formed between the driven roll and the undriven roll, the undriven roll will now be driven by the web, which is in the nip. More accurately, it will be driven by the top layer of web in the nip. The flow of power from the bottom roll will pass to the bottom layer of web, through the adhesive that bonds the two layers, through the top layer of web, and to the top roll. The load that the top undriven roll exerts on the top layer of web will apply a braking effort (or drag) to the top of the web. It is possible that the drag will cause the top layer of web to slip relative to the bottom layer of web. That slippage is delamination. In this case it will be necessary to provide both rolls with drives. That can be done by connecting the top roll to the bottom roll drive through a clutch or by connecting a separate drive to the top roll.

A separate drive on the top roll must be adjusted to supply the frictional losses of the top roll and some of the nip losses. If the top roll drive is adjusted to provide an excess amount of load, power will be added to the top layer of web, and a delamination can occur.

10.2.11 Coating Roll Drives

Coating roll drives also require special attention. Blade coaters on paper webs may require special control to prevent the trailing blade from gouging the drum when the machine is stopped. If the gears and couplings have backlash, the drum can back up when it reaches zero speed. If the blade is still in contact when the drum backs up, the blade can (and has) gouged the drum. To prevent gouging, the drive control must generate a signal as it approaches zero speed. The speed signal is used to retract the coating blade just before the coating drum stops.

Backlash from gears, couplings, and belts is also a problem on coating lines that apply optical quality coatings to a moving web. Backlash can cause chatter marks on the coated web.

10.2.12 Coating Chatter: Sources and Some Solutions

Coating on blade coaters, photographic film coaters, label stock coaters, plastic or metal film coaters, and textile laminators all apply a coating that is judged on the cosmetic quality of the finished web. Coating chatter will result in inferior cosmetic qualities and will cause the customer to return the web to the manufacturer (Figure 10-13). Coating chatter shows up as periodic light and dark coating bands on the web. Coating chatter can be caused by the coating roll, drive motor, control equipment, drive power supply, and power transmission equipment, such as belts, couplings, gear boxes, chains, and universal joints. Chatter can also be caused by adjacent section

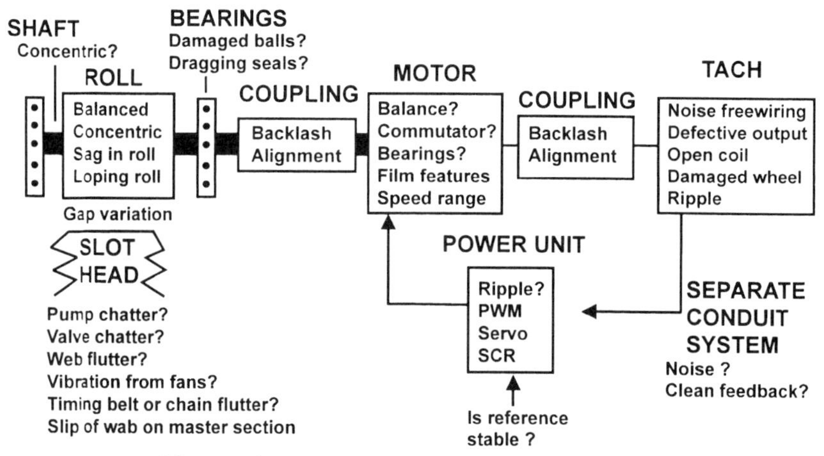

Figure 10-13. Sources of some coating defects.

drives that transmit vibration through the web or by mechanical equipment that transmits vibration to the coating stand through structural members.

Backlash in Timing Belts, Gearboxes, and Geared Couplings

If the coating roll is connected to the drive motor through any type of gear, the profile of the gear teeth may show up as chatter on the coated web, particularly when the drive motor is lightly loaded. Drive motors are frequently lightly loaded on slot coaters. The drive motor must only supply the frictional losses of the coater. Very often, the tension at the coater outlet is increased, which further reduces the load on the coating roll drive. As a matter of fact, the load on the coater may be reduced to the point where it is operating at zero load. With light loads, the torque requirements may shift between motoring and braking. When the load shifts, the gear teeth first drive on the forward side of the gear tooth and then hold back on the back side of the tooth. The changing of the tooth contact must pass through the backlash region. Backlash, "slop," or clearance causes a nonlinear velocity (a perturbation) that results in the coating chatter. Some operators compensate for slop by increasing the tension at ingoing side of the coater. This strategy increases motor load, which keeps the gear in contact with one side of the tooth, thus reducing the effect of backlash. If increasing the tension does not solve the problem, the gears must be replaced.

We have seen chatter occur when heavy-duty timing belts were used to drive an air-knife coating roll. The gearlike teeth in the heavy-duty timing belts were so well defined that the coating stand actually rumbled when the drive was in operation. This particular chatter was eliminated by replacing the heavy-duty timing belts with standard-duty timing belts. In other cases where chatter was a problem, the timing belts were replaced with flat belts. Flat belts seem to provide smoother drive performance than the timing belt. Other engineers have reported that they have reduced chatter by turning the timing belt over and simulating a flat belt surface with the back of the timing belt. New timing belt designs, using a rounded tooth design, have reduced the tendency of timing belts to chatter. We have also seen nonlinear velocity when timing belts, flat belts, or chains were used on *long belt centers.* A long belt center exists when the drive pulley is very far from the driven pulley. Long belt center belts tend to pulsate as they rotate. The use of a belt idler pulley will reduce the amplitude and increase the frequency of the disturbance to a point where it might not be seen as chatter. However, on very critical applications, a direct drive between the motor and the roll may be the best answer to smooth perturbation-free coatings.

Chatter that is the result of gears, gear-type couplings, timing belts, or chains can be eliminated by replacing the "sloppy" devices with zero-backlash devices and direct-drive motors. Care must be taken to ensure that the zero-backlash devices are very stiff so that they do not vibrate during periods of acceleration and deceleration. This vibration is called *torsional vibration*. Let us simply say that an elastic coupling or shaft can begin to vibrate when it acts in concert with the drive control system. The drive control system sees the vibration as a speed error and tries to correct it by accelerating or decelerating the shaft or coupling. The vibration of the coupling or shaft becomes out of phase with the drive system correction, and the result is a system that has gone into torsional vibration and will stay that way until the machine is stopped, until the coupling or shaft is destroyed, or until a stiffer drive transmission device replaces the elastic member.

Remember, a shaft or coupling that seems very stiff and hard to us may appear to be a rubber hose to the drive control system. A coating roll drive needs a control system that is extremely responsive. An elastic power transmission system compromises the system's ideal performance. Long drive shafts, or couplings with rubber inserts, or with leaf springs, or with flexible disks can create problems when applied to high-performance drive systems.

Improperly aligned couplings can cause velocity perturbations that translate into coating variations. Connecting the coating roll directly to the drive motor through a properly designed coupling can eliminate gear reducers, backlash, torsional vibration, and coating chatter.

Chatter Caused by the Roll

Chatter caused by the roll can be due to conditions such as defective bearings, roll imbalance, and a nonconcentric coating roll. Bearings that were installed improperly on the roll, or no longer are within specifications due to normal wear, must be replaced to achieve good coating results.

Roll Unbalance

Those of us who have driven automobiles have seen what can happen when the front tires become unbalanced. The tires will vibrate when we approach certain speeds. The same is true with coating rolls. If the roll is unbalanced, it will vibrate at certain speeds. In addition, we can see the effects of the unbalance as we rotate the roll by hand. As we rotate the roll we will feel a point where we need to exert force to push the roll uphill (applying positive torque). As we continue to rotate the roll we suddenly reach a spot where we must apply a braking effort (apply negative torque) to keep the roll from rotating by itself. This alternating motoring and braking torque requirement is another area that causes chatter when slop exists in the drive system. In addition, the unbalanced roll presents a load that alternately requires a positive torque and a negative torque to rotate the roll. A drive system must alternately supply positive or negative torque if it is to keep the drive rotating smoothly. However, like slop in gear reducers, the unbalance can excite the drive into "torsional" vibration and cause serious damage to power transmission components or to the roll itself. A buildup of dirt or coating solids on the coating roll may also cause roll unbalance. Keeping the roll surface clean will avoid roll unbalance or impressions that occur when a coating lump embosses the web during the coating process.

Nonconcentric Rolls and Drive-Related Problems

A nonconcentric roll can be caused by bent roll journals or by the fact that some rolls will "sag" or deflect if they are kept stationary too long. It is difficult to imagine a metal roll deflecting. Metal rolls will deflect as readily as a rubber hose when a length of hose is suspended between two points. The longer the span, the more noticeable the deflection or sag.

This type of roll is frequently supplied with a "Sunday" or "inching" drive to keep the roll rotating during periods of inactivity. Such inactivity frequently exists on the weekends, during plant shutdowns, or when the machine is idle for maintenance purposes. This drive is important enough to cause many coating mills to assign the plant security staff the responsibility to check the drive as part of routine plant inspection during downtime periods. A slow-speed drive prevents the deflection inherent with some coating rolls if they are stationary for too long a period.

The drive system can also provide a source of noncontinuous velocity that results in uneven coating. To satisfy extremely precise coating needs, the drive motor may require precise balancing, special bearings, and skewed armature slots. Such a motor is said to have "film industry features" to provide smooth rotation. A servo quality motor may be required even to the point of selecting a motor with minimal or no commutator ripple. Modern drive systems advertise perturbation-free performance over a 1000 : 1 or 2000 : 1 speed range. Older drive systems were rated 20 : 1 or 30 : 1. Perturbation-free claims are all relative. See Table 10-3 for comparisons. It is best to find a place to test these claims. Many vendors have test facilities designed to allow customers to test drying equipment, coating equipment, web transport equipment, and drive systems. Drive manufacturers, machinery builders, coating manufacturers, and even customers have pilot coating facilities that are made available to anyone with questions about a new process or equipment.

The output of the power supply to the motor may contain enough ripple to affect coating quality. Single-phase SCR (silicon-controlled rectifier) power supplies have a very high ripple content at lower coating speeds, and this ripple will reduce coating quality. Power supply ripple can be reduced by going to three-phase SCR, pulse-width-modulated (PWM) power supplies, or servo quality power units. Many servo quality drives are designed to operate at extremely low rpm speeds without "cogging" (velocity perturbation or stop–go operation). The recent development of alternating-current (ac) vector drives has given an additional option for precision coaters. However, low-speed operation of these drives should be evaluated for cogging.

Care must be taken in applying *all* types of drives to coating applications. Newer technologies require some research into the advantages and disadvantages of these drives. An understanding of the pros and cons of each drive will result in a good and cost-effective match of technology to the application.

Table 10-3 Comparison of Drive Features

	1 Phase SCR DC	3 Phase SCR DC	DC Servo	Brushless Servo	Vector AC
Motor	DC	DC	DC	Brushless	AC induction servo
X-Proof	Yes	Yes	Yes or purge	Purge	Yes or purge
Micro controller	Least complex	Mid-range complexity	Mid-range complexity	Higher complexity	Highest complexity
	Fewest diagnostics	More diagnostics	More diagnostics	More diagnostics	Most diagnostics
Horsepower	1/4 to 75	3 to 600	1/4 to 30	1/4 to 16	1 to 50
Application	Low power, general applications	Mid–high power, general applications	Low–mid power, general and high performance	Same as DC servo and high speed position applications	Same as brushless in higher HP
	Medium performance	Medium performance	High performance	Highest performance	Very high performance
Speed range	20 : 1	30 : 1	1000 : 1	2000 : 1	1000 : 1
	Cogging at low end	Some cogging at low end			
Gear reduction	Needed	Needed	Little or none needed if enough torque	Little or none needed if enough torque	Little or none need if enough torque
Regeneration	To AC line	To AC line	To snubber or bus	To snubber	To snubber or DC bux
Power factor	Low, 0.68 avg	Low, 0.68 avg	0.95 constant	0.95 constant	0.95 constant
	Higher at higher speeds	Higher at higher speeds			
Efficiency	72–78% at full power	72–78% at full power	Up to 80%, depending on motor and bus	Highest, up to 83%	Up to 80%
	Less at low speeds	Less at low speeds	Fairly constant with speed	Constant with speed	
Speed feedback	Resolver, pulse tach, or analog tach	Resolver, pulse tach, or analog tach	Resolver, pulse tach, or analog tach	Pulse encoder	Pulse encoder
Relative price, Non-X-proof	Lowest	Mid to high	Mid to high	More than 1-phase DC	Highest
		Similar to DC servo	Similar to 3-phase SCR DC	Less than 3-phase SCR or DC servo	
		Economies available in DC servo	Economies available in DC servo		

Feedback Devices

The perturbations in rotational velocity of couplings or belts that drive velocity feedback devices can create "ripple" in the output of the velocity feedback device. If a digital device is used as a velocity feedback, its signal converter may contain enough ripple to influence the drive system velocity. If a digital reference is converted to an analog reference for the drive, the resulting analog output may contain an unacceptable amount of ripple. If the velocity control system responds to the ripple of the feedback or set-point devices, the application of coating may be so nonuniform that it will create cosmetic defects in the finished product. The nonuniformity may cause the quality control group to scrap the product.

It is imperative that the velocity feedback device be connected to the motor with a stiff coupling that is aligned perfectly. It is not uncommon now to use laser alignment equipment. The use of belt-driven tachometers is acceptable for paper machines and for many coating lines, but for the coating rolls on converting lines, an encoder that is direct driven is the most troublefree method of feeding back velocity signals. There are a variety of analog and digital velocity-sensing devices for mounting on motors. Many of these devices show better maintenance characteristics when they are connected directly rather than belt connected to the motor, as the tachometer bearings fail more often on belt-driven units. Many mechanics apply belt drives with excessive tension. We have seen cases where the tachometer shaft is deflected due to belt tension. The tachometer or encoder bearings are not designed to withstand excessive belt tensions and, as a result, will fail in a short amount of time.

Coating Supply System

The coating supply system to the coater can provide sources of coating chatter. Unstable feed valves, positive-displacement pumps, and unstable drives for the pumps can all cause coating variations. Unstable feed valves or feed valve control systems can permit surges that cause variations in coating thickness. It is important to match the characteristics of the coating to the characteristic of the feed valve. Valve curves and a knowledge of the coating fluid characteristics will help to avoid coating problems.

Like gear pumps, positive-displacement pumps may transfer the gear-tooth profile through a slot head to the web. Unstable drives on positive-displacement pumps may cause surges and transfer thickness variations through a slot head to the web. Some coatings may require precise drives for the supply system pumps so that they may operate smoothly over a wide speed range. A pump that will be operated over a wide speed range may require premium-quality motors and power supplies to ensure proper performance over the prescribed speed range.

Transmission of Vibration Through Structural Members to the Coating Stand

If dryer fans are mounted in the ductwork of the machine and the ductwork is supported from the coating stand, the fans can cause the ductwork and coating stand to vibrate. Any vibration at the coating stand or coating head support can cause coating chatter.

The practice of mounting fans in ductwork is evident on lab or test coaters, where space is limited or the machine is portable. If any coating stand is not mechanically isolated from the rest of the machine, vibrations caused by the surrounding machinery will cause the coating stand to vibrate. These vibrations are transmitted to the web and can cause coating chatter. Do not attach support members to the coating stand, but instead, design an isolated coating stand structure that is separated from the rest of the machine and building if that is what is required to eliminate vibration. Elaborate machine isolation pads are available and some consultants specialize in designing buildings and machine isolation systems which reduce vibration to tolerable levels.

10.3 EFFECT OF DRYER DESIGN ON COATING QUALITY

Dryer nozzle design can affect the quality of wet coatings by disturbing the layer of web coating (see also Chapter 9). Impinging air can cause such defects as *railroad tracking*, which occurs when a hole or slot spreads the coating in one location on the web. That spread will show up as a straight line, running in the machine direction, on the dried coating. It is important for the coating plant to test their coatings on different drying systems at the vendor's plant. Test a wide range of coatings rather than just the one you believe might cause a problem.

A great deal of concern always surrounds the effect of flotation dryers on coatings that are on the web and can flow. This concern centers around the question whether these dryers will displace the coating as the web travels through the dryer. Here again, a test at the vendor's facility will uncover any problems that are anticipated and also many problems that were considered to be nonexistent.

10.4 TROUBLESHOOTING "SUDDEN" COATING PROBLEMS

It takes a great deal of patience and familiarity with the machine to identify the source, or sources, of coating defects. Knowledge of both the machine and process are needed to determine whether a mechanical or process problem is causing the defect. A recurring mechanical failure will cause a coating defect. It is important to identify that failure, the frequency of that failure, and to establish a preventive maintenance schedule that anticipates such an event. The implementation of a preventive maintenance schedule will encourage inspection, adjustment, replacement, or cleaning of the equipment in question so as to *minimize* future failures.

Devices such as friction brakes or magnetic particle brakes need periodic maintenance to maintain their performance at acceptable levels. Items such as brake friction linings should be resurfaced or replaced while magnetic particles must be replaced with new magnetic particles. Bearings must be inspected, cleaned, lubricated, and replaced on a routine basis.

Motors, tachometers, and force transducers must be examined and kept free of dirt, worn parts, defective bearings, and worn belts if system performance is to be kept at an optimum level. Brushes for direct-current (dc) motors and tachometers should be examined on a timely basis and should be replaced as recommended by the equipment builder. Knowledge of the machine and process are needed to determine whether a mechanical or process problem is causing the defect.

Awareness of any changes that might have occurred just before a defect was noticed is important. It is extremely important for production personnel to know if machine modifications are planned for a shutdown period. The following activities are usually timed to occur during shutdown periods, much as plant shutdowns during planned summer and winter shutdowns.

1. Change a coating roll or drive roll. If the roll, bearings, or power transmission equipment is not installed properly, coating quality will be affected. It is my belief that this type of change should be performed by experienced personnel familiar with modern alignment instruments and the critical needs of the process. In some ways I lament the loss of the apprenticeship programs that provided training for entry-level trades personnel. We have too often adopted the "let's see what the new guy can do" attitude to the frustration of the production personnel and to the cost to the corporation.
2. Were process air filters changed? If so, are the filter medium and the filter seal compatible with the process? Consider the case where the purchasing department sent out for competitive bids for a filter with inadequate specifications. A vendor said that he could meet the specifications with a less expensive line of filters. The filters were purchased. To our chagrin we discovered that the "identical" filter medium was fine but the filter gasket contained a chemical that would cause serious fogging problems on the photographic film that was coated. We discovered the change in filter brands by accident when an air-handling mechanic complained that he needed more machine downtime to install the new filters. When we examined the new filters, we discovered the difference. Installation of the filters would have resulted in serious financial loss to the company.
3. Was new equipment added to the machine? If so, what effect will the new equipment have on web tension, web guiding, or wrinkle formation? Will the operators be able to thread the web through the section safely and conveniently? New capital projects are too frequently carried out with little involvement on the part of those that are most affected by the changes—the operators.
4. Was maintenance performed on the machine? Were rolls realigned or leveled? If so, who asked for the realignment? Were there tracking or wrinkling problems, or was the realignment a result of preventive maintenance? If the operators had no complaint about web tracking, then the maintenance staff should document out-of-level and misaligned rolls for review with the production department. If the data are consistent with problems that the operating crew sees, proceed to level and align the rolls. It is not uncommon to have the maintenance staff check machine alignment and level once a year or once every two years. This check will not only expose problems that occur when roll bearings are replaced during the year, but it will also expose problems with a new building "settling" after a major capital expansion.
5. Were brakes and clutches rebuilt? If so, will the brakes and clutches require different settings? It is important for the maintenance and production to communicate with each other about work being done during shutdown so that the production group will be aware that something is different. That difference may require changes in operating techniques and control system set points by the operating crew.
6. We have seen cases where fan bearings were replaced during a shutdown but the mechanic forgot to replace the fan drive belts. Production was delayed until

an air-handling mechanic was summoned to install the belts. Luckily, fan pressure was monitored as part of the control strategy. This monitor would not allow the coater to start up without proper fan pressure. Otherwise, it may have been possible to coat the web with improper drying conditions. Significant losses would have occurred without proper monitoring of the fan.

7. Was instrumentation recalibrated during a plant shutdown? Consider the case during a summer shutdown, where an instrument technician discovered that the dryer zone temperature sensors were reading 20° low. Actual temperatures were 20° higher than indicated. The technician recalibrated the sensors so that they read accurately, but he did not tell the operators what he had found. The operators set up the machine with the temperatures approved for the product. Drying was affected and product quality was poor. Drying was identified as a problem but the operating crew was not authorized to make drying set-point changes. The machine was shut down, and a committee was formed that ultimately identified the problem. In the meantime, production was lost at a time when the company was trying to build inventory for end-of-the-year holiday sales. Here again, the operating crew must know what calibration will be performed, what problems were found, and what was done to correct the problem, and what temperature set point is now appropriate. They should also be instructed by the maintenance group as to how to proceed into production with the newly calibrated instrument system. As a matter of routine procedure, the machine should be operated on a test run before going into a production phase. The test should be designed to confirm the satisfactory operation of the machine and to determine if anything has changed in the two-week shutdown. This test reduces the waste that comes from starting production on costly products which use resources that are both expensive and difficult to obtain (gold, silver, or materials that are supplied by vendors who have limited production capabilities).

Can any of these events cause a web coating defect? Absolutely! Solving these problems may require a major troubleshooting effort requiring human resources, vendor involvement, diagnostic equipment, and the cooperation of the management, maintenance, production, and engineering disciplines.

10.4.1 Monitoring Speeds and Tensions of Web

We tend to process webs at specific speeds and at specific levels of web tension. It should be no surprise that we are interested in the speed of the motor, how much torque the motor is applying to the web, and the measured web tension.

Speed Monitoring System

Modern speed monitoring systems consist of a modern microprocessor-based controller that accepts a speed-related signal from each motor-driven section and some sections that are driven by the web. These systems are available with a wide range of alarm and report generation capabilities. Reports can be generated when an event such as a web break occurs. The report will highlight the speed of all sections just prior to the event. Speed monitoring systems are ideal for older coating machines that do not have process computers to provide this function. They may also be ideal for modern coaters when the operator wants a continuous display of speed. Continuous display of speed is important during the startup and debug phases and while trying to solve product defect problems or trying to identify the cause for a web transport problem. A speed monitoring system is also used, after commissioning, to identify drive performance problems with a section speed controller or its tachometer.

Monitoring the Speed of Idler Rolls

Many monitoring systems include a speed signal from a roll that is driven by the web (often called a paper roll or idler roll). If the idler roll slips, the speed signal will be inaccurate. Select a roll with a high degree of wrap to ensure accurate speed indication. Slip can then be detected quickly by monitoring the speed differences between the master drive section and the idler roll. Some monitoring systems provide an alarm or computer output to indicate slippage beyond acceptable limits. Such an alarm is very valuable when an idler roll stops turning and scratches the web as the web is dragged over it. Coating machines with long dryer sections have dozens of idler rolls which may stop turning due to a lack of lubrication or a failed bearing. The monitoring system helps identify the problem.

High-speed machines tend to drive their web-carrying rolls so that they will be accelerated and decelerated in a reasonable time without damaging the web. Undriven web-carrying rolls on high-speed machines coast to a stop so slowly that they represent a hazard to an operator who might be anxious to clean up the machine after a web break. However, a motor that drives the web-carrying rolls will decelerate the roll to a stop fairly quickly and also may have additional braking for an emergency stop mode.

Monitoring for Slip Conditions on High-Speed Coaters

High-speed machines monitor the speed of the master and the tension-controlled sections. If a tension-controlled section is running at speeds that are different from the master drive, we know that the section is slipping. Slippage can result in scratches, wrinkles, mistracking, and web breaks. If all the tension-controlled drives are running faster than the master, we know that the sheet is probably being pulled through the machine—most likely by the rewind. Rewind tension may be high and unwind tension may be low.

If the tension-controlled sections are running slower than the master, we know that the sheet is probably being slowed down by the unwind. Unwind tension may be high and rewind tension may be low. It is important to document acceptable operating parameters for speed and tension for a given coating machine because that is the best way to determine what is acceptable and what is unacceptable when product defects appear. In summary, a speed monitoring system will help on startup and debug, can assist the operator during production runs, will provide a warning if speeds wander outside an acceptable operating envelope, and can be used by operators and maintenance personnel as a problem-solving tool.

Tension Monitoring System

We have said that tension is the force applied to the web between adjacent sections. It is important to know and record the machine operating tension for each product between each section. It is important to document these values on instruments or computers, and it is even more important to know the proper operating tension for each product and process. Routinely examine drive data to establish that normal levels of speed, tension and motor load exist for each product and process.

All tension monitoring systems measure the total tension that is applied to the web (Figure 10-14). If the coating machine processes only one web width, the tension display can be displayed as unit tension (pounds per linear inch) or total tension (total pounds). The operators and engineers can easily see what values of tension are being used to transport the vendor-supplied web. Tension data will be used to select motor size and brake/clutch size when the machine is speeded up or when new machines are purchased. A knowledge of how each product is run may allow us to rebuild the machine to operate at higher process speeds without increasing the size of existing drive equipment.

Replacing an existing drive requires extensive capital investment and downtime with its attendant loss of production. *Rebuilding* an existing drive may require little or no interruption of production time and much less capital cost. Drive rebuilds frequently can be accomplished during scheduled summer and winter shutdowns if existing equipment can be utilized. Monitoring speed, tension, and load operating data provides us with the information that helps us make decisions that improve productivity at the lowest installed cost.

Specifying the Tension Range

The selection of a tension-indicating system begins with a realistic definition of the operating tension range which will be required by the web. If too wide a range of tensions are specified, machinery builders may need to design and implement costly multirange tension control systems which may not be ideal for the actual operating conditions.

The machinery builder may also need to supply robust web-carrying rolls which may have an adverse affect on light tensions that are run for thin films. We have seen cases where the operating tensions were actually $\frac{3}{4}$ to $1\frac{1}{4}$ lb/lin. in. Specifications called for a range of $\frac{1}{4}$ lb/lin. in. to 3 lb/lin. in. We discovered that the actual tension range capability was limited to $\frac{1}{2}$ lb/lin. in. to something in excess of 3 lb/lin. in. The vendor had cheated toward the high side of the tension specification. The components of the tension sensing and correction system would have been better

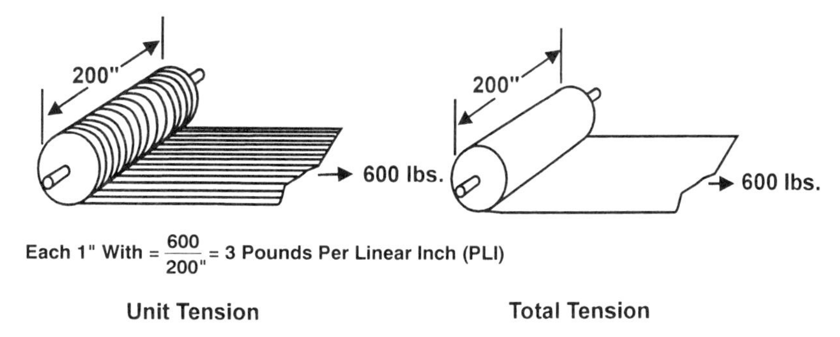

Figure 10-14. Unit tension is the total tension divided by the width of the web.

matched to system requirements if a range of $\frac{1}{4}$ lb/lin. in. to 2 lb/lin. in. had been specified.

All tension measuring systems measure the total tension supplied to the web. As long as the width of the web remains the same, tension can be displayed as unit tension (pounds per linear inch) or as total tension (total pounds). Operators and engineers can use the indicators to balance machine tensions or for troubleshooting. The values of operating tension are used directly to select appropriate brakes, clutches, load cells, dancer roll systems, and drive motors.

If the machine processes products of different widths, tension systems must be able to display the proper tension regardless of web width. For machines that accommodate multiple web widths, it is common to display tension in total pounds. For machines that are blessed with a computer system, the tension is displayed in pounds per linear inch and the process computer menu requires web width as one of the parameters.

Why Tension Control is Needed

1. Tension control provides a valuable tool to the operators, the machinery builder, and the drive vendor.
2. Tension control allows the operator to thread the web through the machine and start production more quickly than if a pure speed controller were used.
3. It corrects for the tension transients that occur when the coating applicator is applied to the web.
4. It allows the machine to process webs of different widths, weight, and caliper.
5. It corrects for the normal variations that occur in paper and plastic webs during the coating operation.
6. Tension control is valuable for the drive vendors who use it to compensate for factors that are within their control, yet may go unnoticed except through changes in web tension. The older analog speed-regulated drives drifted so much that web tensions changed radically during the coating process. Drift was so dramatic that web breaks occurred unless some form of tension control was used.
7. Tension control also compensates for the tension transients that occur when the machine is accelerated and decelerated. The tension control system ensures that machine sections, with different inertias, will successfully accelerate the web to process speed and will then successfully decelerate the web to a stop when the coating run is complete. High-inertia sections may not follow the acceleration/deceleration ramp as closely as the sections adjacent to them. If all sections vary during transient periods, tension changes can be severe enough to cause a web break.

Effects of Wiring Practices on Tension Control

Manufacturers of modern electrical drive systems and computer control systems have definite guidelines for the wiring of power and control devices. They recommend that power device wiring be separated from signal-level wiring. Separation is needed to ensure optimum system performance. Signal wiring is further separated into digital and analog signals. Some system users have chosen to disregard these guidelines and have combined wiring for motors, solenoids, and lighting with signal leads.

When all wiring is combined in today's system, the system performance will be compromised when "electrical noise" interferes with and modifies electrical signals that are used to control the process. We have seen the coating flow to the coating head disrupted when a motor at the winder was jogged. Power cables and signal cables had been combined into one raceway. We saw a signal transient on the flowmeter feedback signal each time the motor was jogged. Coating flow increased momentarily each time the transient occurred. We placed the flowmeter signal cable into a separate raceway and eliminated the electrical "noise" problem.

We also note in Figure 10-15 that the tracking of a coater was seriously influenced when the wiring to the drive motors was combined with the wiring to the drive system tachometers. We discovered that the coater lead section was running slower than the number 4 draw roll. The draw roll pulled the web through the machine at a speed that was about 20 ft/min faster than the master. Coating was being applied with a slot head. Coating coverage was lower and out of specification, because of the slip condition that existed. Web breaks were common on this machine. We recommend that all motor and signal leads be separated in accordance with TAPPI "Recommended Wiring Practices for Automatic Control and Regulator Systems" (number 406-12).

We also discovered that this machine had faulty analog controllers that tended to "drift," causing variations in speed over the course of a production run. We replaced these controllers, added tension control to the machine, and provided the operators with an easy-to-thread and flexible operating unit.

When to Use Pure Speed Control

A block diagram of a basic speed regulator is shown in Figure 10-16 and should be used in the following applications:

1. When web stretch or shrinkage is intolerable during processing. This requirement typically reflects process or product limitations rather than drive control considerations.

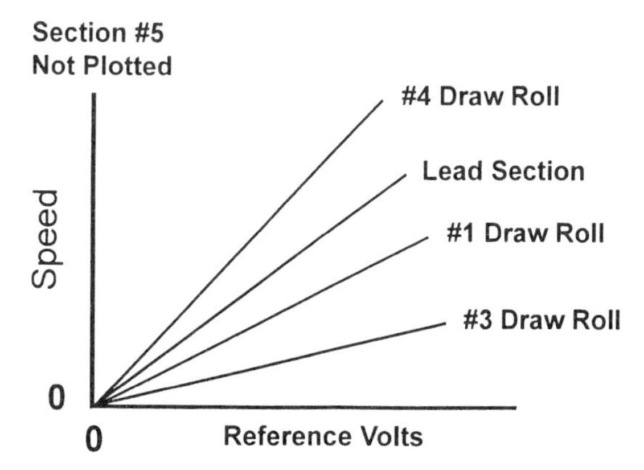

Figure 10-15. Drift, and electrical noise that is due to running the power cables and the signal wires in the same conduit, can greatly affect motor speed.

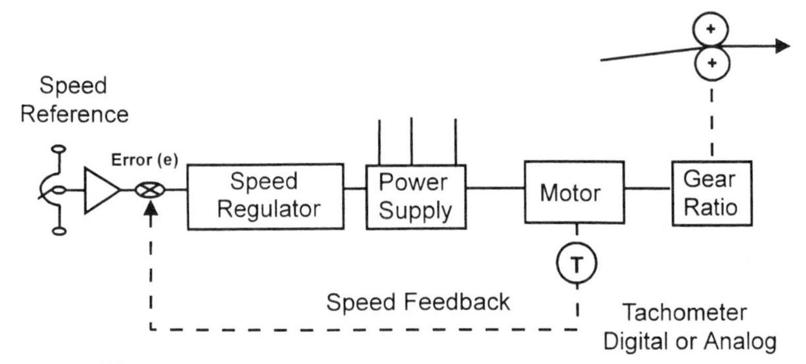

Figure 10-16. Block diagram of a basic speed regulator.

2. When the web is being stretched to a specific ratio between adjacent sections in a process. Tension controllers maintain tension by modifying section speed and torque. If the tension being controlled is higher than the yield point of the pliable web, the tension applied may cause the web to stretch unacceptably, even to the point of a web break. Conversely, if the web tends to shrink, too low a tension may permit unacceptable web shrinkage.
3. When process conditions create a pliable web, and stretch or shrink is undesirable, the attributes of tension control may produce undesirable web deformation or web breaks.
4. To avoid web breaks when the web is so weak that it can barely support its own weight.
5. For the lead or master of the machine.
6. Whenever a process creates or treats a pliable web whose characteristics change with each roll introduced into the machine, applying constant tension can produce variable results. Sometimes a speed regulator may be the only tool capable of producing good results. In any case, an installed tension indicator will help monitor the machine process.
7. If the machine has multiproduct requirements, a drive system should be designed that is capable of running in either tension or speed mode, depending on web and product needs.

Speed regulators compare an incoming speed reference signal to a motor speed feedback signal. If the feedback signal indicates that the motor speed is low, the speed regulator increases power to the motor to accelerate it to the speed reference level. If the feedback signal indicates that the motor speed is too high, the speed regulator will reduce power to the motor to decelerate it to the speed reference level.

There are *common problems that occur at startup* that can be corrected by establishing a disciplined startup and checkout process. Most wiring is checked and verified by the construction technicians before power is turned on, so problems are kept to a minimum. However, an occasional error can occur. During startup and checkout, operate only one section at a time. This will eliminate many problems that can occur when many tachometer cables enter a drive cabinet. Some tachometer cables are not connected or may be connected to the wrong section. Keep in mind how a speed regulator functions to match feedback and reference levels. If there is a difference (error) between the speed reference (set point) and the speed feedback (process signal), the speed regulator will adjust the power to the motor so that the reference and feedback are matched.

1. Frequently the wiring for the tachometers is not connected to the drive controller. When the drive controller tries to match speed reference and speed feedback, the drive speed controller sees no speed feedback signal from the disconnected wiring. The drive speed controller treats this condition as a signal loss, which indicates that the feedback signal is too low. The speed regulator continues to increase the power to the motor in an attempt to match the two signals. The signals cannot be matched if the tachometer is not connected to the speed regulator. Therefore, the speed regulator eventually increases power until the section is running at maximum speed. The section will continue to run at maximum speed until the section is stopped and the problem is corrected. This runaway condition can cause web breaks, web slip, or major machine damage. One way to avoid this problem during startup is to check the wiring at the terminal board manually to confirm that the correct tachometer is connected to the speed controller. Slowly rotate the motor by hand and measure the voltage at the terminals of the speed regulator. Observe if the meter deflects and ensure that the polarity of the tachometer signal is correct. If the section is too difficult to rotate by hand, run the drive at slow speed and determine that the tachometer is connected to the correct speed regulator and that the tachometer voltage is the correct polarity.

2. Tachometer wiring is connected with the incorrect polarity. The results will be identical to the condition where the tachometer is not connected at all. The section will accelerate to maximum speed. Check the tachometer voltage level and polarity at the speed regulator to identify this problem.

3. Tachometer wiring is run in a common wiring raceway. It is a common problem to mistakenly intermix tachometer wiring from adjacent sections incorrectly. Tachometer 1 may be connected to motor 2, and tachometer 2 may be connected to motor 1. The result is that both sections will seem to have defective tachometers. Here again, the section will accelerate to a maximum speed and stay there until the problem is corrected. This problem can be exposed by running only one section at a time. Let us assume that tachometer 2 is connected to motor 1. When motor 1 is run by itself, tachometer 2 will be stationary. No tachometer voltage will be present at the speed regulator and the section will accelerate to maximum speed. Shut the drive off and have the electricians verify the accuracy of all wiring. Once an error is found on wiring, it is common to commit the same error on all sections. This error also occurs when the sections on a machine are assigned names by the machinery *builder* that differ from the names assigned by the drive *vendor*. Section names become even more cloudy if the machine *user* becomes involved in naming equipment while the machine is being installed. For instance, a coating roll may be called a coater roll or the number 2 pinch roll. Pull rolls may be called draw rolls, capstan nip rolls, vacuum rolls, or vacuum tables. In the absence of a clearly defined name for a section, an electrician may assign a name arbitrarily. The electrician has usually won the opportunity to wire a machine on the basis of his or her submission of the lowest bid. The profit will be reduced or may disappear if the electrician attempts to

clarify questions of nomenclature. He or she will work to ensure the maximization of the profit but does not gain profit by taking the time to clear up questions. As a matter of fact, if the mistake can be blamed on poor documentation on the part of the client, the electrician will be paid for corrective work efforts. Such work is frequently corrected on an overtime basis because it occurs when a full operating crew is assembled and waiting to apply coating. The contractor will be paid if incorrect nomenclature results in the need to rewire some equipment. Rework is one way that contractors have to recover profit from a very competitive bid.

4. Infrequently, a tachometer will have a problem with its signal output. We have seen cases where a tachometer has failed totally. When the speed controller tries to match the speed reference with the speed feedback, the defective tachometer output is zero volts. The speed controller will cause the section to accelerate to maximum speed just as though the tachometer was not wired to the speed controller at all. The tachometer must be replaced. The problem can be identified by measuring the tachometer output while the section is at full speed. A zero or very low reading will indicate a defective tachometer. Problems of this type can occur due to poorly seated brushes on the tachometer or failed windings. Replacement of the brushes and seating them properly may solve the problem. If it does not, replace the tachometer.

5. Occasionally, a tachometer will have an intermittent problem. One winding may be defective, which will cause the tachometer speed feedback signal to *decrease* once per revolution. The decrease will be of a definite period and amplitude. The decrease in voltage will indicate that the speed of the tachometer is low. The speed controller will compare the speed reference to the varying-speed feedback and will see a decrease in signal level once every tach revolution. In effect, the tachometer is indicating a speed variation where no variation exists. The speed controller will try to correct for the once-per-revolution variation and will increase power to the drive motor for each speed feedback perturbation. When power increases to the motor, it accelerates to a higher speed. When the motor accelerates, the perturbation passes and the tachometer is now running at too high a speed. The voltage level of the tachometer is too high and the speed controller reduces power to the motor, which slows down. The speed controller is causing section speed to "surge." Surging can cause web breaks, web slippage or machine damage (if the web is strong enough).

Consider the rpm speed of a 3-ft circumference coating roll running at 180 ft/min. The roll will run at 60 rev/min or 1 rev/sec. Consider the ability of modern speed controllers which correct for step changes in speed within $\frac{1}{2}$ sec (or faster for servo drives). Such a controller can cause rapid cyclical changes of power which can cause major damage to the power transmission equipment as the controller tries to correct for intermittent speed feedback variations that incorrectly indicate speed variations.

6. Digital encoders or digital tachometers can have the same difficulties as those of standard tachometers. Wiring can be misdirected, and digital feedback devices can be damaged. If an interface card is used to convert the digital pulse train to an analog signal, the card output can fail or contain enough "ripple" to cause the speed controller to make the section surge. Digital speed feedback devices develop a certain number of pulses per revolution (ppr). Encoders commonly provide 60, 120, 300, or 600 pulses/rev. If precise speed control is not required, 60 pulses/rev is often used for pump and fan applications. Higher pulse rates are used where accuracies of 1% speed control are required over the operating speed range of a coating or laminating

machine. It is the coordination of the speeds between adjacent sections that demands such precision.

Much higher pulse/rev rates are available for positioning motor shafts or when product tolerances demand high accuracy. Such applications are incremental motion machinery or multicolor printing machinery, which require registration of a series of tools to a specific position on the traveling web. Such digital devices are rated in the thousands of pulses per revolution and carry a marker pulse to define the position of a shaft or the position of a spot on the web. These digital devices are frequently incorporated into web defect monitors, which indicate the position of a defect in the roll and the length of the defect.

7. Speed errors that are generated by "electrical noise." We frequently see signal wires run parallel to motor power cables. When the motor is started, stopped, or jogged, the current that is carried in the power cables induces a voltage in the signal cable. The longer the parallel run, the worse the generated noise. Noise that is superimposed upon an existing tachometer signal will cause an intermittent jump in signal voltage. The speed controller will respond to the noise and will momentarily change the speed of the drive motor. The result can be a web break or worse. We have seen an operator jog the machine. That jogging action placed noise on the feedback tachometer of the coater. We saw momentary and intermittent changes in coater roll speed as well as intermittent changes in coating flow to the coating applicator. The solution to the problem was to separate the signal wiring from the power wiring. The problem was identified by monitoring motor loads and feedback signals with a multichannel oscillograph. Once the frequency and amplitude of the noise was related to the "jogging" of the winder, the solution was obvious.

10.4.2 Equipment Lists Avoid Problems and Help Plant Personnel

Misnaming the sections at any time creates a nuisance and extends the startup effort into costly late-night problem-solving sessions. Misnaming can be avoided if the machinery purchaser, the engineering consultant, the machinery builder, the drive manufacturer, and peripheral equipment manufacturers preassign a name to each section and to each piece of equipment during the early stages of the project. It is the user who must name the equipment in accordance with the user's standard production department nomenclature. A knowledgeable user representative should be present during construction to answer questions concerning equipment identification and location. It is common to supply an equipment list that identifies each motor, fan, pump, machine section, drive cabinet, load cell, dancer roll transducer, and instrument.

Each equipment list identifies the equipment as to selected name, commercial name, catalog number, location in the building, and special features. This list is used by the plant maintenance personnel to identify standard components such as drive belts, pulleys, and bearings when maintenance or replacement is required.

10.4.3 The Frequency and Direction of a Speed Perturbation

Review again how a speed controller functions. A speed reference is compared to a speed feedback. The difference between the two signals (called *error*) is used to correct the speed of the driven section. To identify a problem involving a varying speed condition, we must record a number of section parameters. We must record section speed feedback, section motor load (or motor current), and section speed reference. These signals help us establish if disturbances are random or cyclical. Random disturbances may be the noise generated by a motor being jogged. Cyclical disturbances may be generated by a defective bearing or a roll that is unbalanced. If we record the speed reference, the motor load, and the speed feedback, we will have valuable data. The data can be used to determine what is the problem.

When a tachometer is defective, it is helpful to use a second speed-indicating source, similar to a digital speed-indicating system, to confirm if the information from the feedback tachometer is accurate. A defective tachometer output may show a momentary loss in section speed when no such speed change exists. That defective output will cause the section to speed up. The independent tachometer will show that the section is truly speeding up, due to the defective tachometer.

We should also examine the data to see if the speed changes produce motor load changes that are in the same direction or the opposite direction. For instance, is a random speed increase accompanied by an increase in motor load? If that is so, it is what one would expect if something were changing within the drive controller or in the master reference signal. This points to an electrical malfunction within the drive reference circuit. Defects can be in the components that make up the reference network, from signal noise in the reference circuit, or from defects in the controller itself. Is a speed increase accompanied by a decrease in motor load? This phenomenon may be explained by a defect in the feedback circuit. Is the tachometer voltage changing? We expect that a random loss of tachometer feedback will result in a random increase in motor current if the speed controller is to maintain speed at the set-point level. Recording data in any controller network will yield information that allow one to analyze problems. But the data are meaningless unless you understand what can cause the changes that are on the recording oscillograph chart. An understanding of what each signal level does will help you identify a cause-and-effect relationship.

10.4.4 Problems That Occur After the Machine is in Production

All connections on an electrical drive are checked to ensure a tight electrical connection. The connections are checked during a normal summer shutdown when the drive cabinets are vacuumed and cleaned. During one such cleanup, an apprentice electrician was told to tighten every screw in the cabinet. The electrician dutifully tightened every screw in sight, including those marked "System Stability," "Maximum Speed," "Offset," "Bias." Without realizing it, the electrician had adjusted all major parameters of the 12 section drive system. The startup following this preventive maintenance procedure was disastrous. The machine drive was unstable and out of synchronization. The machine had to be shut down and readjusted as though it were the initial startup.

Tachometers are rated to produce an output of 50 V per 1000 rpm *or* 100 V per 1000 rpm. The tachometers are identical in appearance and are designed to mount on the back of a motor. There have been cases where a 100 V/1000 rpm tachometer replaced a 50 V/1000 rpm tachometer. The result is that the 100 V/1000 rpm tachometer will generate twice as much feedback signal at any speed than the 50 V/1000 rpm tachometer. With *twice* the feedback, the section will run at *half-speed.*

1. Belts have been removed from equipment such as fans, pumps, drive rolls, and tachometers.
2. Cables have been disconnected from load cells and dancer roll potentiometers.
3. Drive rolls and coating rolls are replaced with different diameter rolls than those in the machine during startup. If these rolls are driven, the drive speed must be recalibrated to allow synchronization of the new diameter roll with the adjacent sections.

10.5 TYPES OF TENSION CONTROL SYSTEMS

Three types of tension control are applied to drive-control systems.

1. Motor current regulation (Figure 10-17)
2. Swing roll or dancer roll control (Figure 10-18)
3. Force transducer or load cell control (Figure 10-19)

Machine builders and drive manufacturers debate the merits of these methods. Some advocate using dancer rolls in all tension control locations, while others insist that force transducers are superior. All are excellent when used properly.

Current regulation is used wherever it is impossible to locate a dancer or load cell because of process conditions. In particular, this type of control is used where the web is still wet and cannot be touched; usually at the end of a dryer section.

Dancer rolls are frequently used at the unwind and rewind because these sections are most likely to cause a transient during a splice or transfer operation.

Force transducers are used on the interior sections of the machine because they fit into the web path easily.

10.5.1 Motor Current Regulation

Current regulation is simple, easily implemented, and effective. It may be the only way to control web tension if process conditions do not allow a tension sensor to touch the web. Current regulators are used on rewinds with great success and are

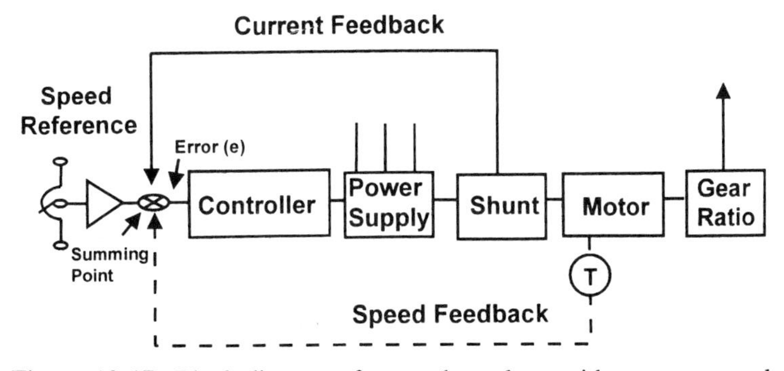

Figure 10-17. Block diagram of a speed regulator with current control.

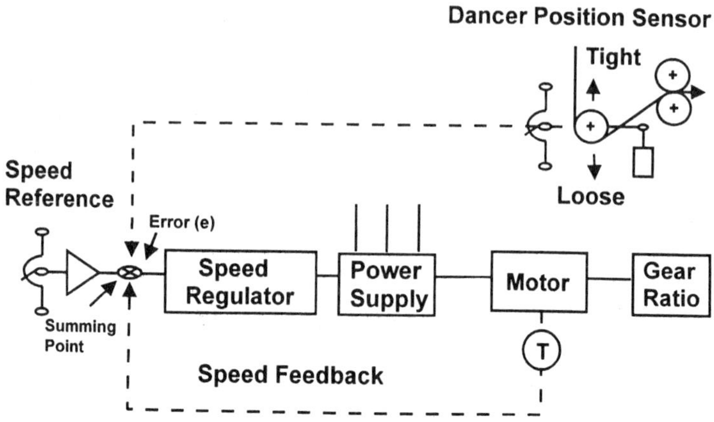

Figure 10-18. Block diagram of a speed regulator with dancer trim.

used wherever a dancer roll or tension sensing roll cannot be designed into the web path.

A current regulator controls the current to a value as set by the operator. It uses total motor current to infer the amount of tension in the web. However, total motor current equates only approximately to actual tension. It is equal to tension load and the losses in the system, but a current regulator cannot differentiate between tension and losses. If load losses in the system are a small part of motor current, a current regulator can effectively control web tension. If load losses are a large part of total motor current, a current regulator should be used only if no other method of tension control is possible.

Current regulators have no direct tension readout, do not provide storage, and cannot differentiate where the load comes from (frictional load that comes from bearings, rotary joints, tension, and cleaning doctors). The current regulator must have compensation circuitry to supply added current during acceleration or when doctors are used on an intermittent basis.

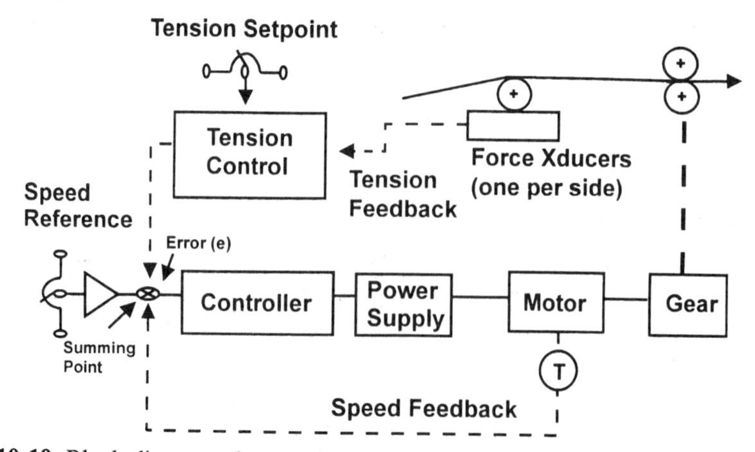

Figure 10-19. Block diagram of a speed regulator with force transducers and tension trim.

The operator must check for tension losses due to defective bearings, cold oil or grease in reducers after a weekend shutdown, or changing friction from steam joints, and must adjust the current to establish tension at the desired level until the problem can be corrected. It is common to start up a machine with higher current during cold startup and then reduce the current as the machine "warms up."

10.5.2 Dancer Roll Control

A dancer roll is a force balance system where the force of the web pulling in one direction is balanced by the loading of the dancer pushing in the other direction. The dancer has the ability to store web to minimize the influence of transients due to acceleration, deceleration, unwind splices, or rewind flying transfers. A disadvantage is that the dancer roll does not have direct tension readout. Tension is derived by calculations that are converted into a family of tension versus air pressure curves.

Drive vendors ask machinery builders to design the dancer loop to store the equivalent of 1 sec of web based on maximum line speed. That amount of storage is not always available, so compromises are made based on factors that depend on the web material and the rates of acceleration/deceleration.

The dancer control system also compensates for step changes in load losses due to doctor loading, nip closings, changes in coater blade loading, and higher-than-normal condensate loads in a dryer by sensing changes in tension and adjusting the section speed to reestablish tension to the set-point value. Dancers also prevent tension changes from occurring by compensating for the diameter change when a driven roll is replaced with a similar roll of a slightly different diameter. When an elastomer roll becomes an ineffective drive section, the roll is replaced or the defective cover is reground. In either case, the roll diameters will be slightly different. Normally, a technician synchronizes the drive for the new roll diameter. Rolls that are replaced in the middle of the night rarely are "synchronized."

Dancer position is precisely monitored by the roll position potentiometer. This system is sometimes called a position regulator.

The design of the dancer system must consider the types of materials that are used, the tension range, and the amount of storage required for good transient response.

Problems with dancers occur when the designers of the system forget the minimum values of operating tension. Cylinders that load the dancers must be friction free or at least have a friction value that is significantly less than the lightest operating tension.

If lightweight and extensible materials are run at low tensions, the dancer design must not allow the stretchy material to cause instability. In some cases, a free loop will be used at "zero" tension on low-tension applications. Where a small amount of tension is allowed, dancer rolls are being constructed from the new lightweight carbon materials to reduce the mass of the roll when it is asked to move during a tension transient.

We have seen dancers where the air cylinder friction prevented use of low operating tensions. Designers have forgotten to consider cylinder friction, the weight of the roll, or the weight of the roll support arms when determining the size of the dancer loading cylinders.

The dancer position sensor is frequently connected to the dancer via a spring-loaded cam-operated mechanism. This mechanism can stick and will provide faulty position information to the controller, causing poor tension control of that section. It is better to positively connect the dancer position sensor with a coupling or a link. The positive connection will ensure the reliable sensing of dancer roll position.

The dancer position sensor is connected to the dancer roll so that the sensor is centered at exactly the midpoint of dancer travel. Thus if the dancer roll has a total of 12 in. of motion, the midpoint of the sensor will be set at 6 in. of travel (the midpoint of dancer travel). If the operator wants the dancer to be offset from the midpoint, electrical adjustments are available to provide the desired offset.

Modern dancer roll control is combined with a speed regulator to provide a machine with a stable control system. The system uses a set point in the drive to establish desired speed. A tachometer feedback is compared to the set point and ensures that the drive is running at the desired speed. The operator adjusts the tension set point to establish the tension level of the drive system. That adjustment is usually a pneumatic pressure control that pressurizes the dancer roll loading cylinder. The dancer roll will move to the midposition. The dancer position sensor will feed back position information, which will cause the drive to impart enough force into the web, which pulls in one direction, to balance the force of the cylinder, which is pulling in the opposite direction. When the dancer is in the middle position the system will be satisfied and tension will be at the set-point level. If the dancer moves from the midpoint a dancer position sensor will provide a signal that will cause the controller to adjust power to the motor so that the dancer roll will return to the midpoint. Typically, dancer control on draw rolls and other constant diameter sections will be set to trim motor speed ±3 to 5%. Dancer rolls for unwinds and rewinds must have enough compensation to control the drive from maximum roll diameter to core size.

Older dancer roll systems allowed the dancer to control the drive motor from 0 to 100% speed. This system was very difficult to stabilize under all operating conditions. Transients such as splices, transfers, acceleration, and deceleration would destabilize the dancer circuits causing slippage, wrinkles, or web breaks. Sections would be bypassed because of these problems. We have had great success modifying old drives to modern standards by converting from 100% dancer control to speed control with dancer trim.

Problems that occur after the machine has been in production are:

1. The dancer roll control position has changed. The dancer pot has been replaced and has been centered improperly. A technician must use a meter to check the center position of the pot versus the center position of the dancer.

2. The dancer position has changed and the pot is centered properly with respect to the dancer midpoint. Someone may have adjusted the pot position to a perceived operating point without asking an operator to verify the position, or there may be a problem with the position amplifier. Center the pot with the existing adjustments or replace the position control amplifier.

3. The dancer roll goes against one stop when the line is run.

a. Check the position sensor to ensure that it centered and is not defective.

b. Check to see if a roll was changed during a shutdown. If a large-diameter roll was replaced by a smaller-diameter roll, the percent change in diameter may exceed the compensation capability of the dancer roll circuit. For example, let's say that we have a 6 in. shell covered with a 1 in. thick rubber cover. That equals an 8 in. roll. If we replace the 8 in. roll with an older roll whose cover has been ground a number of times: say, to 7 in. diameter. The change from 8 in. to 7 in. represents a difference of 12.5%. Most tension control systems will compensate for 3% to 5% or up to 10% maximum. Thus the diameter change will compromise our ability to control tension. If a smaller roll is replaced by a larger roll, the same problem can exist. The larger roll may require more compensation than the system can supply.

4. Dancer control is erratic and "jerky." The dancer does not "float" with the web off the machine but seems to stick in any position to which is moved. Check the loading cylinders to see if friction or binding is occurring. Some cylinders have internal diaphragms that can twist and jam against the inner cylinder wall when air pressure is removed from the cylinder. When binding occurs, tension control is erratic. It is a good idea to design a loading circuit that keeps air pressure in this type of cylinder at all times.

Other pneumatic cylinders have metal scrapers where the piston rod exits the cylinder. The scraper may be binding. Remove the scraper. This same type of pneumatic cylinder may have high-friction chevron packings on the piston. The packings may have dried out and are binding. Replace the high-friction chevron packings with Teflon packings and have the interior barrel chrome plated to reduce friction further. One set of chevron packings can be eliminated to provide a cylinder with minimum friction and drag.

5. New lightweight products do not run well on the machine. The dancers are jerky at low-tension settings. There is a tendency for the web to get tight and then loose between distant drive sections. This points to a dancer system where there is too much friction in the loading system or the cylinders are so large that the cylinders are operating at pressures less than 10 psi. Low air pressures can result in "sticky" cylinders and erratic operation. Also, there may be a conflict between how much force is required to move the web and how much is required for the process. Thinner webs may require low tensions for the process but may require high tensions to motivate web between distant drive points. What may be required is an additional drive point within the machine or a drive that motivates all idler rolls between sections. We have seen cases where there is so much friction in the idler rollers that the tension system "stalls" the web. In effect, the force in the web that comes from tension is less than the force due to the friction in the idler roll bearings. Add impingement dryers to this equation and you will find impressive loops of web forming between idler rolls. The tension in the web is less than the force generated by the velocity of the impingement air on the surface of the web. Once the web starts to hang between rolls we have the added factor of web weight to deal with. Forces due to web weight may exceed the tension set point.

6. Operators complain that a new dancer does not develop enough tension to move heavyweight product through the machine. Tension can be developed on the night shift but not during the day. Check calculations for tension and cylinder sizing. Sometimes the weight of heavy webs is not included in the calculation for the section of a loading cylinder. Wide and thick webs that are processed through towers, with long vertical runs between drive sections, are very heavy. Consider the weight of the web in the selection of the loading cylinder. The cylinder may need to develop enough force to move the weight of the web *and* supply the tension required.

The selection of the loading cylinder size may be marginal. It may be adequate at night when compressed air pressure increases. Air pressure increases at night because less machinery is in operation at night and the demands on the air compressor are reduced.

10.5.3 Force Transducer Control

Force transducer tension control is more expensive electrically but not as expensive mechanically as the dancer roll system. Two force transducers are mounted on a roll that is in the web path. Transducer rolls can be fitted into an existing web path without difficulty. Their signal is summed and sent to the tension regulator, which adjusts motor power output to satisfy the operators tension set point. Unfortunately, some hydraulic and electronic load cells get out of calibration easily. Some types are subject to "drift" and should be selected carefully. It is important to be practical in the selection of tension range to size the transducers properly. The roll weight must be kept within acceptable limits of vendor-recommended *tare-to-signal* ratios. If the roll is too heavy, it will be difficult to separate the signal from the force transducers into the two components: roll weight and tension forces. Force transducers cannot store web and thus cannot absorb transients.

The output of each cell can be displayed to show whether an unbalanced load is being applied by an imperfect web; for example, a cambered web may have a "soft" edge and a "hard" edge. Cambered web tracks through the machine tight on one side of the machine and loose on the other. The tight side exerts more force on its transducer than the loose side. Additional rolls may be supplied to fix the wrap angle of the web over the roll to ensure that the force being measured is always directly proportional to web tension. It is important to maximize the force vector that is measured due to web tension and to ensure that the tare-to-signal ratio is compatible with the transducer manufacturer's specifications. It is also important to keep the wrap angle constant on tension-sensing rolls located in a multiproduct machine where multiple threading paths exist. If the wrap angle is different for each web path, the web tension feedback will be incorrect for all but one web path.

Because the force transducer measures web tension directly, it ignores many of the load losses that bother the current-regulating system. However, rolls mounted on transducers must be balanced very precisely. Dynamic roll balance is especially important on rolls used as tension feedback devices for tension control systems.

Unbalanced tension-sensing rolls falsely indicate a tension variation on each revolution when none exists. Adverse effects of tension roll unbalance can be reduced by using the viscous damping systems that are part of the force transducer product line.

Force transducers are now applied to all sections of the machine, from the unwind to the rewind. The fast response capability of modern drive systems makes force transducers effective in applications that once required the storage capability of dancer rolls. The drive vendor, machinery builder, and user must integrate their efforts to ensure that the electromechanical design of the drive system maximizes the response capability of the modern drive system.

The user must ensure that the process can withstand the fast response times of modern drives. For example, softly wound jumbo rolls that are introduced to the machine may telescope during acceleration or during the splicing cycle. Process demands will determine whether dancer rolls or force transducers are the control of choice.

For narrow webs, up to 12 in. wide, single-transducer designs are available. The sensing roll can be cantilevered or mounted directly on the load cell frame. The force transducer system performance can be optimized by selecting a realistic range of operating tensions. The force transducer produces a signal when a tension force is applied to it. The total deflection of the transducer is about 5 thousandths of an inch (0.005 in.) when the maximum rated force is applied to the cell. There is a tendency to specify a range of tensions that is wider than the actual operating tensions. If the actual operating tension is significantly lower than the rating of the cell, the cell will be operating with a very narrow range of deflection. The signal from the cell will need to be amplified to provide a usable signal, but any error or drift that is inherent in the system is also amplified. This drift tends to cause the web to become excessively tight or slack during a production run. We have recalibrated force transducers once a day to keep the web under control. We have found that temperature, cell design, and amplifier design all created significant drift problems when the transducer was too large for the application. If tension meters indicate tension when no web is on the machine, it is time to recalibrate the tension system.

Calibration of the system begins by setting the signal and meter reading to zero, with no web on the sensing roll. It is wise to check the tension system calibration every week to see if drift is a problem. If drift is discovered, check the actual operating tensions versus the specified tensions. Ask the manufacturer of the force transducers and amplifiers if their modern components have been redesigned to minimize drift. Record the signal from the force transducers and the amplifier. Determine if one or the other is drifting. If the cell output is constant and the output of the amplifier is wandering, you have amplifier drift. If the force transducer output wanders with no web on the machine, the force transducers are drifting. We have seen both components drift in the past. The vendor is in a position to tell you what has been done to minimize the problem.

We have also seen some load cells show a propensity to provide a signal offset on an intermittent basis after a transient occurred on the machine. These cells had a mechanical flaw that would cause the cell to bind and to show an offset in the signal output after a transient. If the operator leaned against the roll when threading the machine, an offset would appear. The offset exceeded the value of tension that we were trying to control. The offset might disappear when the next transient occurred. The cells controlled a winder and affected the quality of the wound roll, causing softly wound rolls that would telescope. These cells were manufactured during the period that lasted from 1987 to 1990. We told the vendor what we were seeing and he acknowledged the problem and advised that they had redesigned their cells. We replaced the load cells after we:

1. Reduced the size of the force transducers to reflect the actual tensions being run. We found that the actual operating tensions were less than specified and that the load cells were sized more for the weight of the roll than tension control.
2. Reduced the weight of the sensing roll so that the ratio of the roll weight to the tension force was more desirable (tare-to-signal ratio). Prior to the roll weight change, the force due to roll weight was 30 times the force due to web ten-

sion. Force transducers are wired to the tension amplifier with a small-gauge shielded cable. When this cable fractures, the signal will be connected and disconnected intermittently from the tension amplifier. Consider the separate case of an unwind that would suddenly increase tension to a maximum level, causing web deformation or web breaks on a 17-mil-thick aluminum foil. We discovered that the tension signal would suddenly drop to zero and then reappear. Tension went to maximum (15 lb/lin. in.) and drop to zero. The problems were solved by replacing the cells and the wiring. Operation improved, but we noted that serious signal drift existed. The amplifiers were replaced and the drift problem was eliminated.

Sizing Strain Gauge Load Cells

The size of strain gauge load cells depends on the total force applied to the load cell. The total force consists of the tension load and the effect of the weight of the roll. Some vendors will take the tension force vector and double it to ensure that the cell will not be damaged by high-tension transients or by running tensions that are higher than specified. This practice may lead to the selection of cells whose rating dwarfs the force applied by operating tensions. If the user exaggerates the maximum operating tension and the strain gauge manufacturer doubles the exaggerated value, the load cell will be too large. In effect, we are weighing a rabbit on a scale designed to weigh an elephant. Here again, the signal available can be amplified. However, this will also amplify the offsets, drift, errors, and signal distortion from all sources. If an unbalanced sensing roll is bouncing on the force transducers, that signal will also be amplified.

Vendors are reluctant to tell you that you may be running into trouble with a wide-ranging tension specification. They are afraid to lose the job to their competition. The message is that you should know your process and the process operating conditions, and work with the vendor to select the best components for your operation.

10.6 SECTION CHARACTERISTICS THAT ENSURE GOOD WEB TRANSPORT CONTROL

10.6.1 Master or Lead Section

The master section on any process line will set the speed of the web through the machine. It does so by having such a positive grip on the web that it moves 1 ft of web for every foot of roll surface travel. If the grip is broken, the section loses its ability to control web velocity and the web will slip past the section. If the master section is the capstan after the unwind, it will draw 1 ft of web from the unwind and deliver 1 ft of web to the following section. Sections that positively control the web are:

1. The unwind or rewind (surface or center-driven reel)
2. A nip section (if the nip is closed)
3. A calender stack
4. A blade coater
5. Impregnating coaters

We should qualify some of these sections by saying that blade coaters or nipped draw rolls seem to slip at speeds above 3500 ft/min. The newly designed short-dwell blade coaters seem less able to control web transport than the blade coaters, which have high web wrap angles on the coating drum.

Nipped sections, calenders, blade coaters, and surface-wound reels have been used as master sections because of their ability to control web transport. In cases where a blade coater slipped at high speeds, the master function was shifted to other sections. A good lead section must maintain its grip on the web at all speeds up to the maximum design speed even when subjected to transients during acceleration, deceleration, or at any stable speed. The lead section must be capable of adding torque (motoring) or holding back torque (regenerating) to the web in order to control web speed.

Defects That Occur When the Master Section Slips

If the master section is unable to control web velocity and the web slips, defects such as a thick or thin coating, overdrying or underdrying, web scratches, coating scratches, and poorly wound rolls can be expected. Web breaks are common when web slip causes tension control systems to become unstable. Worse yet, it may be impossible to track the web as it slips through the machine longitudinally and laterally.

When slip exists, the machine may operate at lower speeds, but transients will cause the web to slip from side to side and eventually break. If slip is bad enough, it may not be possible to perform tasks as simple as threading a machine.

In summary, the master section in the machine must be able to transmit the torque from the motor, through the roll, and into the web without letting the web slip. The total drive system for the master includes the drive motor, the drive shaft, the drive roll, and even the web itself. The drive for the section does not end at the motor shaft.

10.6.2 Sections with Limited Ability to Control Web Transport or Tension

A section that will control tension needs to have the same characteristic as the master or lead section. *It must not slip.* Some sections seem to perform well at slow speeds and then begin to slip as operating speeds increase. Slip begins to occur when the boundary layer of air that is carried along by the web begins to provide a pressure pad between the web and the driving surface. That pressure pad acts as a lubricant that causes the section to slip. Sections that tend to slip are:

1. Unnipped nip rolls and vacuum sections with small wrap or low vacuum.
2. Single elastomer rolls at higher speeds.
3. Controlled gap coaters (impregnating coaters, gate roll coaters).
4. An unfelted dryer section (dryers slip above 600 ft/min. Tension control is needed for acceleration and deceleration).
5. Felted dryers that run above 3000 ft/min with only one drive. Dryer drives are now being reconfigured into two components: a felt drive and a dryer drive. As such, the total section load is measured and split between the two component drive sections. Load split is determined by the best operating results.
6. Unfelted single dryers that are backups for high-velocity impingement dryers.

Dealing with Sections That Have Limited Ability to Control Tension

Sections with limited ability to control tension must be limited as to what they are asked to do and by defining their true operating parameters. Select a practical tension range for the machine. If all web is run at 1 lb/lin. in., a range of $\frac{1}{2}$ to 2 lb/lin. in. instead of $\frac{1}{2}$ to 3 lb/lin. in. may be more practical. Too large a range will have the following effects:

1. It will eliminate current regulators. (Current regulators can accommodate a limited range of current.)
2. It will eliminate force transducers. (These may have marginal value; selection may be difficult if the transducer is to satisfy the highest and lowest tension and still provide adequate transducer overload capacity.)
3. It may compromise the ability of the dancer roll to handle the high end or low end of the tension range. Dual dancer loading systems are used to provide a high and low range of tension. These systems have met with mixed success and enthusiasm at the mill level.

Generally, tension control specifications call for a tension control system to provide ±10% tension trim, separately adjustable, for each section. Ten percent tension trim may cause sections to become unstable and lose control of the web. Experience in the paper industry and some converting plants indicates that the values of trim that are listed below are adequate for machines whose speed differential from the unwind to the windup is less that 1 ft/min.

Draw rolls	3 to 5% usually
Blade coaters	3 to 5%
Single unfelted dryers	0.5 to 1%
Unfelted dryer nests	1 to 3%
Felted dryer nests	1 to 3%
Calender	1 to 5%

Tension trim values must exceed any drift that exists in an analog system. If the drift of the drive system exceeds the tension trim, tension control will be unpredictable. Typically, there is too much tension trim incorporated in the dryer tension regulator, and this causes web control problems during acceleration and deceleration.

At line speeds above 3500 ft/min, it was common to see all sections slipping to some extent. The web tends to track poorly when this occurs and the only tool that the operator has to correct tracking is to increase web tensions at the windup. Significant increases in windup tension can contribute to increased web breaks even as the increased tension improves web tracking. If all sections are slipping, the web is not being supported or controlled by the interior sections of the machine. Only the windup and unwind have control of the web. That represents a long and unsupported draw condition. If unwind tension is normal (say 2 lb/lin. in.) and the rewind tension is increased to 10 to 15 lb/lin. in., one can expect the web to break away and accelerate to a speed that is limited only by the range of windup tension control. One can visualize this phenomena if you consider that it takes almost no effort to keep a 1-ft span of rope in a taut condition. However, if the span of rope is 100 ft long, it takes significant effort to keep the rope taut. Web transport improvements are seen as felted dryer drives are retrofitted with the modern dryer drive concept (i.e., a felt drive and a dryer drive).

On machines where dryer drives have been provided with the modern dryer/felt drive combination windup tensions have been decreased to predictable values. Modern dryer drives become an effective master section within the machine that sets sheet speed and allows adjacent sections to perform properly. When sections within the machine control the web, web tracking and web control become reliable. If single elastomer rolls slip, then vacuum rolls or vacuum tables are being used as replacements.

It is important that large-inertia sections such as dryers and calenders be limited as to how much tension trim will be provided by the tension control system. If these large sections break away and slip, it is questionable if the tension system will be able to limit the transient that will occur when the section stops slipping. We have seen dryer sections slipping at 10% above line speed during a deceleration period. Web breaks occurred when the sheet adhered to the dryer surface as the machine decelerated to 500 ft/min.

Sections That Always Slip or Almost Always Slip

There are times when simply reducing the range of tension trim is not enough to allow a tension-controlled section with marginal tension control capabilities to regain control of the web.

If the section always slips, try using only speed control to control the section. If the section is speed controlled, it will always be close to web speed and will not scratch the web even if a slight difference in speed exists. If the section does not always slip but breaks away easily, we can revise the controller that we are using to be less precise.

All modern controllers use operational amplifiers as the heart of the control system. These amplifiers are more precise than the old proportional amplifiers, but the integrating attribute of the amplifiers may never allow these marginal sections to perform properly. There is success when tension controllers are converted from integral to proportional control.

The softer characteristic of the proportional controller together with the tension trim limit has allowed sections with marginal tension control capabilities to control tension. For high-speed coating lines, the combination of proportional control with tension trim limit may be particularly effective for the older dryer drum drives (dryers that do not have a separate felt drive). On any line where nipped rolls such as coaters and pull rolls tend to break away easily, the combination of proportional control and tension trim limit may provide stable performance to a machine that cannot otherwise support the demands that an integrating controller places on it. Proportional control and tension trim limit may be the control of choice when operating personnel do not have an active system for inspecting the covers of elastomer rolls and for correcting for the reduction in coefficient of friction that normally occurs with time, temperature, or abrasion. The goal is to have a machine that runs consistently, and if a machine performs better with the less precise proportional system than it does with integral control, use the proportional control.

10.7 DRIVE FEATURES THAT GIVE THE OPERATOR ADDITIONAL TOOLS

Jog forward and jog reverse capability at each section helps the operator when the web is hand threaded through the machine. Jog forward and jog reverse on a centerwind allows the operator to peel off defective web easily and to obtain a sample of web from each roll for a routine quality check.

Ammeters and tension meters enable the operator to determine if the machine tensions are balanced and if operating section loads are normal. High section loads may indicate an impending failure in the system. Ammeters and tension meters also show the frequency and direction of any tension changes and load changes. This helps the operating staff to determine where a problem exists.

An independent speed monitoring system can be used to check the performance of a drive when its own system components are defective. A hand-held tachometer permits rapid checks of section speeds when problems occur.

10.7.1 Tools That Are Used for Troubleshooting Drive Problems

1. *Multichannel Oscillograph.* A six-channel recorder is a popular device that allows the technician to view speed, tension, and load data while trying to determine the source of a problem.
2. *Digital Multimeter and Analog Multimeter.* This is to be used for the preliminary troubleshooting effort. It allows the technician to look at speed, tension, and controller parameters quickly.
3. *Oscilloscope.* This is used to look at digital data to determine if the digital systems are operating properly.
4. *Digital-to-Analog Converters.* These allow the technician to record speed data from digital speed feedback devices.

10.8 SUMMARY

Good machine performance can be achieved only when the sheet coming to the coater is reasonably uniform and consistent. A uniform web will result in fewer web breaks and more uniform coating and drying. Each mill must establish web specifications that yield the best finished product. Close attention must be paid to holes in the web, edge cracks, and machine splices. Experience has shown that machine efficiency is best where web specifications are realistic and are policed. In this chapter we discuss some fundamental ideas of web transport and some problems that exist within the machine.

The operators and technicians must understand the needs of the machine so that they can contribute to the solution of problems when problems occur. Understanding begins by knowing the established operating conditions and techniques. Operators and technicians should be able to use devices such as ammeters, voltmeters, and speed indicators to determine if the machine is operating properly. Improper readings should signal a need to investigate further with the use of record oscillographs and other instrumentation. Characterization and documentation of an existing machine's parameters is one way to determine when a process is performing properly or is in trouble. Ignoring the data supplied by operators devices will only lead to frustration when a problem occurs.

CHAPTER XI

Coating Defects Catalog

Easy access to the results of previously solved defect problems is needed to reduce the time required to solve current problems. This is particularly true of the analytical data, which constitute a major part of the time and funds expended in a problem-solving cycle. These data are very characteristic of specific defects. If the complete data from a previous cycle is available, it may be possible to identify a defect just by comparing the initial microscope pictures with previous pictures. If the pictures are similar, it is likely that the defects are the same and the same solutions might work. The analytical time and cost for the second stage can also be reduced by running only those tests that gave the best results in the previous cycles. It is therefore recommended that a readily available coating defect catalog be established which contains this information in summary form. This catalog is in addition to the final closing report for any one defect. It is intended to make the analytical information readily available for the problem-solving team. It should be designed to focus on the characterization data and the solution that was found. A principal reason for having this catalog is that one picture is worth a thousand words. Comparing defect images is much better than reading about them. Also, it is typical in most coaters for certain types of defects to repeat themselves and to have similar causes. Access to this historical knowledge will help isolate the cause and so solve the problem rapidly. It may also suggest tests to run and new questions to ask about the current defect.

The ideal format for this coating defects catalog is a computer database in which each individual record can contain both text and pictures. The contents should be easy to search so that specific information can easily be extracted. Thus the search could be for all mention of chatter, bubbles, and so on, and only those records would be displayed. This is much easier and more effective than scanning pages in a loose-leaf notebook. In addition, it is simple to add new material to the database, and it is easy and inexpensive to copy and distribute files from the database, or just selected portions of a particular file.

There are a wide variety of commercially available, suitable database programs that function on personal computers or on larger centralized computers. The images from a digital image analysis system (Cohen and Grotovsky, 1993) are in computer format and can easily be incorporated into most databases, along with the other analytical data.

The defects catalog can be as simple or as complex as the user desires; however, it is our experience that a simple catalog with some basic information is very effective because, if it is simple, it will be used. A complex system will be more difficult to use and it will be more difficult to enter new information. As a result, it will not be used often. For the system to be effective, it must be up to date with data on a wide variety of defects. A simple, easy-to-use system will be used, will grow, and will be more and more useful.

Another valuable function for the catalog is as a training aid for personnel. Part of the job for coating operating personnel is to eliminate defects while operating the coater. After a year or so of operation, several types of defects will have been found along with the methods to eliminate them. Instead of forming a problem-solving team, the operators will be able to use past history—their experience—to solve some defects problem. As experience is gained with more products, the operators will learn to solve the problems that have occurred previously. This experience is available only as long as the same people work on the coater. However, as people change jobs, this experience is lost and new operators have to be trained. This defects catalog should be used as part of the training procedure and will be a valuable reference while experience is being gained.

The catalog should contain a single record for each defect cycle and for each occurrence of a defect. The record should briefly describe the defect, contain pictures of the defect, name the defect and give its source, supply the chemical analytical data, and summarize how the problem was solved. There should also be a reference to more detailed information, such as the report from the problem-solving team or to generalized reports. The record can also have the names of local experts who can be called on for help. The size of a record depends on the needs and preferences of the user. It can be as short as two or three computer screens, or it can be as long as several pages. A template and directions for adding new information should be included in the system. Each occurrence of the defect should have its own record. Defects are statistical in nature and several occurrences are needed to get the true indication of what is happening. A record does not have to be complete to be entered. It is often useful to enter what is known about a defect and what was done about it. A simple record on chatter in a specific product, giving frequency, a record of coater data, and a note indicating that a bearing was changed in the coating roll, may help someone the next time chatter is seen.

To search this type of database, one can specify key words of interest, such as bubbles, chatter, streaks, spots, and so on, or the product type, and the computer will search all records and display those that meet the specified criteria.

It is also possible to classify the defects, such as was done in Chapter 2, and add text that contains these classification terms. The records could then be searched and all records displayed under this classification. This is useful for general types of defects when an exact match is not found. There are many ways to classify defects and there is no agreed upon naming convention. This complicates the classification. Table 11-1 gives a basic classification scheme that one of us uses and finds helpful. It classifies defects by process sources, size, and type and is very flexible to meet specific process needs.

Table 11-1 Defect Classification Scheme

Attribute	Characteristic Terms		
Defect source	Support		
	Solution preparation		
	Solution delivery		
	Web transport		
	Coating		
	Drying		
Defect size	Large	seen with the unaided eye	
	Macroscopic	seen with a 10× magnifier	
	Microscopic	seen with a microscope	
Defect type	Continuous in the machine direction		
	Continuous in the transverse direction		
	Spot		
	Gross distortion		
	Subtle		
Examples			
A bubble might be:	Solution delivery	Large	Spot
Wrinkles might be:	Web transport	Large	Gross distortion
Chatter might be:	Coating	Large	Transverse direction

11.1 DEFECTS CATALOG: EXAMPLES OF ENTRIES

Some examples of specific entries in a defects catalog will illustrate the concept. Since coaters and the products coated are unique, the specific defects and the methods to eliminate them tend to be unique. Therefore, each coater should have its own catalog. A series of typical records will be presented to show some different types of catalog entries that can be created. The focus will be on the type of information that a record can contain.

Again, the basic requirement for selecting a specific computer database is the ability to store both text and graphics, and the ability to search for key words. Most databases are flexible and will permit the user to design a form to match their needs. Figures 11-1 and 11-2 show the text and graphics for a flake defect from a Macintosh-based defects catalog.

In entering the graphics into a specific record, the preferred picture is an actual micrograph of the defect. When this is not possible, it is acceptable to enter a sketch or drawing. This may be necessary when the defect is subtle, such as mottle or a light chatter, or covers a large area such as wrinkles, where micrographs will not show the defect. A sketch is better than not having a picture. This was done for repeating spots in Figure 11-3, which is good example of a basic entry. The defect is described and sketched, and a calculation is shown that can help locate the source of the problem. The entry suggests that when a repeating spot is found, it probably comes from a contaminated roll. Thus calculating the roll diameter with a circumference equal to

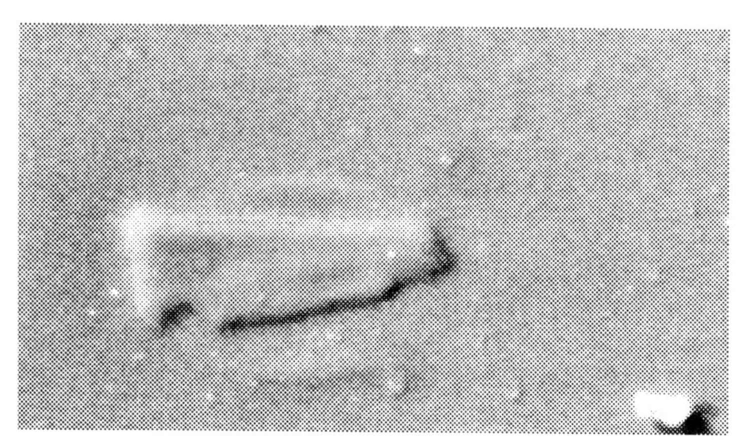

Figure 11-1. Defects catalog micrograph of a base flake. Product: 824-089 end (raw stock). Magnification: 50×. Lighting: oblique; #22-0.

the repeat spacing helps locate the roll. Other examples of basic entries are shown in Figures 11-4 to 11-8.

A catalog entry can also be far more descriptive and can summarize information from a whole series of defect cycles. This should be done when sufficient data have been accumulated so that the pattern and mechanisms of formation are understood. It is also useful to summarize the information when there are many examples of a particular defect. It is much easier to view one summary record than to examine a number of records. In the next sections the entries for bubbles and chatter will be presented.

Defect number:	#22-O, R, T
Defect type:	Base flake
Product number:	824-089
Magnification:	50×
Coating type:	Raw stock
Date of entry:	11/5/92
Immediate action:	None
Corrective action:	If problem persists, the base could be dirty. A decision would have to be made either to stop using this base roll series or to use base cleaning measures, such as a tacky roller, and make sure that the vacuum system is working.

Remarks: It is important that you use all three lighting techniques in order to properly identify this type of defect. Using transmitted light will usually produce the best picture for analyzing base flakes. This is because it will show the actual outline of the flake. Oblique and reflected light will show the flake being coated by the emulsion. You may also see this defect on gel coatings.

Figure 11-2. Defects catalog text for the base flake of Figure 11-1.

Classification terms: Coating, drying, web transport, macroscopic

Description:

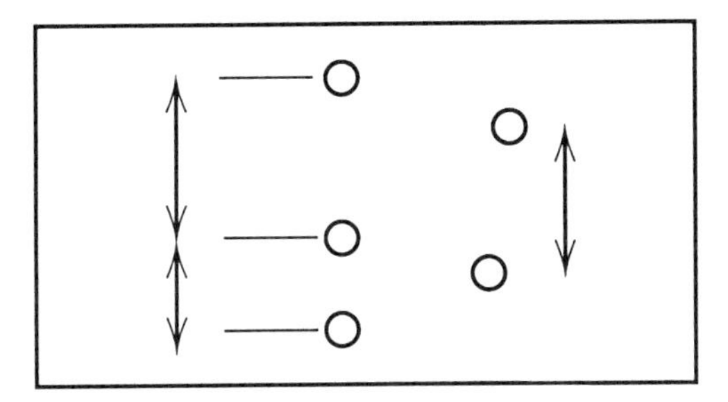

Measure the repeat distance:

Repeating spots with a constant repeating distance between them may result from dirt on a roll transferring to the coating. If the dirt remains at the same place on the roll, the distance between spots will be the circumference of the dirty roll. When this is suspected, measure the distance between several different spots and calculate the average. Divide this number by π, 3.14, to estimate the roll diameter in the same units. This can then be compared with a table of coater roll diameters and circumferences, which will indicate which roll to check.

Roll Diameter (in.)	Roll Circumference (in.)	Roll Type
3	10.2	Idler rolls
4	12.6	
6	18.5	Vacuum roll
8	25.1	
10	31.4	Coating roll

Then check and clean the rolls thoroughly to remove debris-causing spots. Repeating spots can come from any roll in the coater or the unwind or rewind which comes in contact with the uncoated or coated base.

Figure 11-3. Description of repeating spots.

11.1.1 Bubbles

Classification : Coating solution preparation and delivery
Large to macroscopic
Spot

Bubbles are defined as "a small body of gas within a liquid," "a thin film of liquid with air or gas," or "a globule within a solid." In coating solutions air can become trapped leading to bubbles. These then can stay with the liquid, end up in the coating, and lead to defects. Typically, in the dryer the skin that forms the bubble and contains the air breaks and leads to a clear round spot that is called a bubble. Bubbles are very common in aqueous solutions that contain surfactants to increase the wettability on the support. Bubbles are particularly common in low-viscosity solutions but do occur

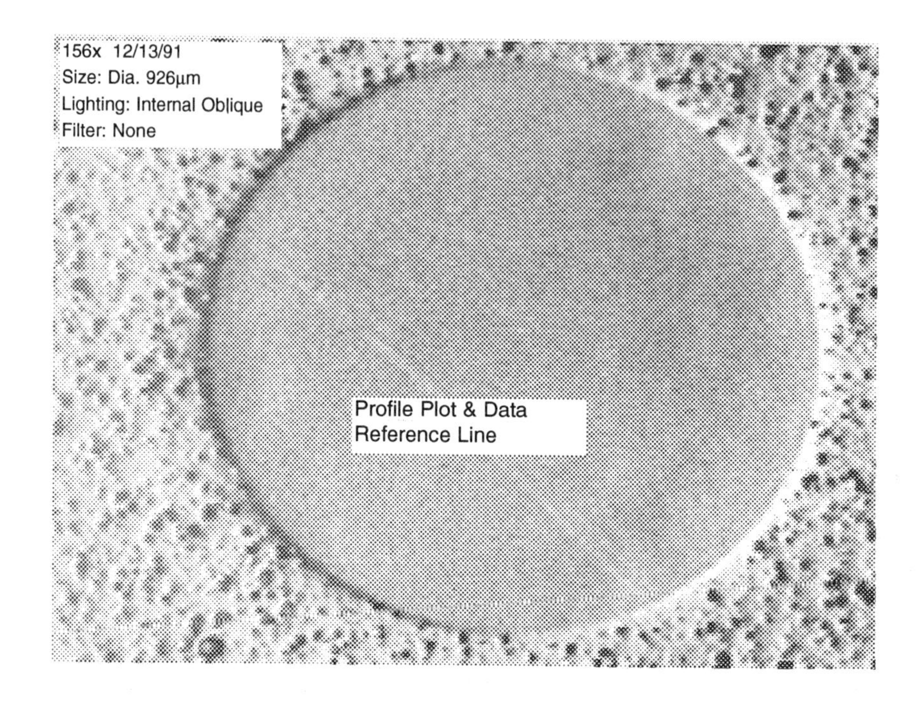

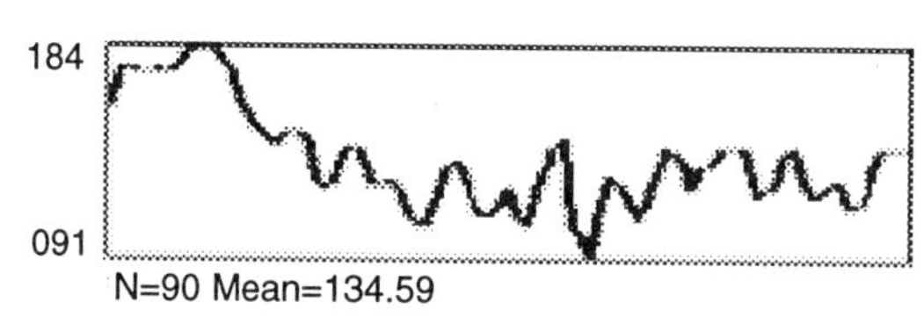

Figure 11-4. Description of a repellency spot on the base. Classification terms: support, large, spot. Cause: oil contamination on base. Remedy: check web path prior to applicator for spray or dripping; use new support role. From E. D. Cohen and R. Grotovsky, *J. Imag. Sci. Technol.*, **37**, 133–148 (1993), with permission of the Society of Imaging Science and Technology.

over a wide range of viscosities and are more difficult to remove at higher viscosities. They are an ongoing problem in most coating applicators. Examples of bubbles are shown in Figures 11-9 and 11-10.

Most bubbles can easily be seen by eye but are analyzed by microscopy to show up the contrast between the normal coating and the thin coating in the bubble. Bubbles can come in a variety of shapes and many specific forms. They vary in size from 50 to 1200 μm. All share the basic characteristic of a thin coating in the center where the gas was trapped, and most show the collapse of the upper skin when the bubble bursts. Bubbles tend to be round because of surface tension.

The differences in appearance arise from the properties of the coating, the size of the bubble, the location in the wet coating, and the temperature of the dryer. If the coating is weak, as the air heats up and expands in the dryer, the bubble will break and the solution will drain down to surround the clear spot. If the coating is strong or forms a solid gel, the bubble may remain intact and not break.

When bubbles are discovered, since they represent air trapped in the coating liquid,

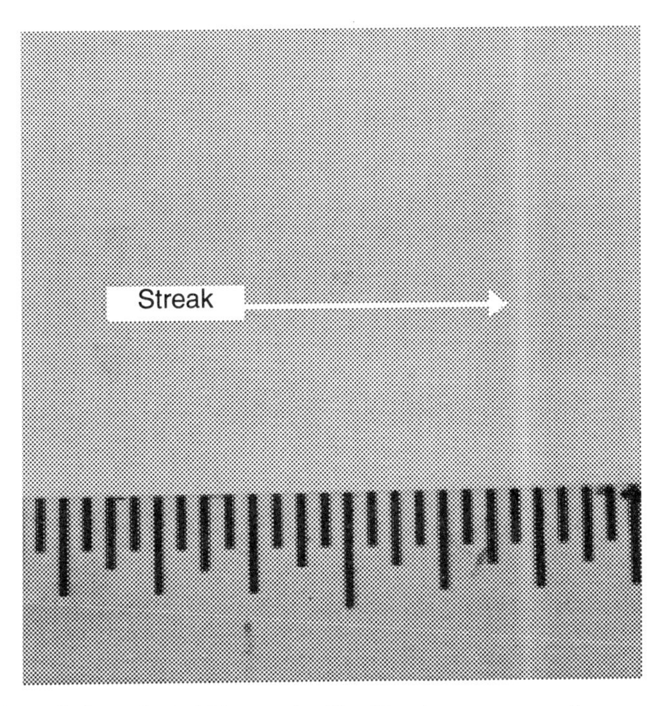

Figure 11-5. Description of a bubble streak. Classification terms: coating, macroscopic, continuous machine direction. Possible cause: bubbles trapped in coating bed. Remedy: remove bubbles with pointed stick; widen applicator coating roll gap; check for bubble sources in feed lines.

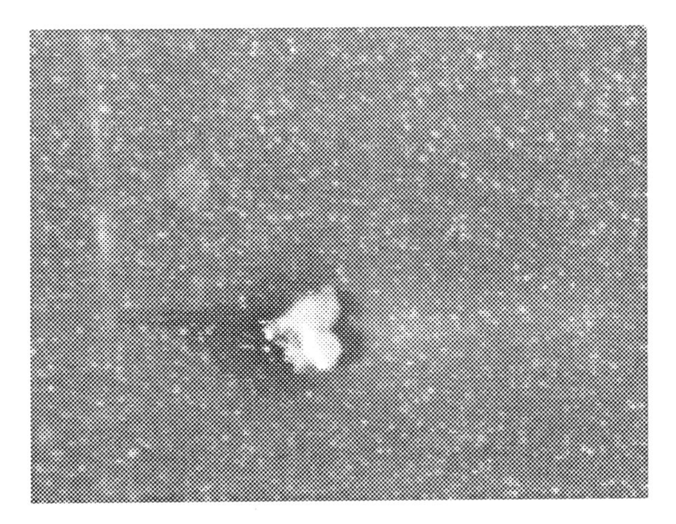

Figure 11-6. Description of a gel slug. Classification terms: coating, macroscopic, spot. These are typically seen at the start of a production run and will disappear within 200 to 300 ft. If they persist shut down, clean lines, and switch to a new batch of coating solution. A smaller filter can be used on the current batch.

Figure 11-7. Description of a bubble with a tail. Classification terms: solution delivery, large, spot. Magnification: macro lens. Lighting: light table; #9-LT. These bubbles were seen in a new product experiment. A check of the solution delivery system indicated that the debubbling device was not functioning. This was corrected and the bubbles disappeared.

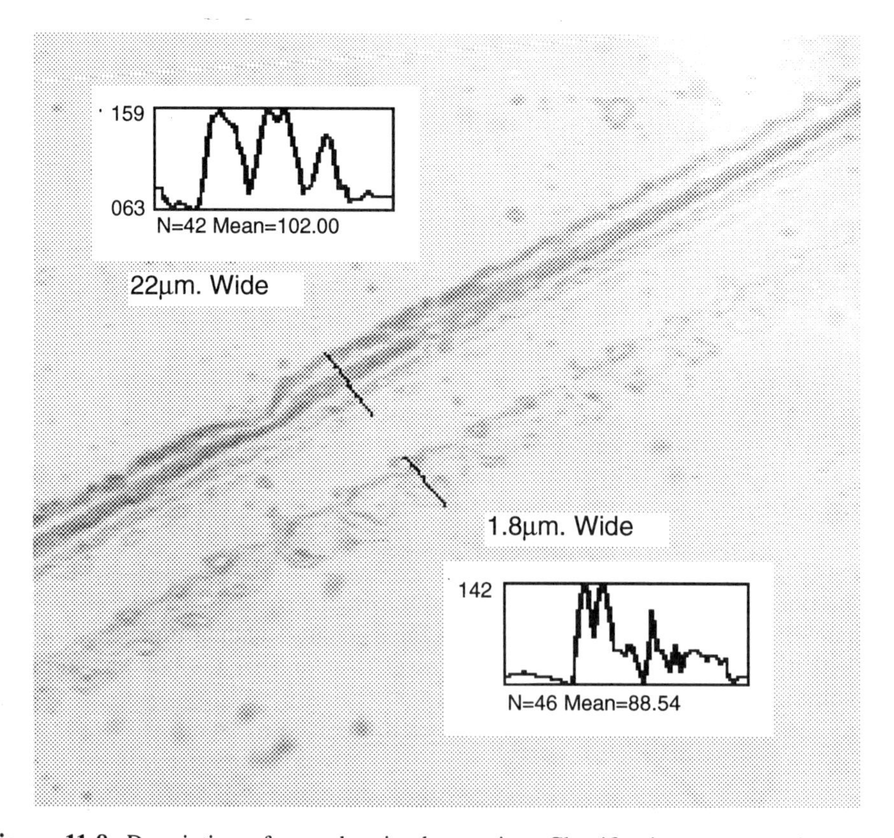

Figure 11-8. Description of scratches in the coating. Classification terms: web transport, macroscopic, continuous machine direction. Magnification: 618×. Lighting: reflected light polarized; no enhancement. These scratches were not in the coating after the dryer but were in the final wound product roll. They were caused by a contamination band dried on the surface of the roll before the winder which was scratching the product.

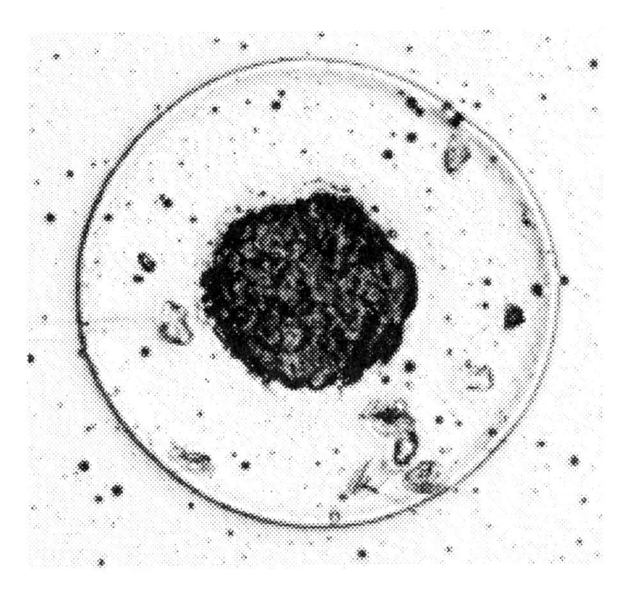

Figure 11-9. Description of bacteria-induced bubbles. Classification terms: solution delivery, macroscopic, spot. Magnification: 50×. Lighting: reflected; #13-R. Although appearing to be bubbles, these spots were caused by bacteria that had accumulated in the solution delivery lines. The black center consists of bacteria, which are also present in some of the bubbles. In some samples the bacteria have exuded from the bubble. To eliminate this problem clean and or replace all feed lines and clean the delivery pump.

the process has to be checked for all potential sources of air. This starts with the kettle used to prepare the coating. Examine the process during mixing and check for foam and air entrapment. Check that the agitator is turning at the standard speed, as too high an agitation rate will entrain air. During coating the agitation rate, if an agitator is used, should be the minimum required to maintain the dispersion homogeneous.

Bubbles are lighter than liquid and so will tend to rise to the top of the vessel. Thus the coating solution should be drawn off the bottom. Also, after preparation the liquid should stand for some fixed time to allow the bubbles to rise to the surface. A vacuum pulled on the vessel during this period causes the bubbles to expand and thus rise much faster to the surface. Check that the standard standing time was followed. Check that the withdrawal point of the vessel is at or near the bottom, and of all reservoirs is below the surface.

The lines that lead from the mixing vessel to the feed reservoir and from there to the coating stand can contain air. These lines should not enter below the surface of the next vessel. The lines should be checked to make sure that there are no leaks and that all connections are tight, so that air cannot leak in as it can if the pressure in the line goes below atmospheric. The lines should not have any points where bubbles can accumulate. When solution is first introduced it should be recirculated back to the first vessel to clear the lines. Pulsing the lines—having an intermittent high flow rate—helps clear them of air.

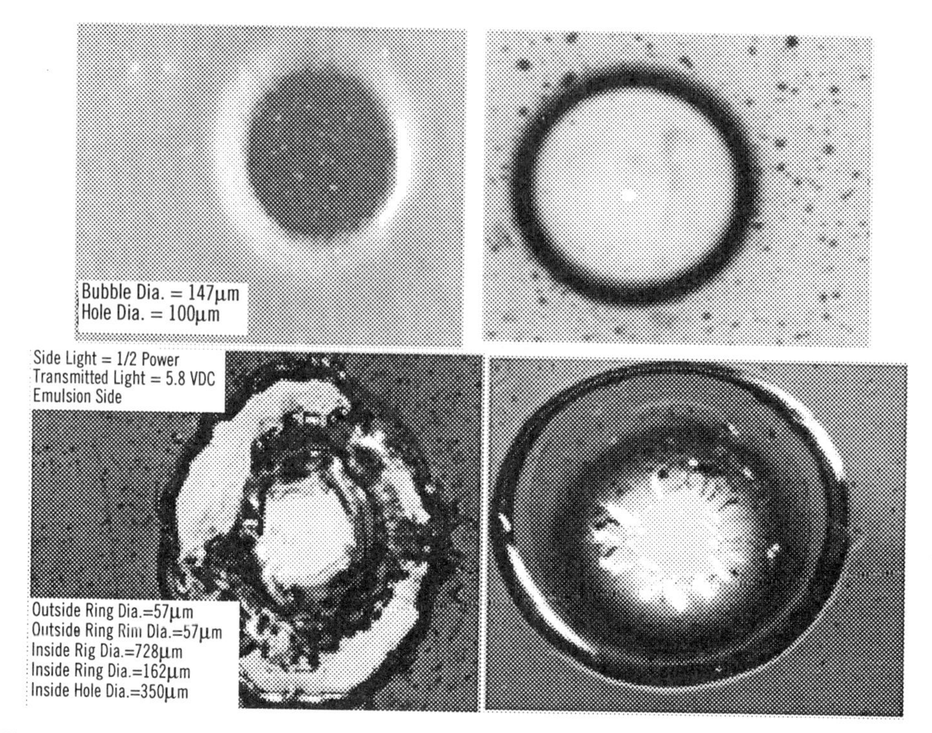

Figure 11-10. Examples of bubbles: (*a*) standard bubble; (*b*) unbroken bubble; (*c*) volcano type; (*d*) volcano type.

A pump is often used to feed the coating liquid to the coating stand. Air can be sucked into the pump on the suction side if the connections are loose or the pump seal is worn. Also, cavitation can occur at the vanes of the pump to form bubbles. Cavitation is always bad, as it causes erosion of the pump vanes. If dissolved air is present, these bubbles will contain air which may not redissolve.

During drying, dissolved air can come out of solution to form bubbles. The solubility of any gas decreases with increasing temperature. At the higher temperatures of the dryer, air can come out of solution. It is desirable to have as little dissolved air as possible. It is sometimes helpful to reduce the level of dissolved air by pulling a vacuum on the coating vessel before coating.

Air can also be introduced under the coating during the coating operation. This very thin sheet of entrained air then breaks up into fine bubbles. These often then dissolve in the coating liquid. At higher coating speeds more air is drawn in to form larger bubbles, which may remain in the coating. This problem is normally observed only when the coating speed is increased.

11.1.2 Chatter

Classification : Coating
Large
Continuous transverse direction

Chatter is the term used to describe coverage variations that appear as uniform bands across the width of the web. These bands are repetitive and have a frequency that can be related to the cause of the chatter. The coverage variations can be varied from the subtle, which are difficult to detect, to the gross, where there is no coating at all in the low-coverage regions. Chatter can be caused by mechanical sources, or it can arise from the hydrodynamics of the system.

Mechanical chatter can be caused by vibrations from the building, pumps, coating machine, drives, and so on, which are transmitted to the coating stand or to the web. Chatter can be caused by fluctuations in drive speed and by fluctuations in liquid feed to a coating die. It can be caused by pressure fluctuations in the bead vacuum system, if one is used. It can be caused by flutter in the web. It can be caused by the coating roll being out of round, or with a bad bearing, causing a fluctuating coating gap. Chatter results in coverage oscillations, and thus means that there is either a fluctuation in the web speed or a fluctuation in the flow rate onto the web. The basic troubleshooting approach is to determine the source of mechanical instability that causes the chatter. The first step is to determine the frequency of the chatter. The frequency is found from the web speed and the measured wavelength or spacing of the chatter marks. Thus

$$\text{chatter frequency}, \frac{\text{cycles}}{\text{s}} = \frac{\text{web speed}, \dfrac{\text{length}}{\text{s}}}{\text{wavelength}, \dfrac{\text{length}}{\text{cycle}}}$$

If speed is in feet per second, the wavelength or spacing should be in feet. Often, the repeating pattern can vary from being close together, perhaps 1 cm apart, to far apart, perhaps 1 m or more. Several measurements from widely spaced samples should be taken for the wavelength.

The chatter frequency should then be related, if possible, to some mechanical characteristic of the coater. For example, a frequency of 60 Hz, the frequency of alternating current in the United States, suggests an electrical cause, such as pump motors or drive motors. Lower frequencies suggest mechanical vibrations. Higher frequencies are sometimes from rolls. The length of a chatter pattern on the web should be converted into a diameter to see if it matches a roll diameter in the system. A poor bearing, especially in the coating roll, is sometimes the cause of chatter. A measurement that helps in troubleshooting is to run the coater over a range of speeds and measure the repeating distance and the chatter frequency as a function of speed. This will give some indication of the possible sources of vibration. Thus, if a bad bearing on a roll is causing the chatter, the linear frequency would not change.

Ideally, the next step is to have a vibration analyzer to determine the vibration frequencies at the coating stand and in the coating machine. This equipment can determine unusual vibrations in the rolls that drive the coater, in the idler rolls that transport web, and in the unwind and rewind equipment. The measured vibration frequencies at different positions can be compared with the calculated chatter frequency and also with historical data to determine changes.

Some other items to consider are:

1. Check the bearings in the coating roll, the transfer rolls, and web support rolls, to be sure that they are turning freely and are not binding. Replace as needed.
2. Check the rolls to see whether they are round and rotating uniformly. With the coater locked out so that it will not run, turn the rolls to check that they are rotating freely.
3. Check the pumps by hand for any unusual vibrations. Observe the lines for any pulsations.
4. Check the web tension and ensure that it is uniform across the web. Try varying the tension.
5. Check the web speed indicator while running for speed fluctuations.
6. Check the support at the unwind stand for roll uniformity and straightness. A poorly wound roll can vibrate and send pulsations through the web.
7. Try a different roll of web.
8. Check the coating stand to ensure that it is assembled correctly and that all parts fit as designed.
9. Check the heating and ventilating equipment in the dryer and in the building for any unusual vibrations.
10. Check the bead vacuum system, if used, for high-frequency pressure fluctuations.

If no mechanical causes are found, the chatter may be caused by the system hydrodynamics. This usually occurs near the limits of coatability. Try coating at a tighter gap, or try changing the bead vacuum. Also try coating at a lower speed.

REFERENCES

Cohen, E. D., and R. Grotovsky, "Application of digital image analysis techniques to photographic film defect test method development," *J. Imag. Sci. Technol.* **37**, 133–148 (1993).

Index